PO-3381   27396-1001

INTRODUCTION
TO
SOL-GEL PROCESSING

THE KLUWER INTERNATIONAL SERIES
in
SOL-GEL PROCESSING:
TECHNOLOGY AND APPLICATIONS

Consulting Editor
Lisa Klein
Rutgers, the State University of New Jersey

INTRODUCTION TO SOL-GEL PROCESSING

by

Alain C. Pierre
Université Claude-Bernard-Lyon I

KLUWER ACADEMIC PUBLISHERS
Boston / Dordrecht / London

Distributors for North America:
Kluwer Academic Publishers
101 Philip Drive
Assinippi Park
Norwell, Massachusetts 02061 USA

Distributors for all other countries:
Kluwer Academic Publishers Group
Distribution Centre
Post Office Box 322
3300 AH Dordrecht, THE NETHERLANDS

Library of Congress Cataloging-in-Publication Data

A C.I.P. Catalogue record for this book is available
from the Library of Congress.

To Marie-Claude
David and Valérie

Kaolinite gel network. From K. Ma and A. Pierre - Unpublished photograph.

TABLE OF CONTENT

PREFACE

BACKGROUND

Sol-gel processing has been known for a long time; the first silica gels were made in 1845 by M. Ebelmen at the "Manufacture de Céramiques de Sèvres" in France. However this processing technique has known a very important development during the last two decades. Several series of international conferences have addressed this theme, in particular: "International Workshops on Glasses and Ceramics from Gels" and conferences on "Better Ceramics Through Chemistry", "Ultrastructure Processing of Ceramics, Glasses and Composites", "Sol-Gel Optics", "Hybrid Organic-Inorganic Materials", "Aerogels". This list does not include many specialized sessions in international conferences devoted to materials.

Sol-gel processes have brought a new view in the domain of glass and ceramics fabrication and they have enlightened the importance of chemistry along the complete fabrication lines of materials, from initial chemical precursors to the final products. The basic idea is to progressively create an oxide network by polymerization reactions of chemical precursors dissolved in a liquid medium
The importance of sol-gel processing and its complexity is such that it deserves books and excellent ones, in particular on "Sol-Gel Technology for Thin Films, Fibers, Preforms, Electronics and Specialty Shapes", edited by L. Klein (Noyes, Park Ridge, N.Y.,1985) and on "Sol-Gel Science. The Physics and Chemistry of Sol-Gel Processes" by C.J. Brinker and G.W. Scherer (Academic Press, N.Y.,1990), have appeared. The latter book definitely constitutes an outstanding and complete reference on the subject which researchers are advised to consult first. Certainly, new informations on sol-gel have been discovered since its publication. However, the aim of the present book is different.

SCOPE

The subject needs now to be taught in universities, in courses on inorganic chemistry, surface science and ceramics processing or combinations of them and books less extensive on research developments than those quoted and more concise on the main basic theories, are needed. The present volume is an attempt in this direction. It is the exact content of lectures which were progressively developed and taught, both at undergraduate and graduate levels, first at the University of Bordeaux in 1987 in France, then at the University of Alberta in Edmonton, Canada and since 1995 at the University of Lyon in France. A first shorter version was published in 1992 in French by the printing company Septima, in Paris.

The book is organized in such a way that each chapter actually corresponds to one of the main chronological order steps involved in sol-gel processing. The detailed developments in many sections are a consequence of the input of students. Other clarifications are certainly needed, as new ones appear necessary each year during the course of a teaching session and I would like to apologize for this to the readers. I will be very grateful to all users of this book if they could communicate their remarks and suggestions for further possible versions dedicated to teaching.

ACKNOWLEDGMENTS
This book would not have been written without the strong initial encouragement of my colleagues at the University of Bordeaux: J. Etourneau, M. Onillon, J. Portier and B. Tanguy. I would like to sincerely thanks them, as well those who helped me at some stage, in particular E. Matijevic of Clarkson University, S. Sakka from the University of Kyoto and mainly J. Livage of the University of Paris-VI .

I would like especially to express my deep gratitude to L. Klein, Professor at Rutgers University, for her support and reviewing a first draft of the manuscript. This book would not have existed without her help. The technical support of the Kluwer printing company was very helpful and I am very pleased to acknowledge their help.

Challenging discussions with many students over quite a few years led to drastically modify the content of many sections. This book is in part their fruit, indirectly, and I am very thankful to them. One of the main contributors, on all the chemical aspects of sol-gel, was my own daughter Valérie, student in engineering chemistry at the French "Grande Ecole" "Chimie , Physique, Electronique", of the University of Lyon-I; I am very much indebted to her.

GENERAL INTRODUCTION

1.1 - SHORT HISTORY

Sols and gels are two forms of matter that have been known to exist naturally for a long time. They include various materials such as ink, clays, and a number of other substances such as the eye vitrea, blood, serum, and milk [1]. Sols and gels have arose scientific interests for a long time. The oldest sols prepared in a laboratory were synthesized with gold by Faraday in 1853. They are still stable nowadays[2].

It was not until 1861 that Graham founded colloidal science. Since then the study of ceramic colloidal sols has been slowly progressing so that the manner in which the sols forms and their sensibility to a number of different factors are now starting to be understood. It is only since recently that we can control the size of particles with inorganic salts [2,3]. Similarly, it has been necessary to achieve a good understanding of the nature of sols and of the laws explaining their behaviors before synthesizing well-defined particle dispersions. A major contribution in the understanding of sols chemistry came from the DLVO or electrostatic theory. This theory was particularly the first one to distinguish a precipitate from a stable colloidal suspension.

As for gels, apart from some silica gels which exist naturally [4], their synthesis has only been achieved since the nineteenth century. Ebelmen [5-7] produced the first silica gels in 1846 and Cossa [8] synthesized the first alumina gels in 1870. Since then, all kinds of inorganic gels have been progressively synthesized and various techniques such as the supercritical drying of Kistler have been introduced [9]. This latter led to the production of the first aerogels of silica, alumina, zirconia, stannic and tungsten oxides. Still in ceramics but outside the field of oxides, Stock and Somieski [10] synthesized the first silazanes, the precursors to Si_3N_4. The most recent techniques include the use surfactants, which allows the creation of porous structures with regular hexagonal packing of cylindrical pores [11] and other new hybrid of organic-inorganic materials [12].

Some gels have been synthesized only recently. The first borate gel, for instance, has been produced in 1984 by Tohge et al [13]. The number of publications on "gelatinous" and "colloidal" forms of ceramics is, however, steadily increasing since the beginning of colloid science and some of the properties and structure of gels are now starting to be elucidated. For instance, the first historical details have been summarized by Iler [4] for silica and by Gitzen [14] for alumina.

Flory [15] then set the theoretical basis needed in order to understand the network structure of a gel and the kinetics of its formation and gelation. His original work addressed only organic gels but was soon extended to inorganic ones. Shortly after, the theory of percolation of Hammersley [16] explained the gelation process as a special case of the critical phenomena occurring in physics. The lace geometry that characterizes a gel network, and the porous ceramic obtained at the beginning of densification were then described by the fractal geometry introduced by Mandelbrot [17].

For a long time sols and gels were only of pure scientific interests. However, their extremely high specific area made them increasingly interesting in the field of catalysis. By the beginning of the 1960's, following the works on clays [18] and nuclear fuel oxides in England, and those of Matijevic [19] on the production of colloidal particles with controlled size and shape, new ceramics were produced for more technical use.

Since the 1970's, an increasing number of publications in the field of gels and inorganic colloids has been published. This led to an increasing number of potentially interesting applications in high technology ceramics; some of them have already been commercialized. The new sol-gels processes can now fulfill at least to some extent the current need for new and better products [20,21] High purity submicron powders, nuclear fuels, electronic and ionic conductors and magnetic materials can now be produced by sol-gel techniques. Sol-gel processing is also very useful and important when the production of reproducible homogeneous complex ceramic materials is necessary. As a matter of fact, this technique is often considered as the route to "better ceramics through chemistry." However the fundamental aspects of physics are also very important and must not be neglected. Those aspects include particularly the DLVO theory, the percolation approach of gelation and more generally, the "ultrastructure" design (i.e., at the scale of very small aggregates of atoms) of any property that it be mechanical or magnetic.

1.2 - SOLS, GELS AND GELATION

SOLS

A sol is a stable suspension of colloidal solid particles within a liquid [22]. For a sol to exist, the solid particles, denser than the surrounding liquid, must be small enough for the forces responsible of dispersion to be greater than those of gravity. Moreover, these particles must include a number of atoms macroscopically significant. In fact, if the particles were too small, then it would be more accurate to speak of molecules in solution.

Originally, "colloidal" only referred to macroscopic particles that could not pass through a dialysis membrane. This definition, however, did not give limited and accurate values of the size range of the particles concerned. Practically, particles in a colloidal sol must have a size comprised between 2nm and 0.2μm; this corresponds to 10^3 to 10^9 atoms per particle [1]. Particles in this proper size range can be divided into three categories. They can be either composed of subdivided parts of bulk matter (example: small particles of α-alumina), of real macromolecules that are big enough to be colloidal (such as proteins), or of small particles that can be considered both macromolecules and as tiny parts of macroscopic matter (such as lacey particles). In the case of subdivided parts of bulk material, there is two thermodynamic phases and the sol is considered as lyophobic, or hydrophobic if water is the main solvent used. In the case of real macromolecules, there is thermodynamically only one phase and we then speak of a lyophilic or hydrophilic solution. The small lacey particles are more difficult to characterize and occurs, for instance, with some colloidal forms of silica. The solvent used to disperse stably the colloidal particles of a sol is often either pure water or a solution composed mostly of water. Nevertheless, other solvents such as alcohol can also be used.

GELS
A gel is an porous 3-dimensionally interconnected solid network that expands in a stable fashion throughout a liquid medium and is only limited by the size of the container. If the solid network is made of colloidal sol particles the gel is said to be colloidal. If the solid network is made of sub-colloidal chemical units then the gel is polymeric. A polymer, as defined by Flory [23], is a group of molecules whose structure can be generated through "repetition of one or a few elementary units." There is a great diversity of inorganic sols and gels and several classifications have already been proposed, the most complete of all is again the one given by Flory [24].
The nature of gels depends on the coexistence between the solid network and the liquid medium. Several remarks therefore have to be made on this equilibrium. The liquid is present between the mesh of solid network that composes the gel; it does not flow out spontaneously and is in thermodynamic equilibrium with the solid network. If the liquid is mostly composed of water, and if that aqueous phase is the one present in greatest proportion, then the corresponding gel is an aquagel (or hydrogel). An aquagel is a soft material that can be easily cut with a knife. If the liquid phase is largely composed of an alcohol then the gel is an alcogel. Finally, if most of the liquid is removed, then the brittle solid obtained is either called a xerogel or an aerogel, depending on the drying method (they can also be called more simply dry gels).

GELATION
A gel forms when the homogenous dispersion present in the initial sol rigidifies. This process, called gelation, prevents the development of inhomogeneities within the material. A sol, or a solution, can be transformed into a colloidal (or polymeric) gel by going through what is called a gel-point [25]. Practically, it is at this point that the sol abruptly changes from a viscous liquid state to a solid phase called the gel. This gel-point and the properties of sols and gels near this point are now better characterized within the framework the new "theory of critical phenomena".

GELATINOUS PRECIPITATES
The intermediate between the gel and the precipitate is called gel-precipitate. Its definition is usually ambiguous and in order to clarify it you must consider that a

precipitate is the result of the formation of separate dense aggregates that in some precise conditions can no longer disperse stably in a liquid. These aggregates have a texture that can be as well densely packed as quite open. In fact, the conditions for precipitation can be easily combined with those required for the formation of infinite 3-dimensional porous networks in order to form gelous precipitates with identical types of links.

SOL-GEL PROCESSES

Many definitions of sol-gel processes exist. Dislich [26], for instance, considers that the sol-gel procedure only takes into account multicomponent oxides that are homogeneous at the atomic level. This definition therefore does not include the colloidal coprecipitates of hydroxides and oxyhydrates since they become homogeneous only by reaction at high temperature. In fact, here the term "sol-gel" is restricted to the gels synthesized from alkoxides. Segal [27], on the other hand, defines the "sol-gel" process as the production of inorganic oxides, either from colloidal dispersion or from metal alkoxides. The sol-gel process, however, no longer includes solely oxides but also some other components such as nitrides and sulfides that have been used recently in the synthesis of hybrid organic-inorganic materials.

We will therefore consider in this book a more general definition of sol-gel processes. This definition state that a sol-gel process is a colloidal route used to synthesize ceramics with an intermediate stage including a sol and/or a gel state.

1.3 - OUTLINE OF SOL-GEL PROCESSING

Many variations can be brought to the sol-gel synthesis of ceramics. In fact, sol-gel processing does not only designate a unique technique, but a very broad type of procedures that centralizes around a single scheme as presented in figure 1.3-1.

The first step of any sol-gel process always consists in selecting the precursors of the wanted materials. It is the precursor that, by its chemistry, lead the reaction towards the formation of either colloidal particles or polymeric gels. When the future material is composed of several components (i.e., when several oxides are mixed together), then the use of a combination of different precursors and procedures enhances different chemical synthesis and hence different products. Recently, new types of materials called organic-inorganic hybrids have been created using this technique. The colloidal particles obtained can then be precipitated and treated according to one of the conventional processing techniques, such as cold pressing, hot-pressing and sintering, in order to produce the desired ceramic. The colloidal particles can also be dispersed into a stable sol before being transformed into a gel.

Sols and gels can also be spinned into fibers or transformed through one of other various techniques into a coating material. If a gel powder for melting purposes is required, than the gel may be dried and no special cares need to be given to fracturation. On the other hand, controlled gelation and drying leads to the formation of monosized droplets which can reach several hundred micrometers in diameter.

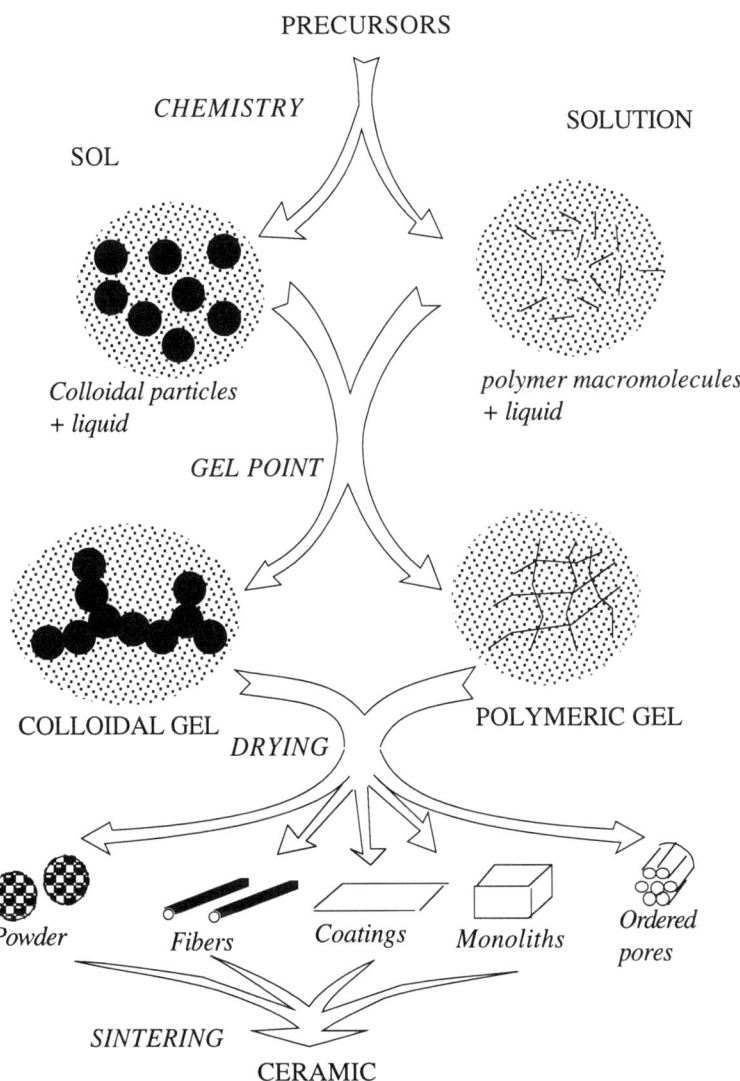

Figure 1.3-1 - Simplified chart of sol-gel processes.

It is also possible to synthesize small ceramic parts by using, after a simple sintering, an all-monolithic process that avoids fracturation. Furthermore, the size of pores can also be controlled by supercritical drying or by the use of surfactants. This is particularly useful in the field of catalysis. In sol-gel processing we do not only have a unique route, but rather a map with several outstanding routes and many by-ways that are far from being fully explored yet.

1.4 - RECENT DEVELOPMENTS

The applications of sol-gel processing are still not very widespread. They are, however, in expansion and the potentiality is still quite large. The most widespread use of sol-gel is in coatings devices such as in antireflective coatings with index gradation; optical or infra-red absorbing coatings; electrically conductive coatings; and, finally, coatings that protects against scratch, oxidation and erosion on all types of materials. Research in the field of new ceramic fibers and mainly in the application of composite materials is also under way and will certainly result in the production of quite interesting materials. Outstanding success has already been met experimentally in advanced fine-grains ceramics with ferroelectric, dielectric, piezoelectric, optical or electrooptical properties. This success is due to the feasibility of better ultrafine ceramic powders of a new type [28]. New interesting glasses and glass-ceramics can now be synthesized from sol-gel monoliths as well as by the sintering or melting of sol-gel powders. Those glasses could not be obtained by any conventional processing. Moreover, sols and gels of ceramics have specific and particular properties that make them quite interesting in some new specific applications, as in the coatings of photographic films. The two most noteworthy recent achievements are the design of new organic-inorganic sol-gels used in the embedment of photochromic and laser dyes [12], and the synthesis of ordered pore structures from micellar surfactant solutions [11]. Sol-gel processing is therefore the newest technique that must be considered in the synthesis of ceramics.

1.5 - ADVANTAGES AND LIMITATIONS OF SOL-GEL PROCESSING

There are many advantages to sol-gel processing. It not only allows for materials to have any oxide composition, but it also permits the production of new hybrid organic-inorganic materials which do not exist naturally. Sol-gel processing also has many other significant advantages. Very pure products are obtained by simply purifying the precursors either by distillation, crystallization, or electrolysis. Moreover, the chemical processes of the first steps are always carried out at low temperatures. By comparison with the classical high temperature synthesis of conventional ceramics, this minimizes considerably the chemical interactions between the material and the container walls. Another advantage is the association of the solid colloidal state with a liquid medium, thus avoiding any pollution by the eventual dispersion of dust. This explains why the biggest industrial application of sol-gel synthesis of ceramics is for nuclear fuels; pollution is very critical in this domain.

There are other more fundamental advantages to sol-gel processing. For instance, the kinetics of the various chemical reactions can be easily controlled by the low processing temperatures and by the often dilute conditions. The nucleation and growth of the primary colloidal particles can also be controlled in order to give particles with a given shape, size, and size distribution. This size domain is of course largely submicronic. However, in order to do that, the sintering temperatures of both the amorphous and the crystalline material must be lowered of several hundred degrees Celsius [29]. Apart from this energetic disadvantage, this cooling may be critical for some applications. This is the case of ceramic-ceramic composites for which the processing temperature must be lowered below the temperature needed for the fiber-matrix reaction. This is also the case of hybrid organic-inorganic materials.

The structure of sol-gel ceramics can be as easily controlled as the size of the particles. An amorphous or semi-vitreous state is, for example, quite easy to obtain and many new types of glasses that were not feasible by the conventional quench techniques can now be synthesized. As a matter of fact, the two problems that the conventional quench techniques encountered, that is the very high fluidity of the liquidus of the material that required a very fast cooling rate for the glass to form and the existence of a miscibility gap in some domains of composition making it impossible to obtain a homogeneous glass, were completely solved by the sol-gel process. The distribution of the pores and crystalline or amorphous phases can also be tailored to a large extent. This is particularly true if the phase present at grain boundaries can be either controlled or eliminated. Eliminating this phase can be important in some cases, as for the ionic conductors composed of sodium and silicon (Nasicon) for which the segregation of ZrO_2 at grain boundaries must be avoided.

Sol-gel processing offers the most outstanding advantages for mixed oxides systems in which the chemical homogeneity of the various elements can be controlled down to the atomic level. This is the case of the Lead Lanthanum Zirconium Titanates (PLZT) [30]. This condition is, however, virtually impossible to achieve when solid powders are mechanically mixed such as in the conventional processes; and hence the optical transparencies obtained are, in this case, not as high as those obtained by sol-gel processing.

The greatest limitation to the synthesis of ceramic by sol-gel processing is still the cost of the precursors, and especially that of alkoxides. Most of these alkoxides are nonetheless quite easy to make; especially if they do not tend to polymerize. A few of them such as Zr and Ti are even used industrially by the Schott company for coating applications [26] and are thus quite affordable. Moreover, alkoxides can also be mixed with much cheaper metal salts provided that a purification step is included in the procedure.

The sol-gel synthesis of ceramics will never be able to compete for the mass production of some large scale materials such as window glass for which the conventional processes can rely on much cheaper raw materials. Sol-gel processing becomes, however, much more interesting for highly advanced ceramics.

Sol-gel processing is not the only chemical route that leads to better ceramics. Another procedure includes the use of organometallic compounds in which an organic group is directly bonded to a metal without any intermediate like oxygen. The chemistry of organometallics essentially concerns the synthesis of complex polymers and especially those containing carbides and nitrides. If the chemistry is carried out in a liquid medium, then it can be considered as a sol-gel process. Precipitation and co-precipitation techniques are also used and are sometimes even considered as a side branch of sol-gel processing. The chemical reactions concerned in this case are the same as those occurring in sol-gel synthesis; they often lead to the production of colloidal particles, but can also be dispersed into a stable sol.

Two other techniques different from sol-gel processing are:

- The thermal decomposition of precursors in the vapor phase [20]. This technique is also known as chemical vapor decomposition (C.V.D). Some of those methods use quite sophisticated nucleation and growth procedures such as heating by laser. As in sol-gel processing, very pure products can be obtained if certain precursors such as, again, alkoxides are used.

- The hydrothermal process. It is carried out on solutions in an autoclave and at higher temperatures than those required in sol-gel processing. This technique also produces crystals with a size ranging in the micrometer and beyond [31]. Hydrothermal synthesis is not within the scope of this book, although the wet chemistry aspect of it is essentially the same as that of sol-gel processings.

1.6 - ORGANIZATION OF THE BOOK

This book aims at offering a synthetic view of the various aspects and interest of the materials produced by sol-gel processing. These aspects are presented according to the classical scheme of a typical sol-gel process as summarized in figure 1.3-1. The liquid medium chemistry is first discussed in Chapter 2, then the colloidal particles in Chapter 3, followed by the phenomenon of gelation (Chapter 4) and the gels themselves and their properties (Chapter 5). A presentation of two recent types of sol-gel derived materials follows in Chapter 6. The phase transformations of gels to higher temperature forms and their sintering are then presented in Chapters 7 and 8. The different applications are finally discussed in Chapter 9.

You will note that at all levels, that it be molecules in an aqueous medium or porous ceramics under sintering, there is always a competition between the formation of dense structures and open ones. This book is a journey through the various stages of this competition.

1.7 - REFERENCES

1 - Livage J., Lemerle J., Ann. Rev. Mater. Sci. 12 (1982) 103-122.

2 - Matijevic E., "Monodisperse colloids (Preparation, Properties and Applications), and interactions in mixed colloidal systems (heterocoagulation, adhesion and microflotation)". Conference presented at the Université de Bordeaux I, France, 9-10 June 1987.

3 - Roy R., J. Amer. Ceram. Soc. 39 (1956) 145.

4 - Iler R.K., "The Chemistry of Silica", Wiley, New-York (1979).

5 - Ebelmen M., Ann. Chim. Phys, 15 (1845) 319

6 - Ebelmen M., Ann. Chim. Phys, 16 (1846) 129

7 - Ebelmen M., Compte Rendus de l'Acd. des Sciences 25 (1947) 854.

8 - Cossa A., Il Nuovo Cimento 3 (1870) 228-230.

9 - Kistler S.S., Nature, 127 (1931) 741.

10 - Stock A., Somieski K., Ber. dt. Chem. Ges. 54 (1921) 740-758.

11 - Corma A., Chem. Rev. 97 (1997) 2373-2419.

12 - Sanchez C., Ribot F., New J. Chem., 18 (1994) 1007-1047.

13 - Tohge N., Moore G.S., Mackenzie J.D., J. Non-Cryst. Solids 63 (1984) 95-103.

14 - Gitzen W.H., "Alumina as a Ceramic Material", the American Ceramic Society, Columbus, Ohio (1970).

15 - Flory P.J., J. Am. Chem. Soc. 63 (1941) 3083- 3100.

16 - Hammersley J.M., Proc. Cambridge Phil. Soc. 53 (1957) 642-645.

17 - Mandelbrot B.B., "Fractals: Form, Chances and Dimensions", Freeman, San Francisco (1977).

18 - Ford R.W., "Drying", Institute of ceramics, Textbook series; MacLaren and Sons, London, England (1964).

19 - Matijevic E., Acc. Chem. Res. 14 (1981) 22-29.

20 - Mazdiyasni K.S., Lynch C.T., Smith II J.S., J. Am. Ceram. Soc. 48 (1965) 372-375.

21 - Wheat T.A., J. Canad. Ceram. Soc. 46 (1977) 11-18.

22 - Hiemenz P.C., "Principles of Colloid and Surface Chemsitry", Marcel Dekker, New-York (1977).

23- Flory J.P., Disc. Faraday soc., 57 (1974) 7-8.

24 - Flory J.P., "Principles of Polymer Chemistry", Cornell University Press, Ithaca, New-York (1953).

25 - Seyferth D., Wiseman G.H., in "Ultrastructure Processing of ceramics, Glasses and Composites", Edited by Hench L.L. andUlrich D.R., Wiley, New-York (1984) 265-271.

26 - Dislich H., J. of Non-Crystalline Solids 57 (1983) 371-388.

27 - Segal D.L., J. Non- Crystalline Solids 63 (1984) 183-191.

28 - Mazdiyasni K.S., Ceramics International 8 (1982) 42-56.

29 - Zelinski B.J.J., Uhlmann D.R., J. Phys. Chem. Solids 45 (1984) 1069-1090.

30 - Haertling G.H., Land C.E., Ferroelectrics 3 (1972) 269-280.

31 - Sapieszko R.S., Matijevic E., Corrosion 36 (1980) 522-530.

THE CHEMISTRY OF PRECURSORS SOLUTIONS

2.1 - INTRODUCTION

If inorganic sols and gels can be obtained by various methods, they are often directly synthesized from chemical reactants dissolved in a liquid medium. The chemical reactant which contain the cation M present in the final inorganic sol or gel is called the chemical precursor. Its chemical transformations are complex and involve a competition at the molecular level between the reaction responsible for the formation of open structures and the one leading to dense solid. Those same reactions are responsible for the controlled dispersion of dense colloidal particles in a sol or their agglomeration into a gel. They are therefore important and will be viewed in a separate chapter.

All types of precursors can be used, provided they are miscible. Two main groups are therefore distinguished: the metallic salts and the alkoxides. The general formula of a metallic salt is M_mX_n where M is the metal, X an anionic group, and m and n stoichiometric coefficients; an example is aluminum chloride, $AlCl_3$. As for alkoxides, their general formula is $M(OR)_n$, which indicates that they are a combination of a cation M with n alcohol groups ROH; an example is aluminum ethoxide, $Al(OC_2H_5)_3$. Since the solution chemistry of these two groups are quite different, the choice of the solvent as either water or an organic liquid depends on the precursor. The ceramics synthesized by sol-gel, for example, are mainly oxides; water is therefore always present as a major reactant in order to transform the precursor. Hence, the electronic properties of the water molecule are important in the transformation process of the sol-gel precursor.

Precursors, other than metal salts and alkoxides are starting to be studied. This is the case of organometallic compounds in which a metal M is directly linked to a carbon atom. The synthesis of non-oxide ceramics, such as carbides, nitrides or sulfides directly by the sol-gel route is another example. These fields deserves therefore a special consideration and will be addressed in the last part of this chapter.

Some materials such as carbides and nitrides are difficult to synthesize directly in a liquid medium by chemical reactions. In this case, the technique first involves forming primary powders by thermal decomposition of salts or by grinding bulk materials. In a second step, these primary particles are dispersed in a liquid medium so as to control their agglomeration further on. The phenomenon then involved are the same as in the gelation of colloidal particles directly obtained in a liquid medium.

An aerosol is obtained by nucleation and growth from a vapor phase. The chemical reactions are also similar to those present in a liquid medium.

2.2 -   SOLVENTS

It is necessary to consider water separately from the other solvents

WATER
The water molecule is, in the Lewis representation, V-shaped (Fig.2.2-1). The oxygen atom is surrounded by four electron pairs; and there is one covalent bond with each of the two hydrogen atoms and two unshared electron pairs. According to the VSEPR model, the oxygen therefore occupies the center position of a tetrahedron and the two hydrogen atoms occupy two of its corners. The angle θ = HOH is 104.5° in the gaseous state, and varies from 118° to 120° in the liquid state.

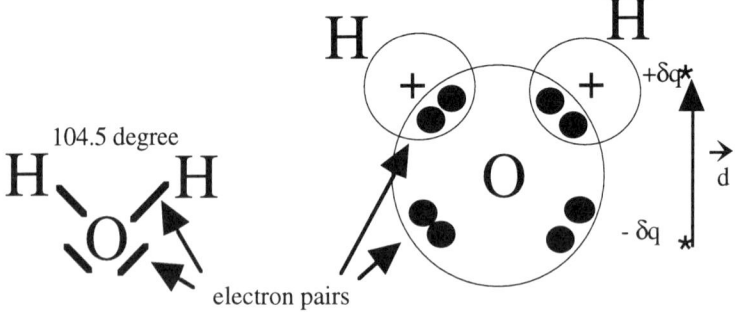

Figure 2.2-1 - Lewis representation of a water molecule

Such a configuration provides the permanent polarization of the water molecule, with a dipole moment defined as

$$\vec{\mu} = (\delta q) \ \vec{d} \qquad\qquad (2.2\text{-}1)$$

In the physical representation, the distance vector $\vec{d}$ originates at the center of the negative charges ($-\delta q$) and ends at the center of the positive charges ($+\delta q$). The magnitude of this dipole moment is therefore μ = 1.85 D (1 D = 1 Debye =   3.336 x 10^{-30} C.m) [1].
The water molecule belongs to the symmetry group C_{2v} [2]. The following theoretical calculations of the molecular orbitals of H_2O are made by linear combination of the 2s and 2p atomic orbitals of the oxygen atom and the 1s orbitals of the two hydrogen atoms. The electrons of the water molecule are therefore placed on the four molecular orbitals which have the following wave functions and energies:

$$\Psi_W(2a_1) = 0.85\Psi_O(2s) + 0.13\Psi_O(2p_z) + 0.81(\Psi_{HA}(1s) + \Psi_{HB}(1s)) \qquad E = -36 \text{ eV}$$
$$(2.2\text{-}2)$$

$$\Psi_W(1b_2) = 0.4\Psi_O(2p_y) \qquad\qquad +0.78(\Psi_{HA}(1s) - \Psi_{HB}(1s)) \qquad E = -19 \text{ eV}$$
$$(2.2\text{-}3)$$

$$\Psi_W(3a_1) = 0.46\Psi_O(2s) - 0.83\Psi_O(2p_z) - 0.33(\Psi_{HA}(1s) + \Psi_{HB}(1s)) \qquad E = -14 \text{ eV}$$
$$(2.2\text{-}4)$$

$$\Psi_W(1b_1) = \Psi_O(2p_x) \qquad\qquad\qquad\qquad\qquad\qquad\qquad\qquad E = -12 \text{ eV}$$
$$(2.2\text{-}5)$$

In these equations Ψ_W represents the molecular orbitals of the water molecule, while Ψ_O is the atomic orbitals of the oxygen atom, and Ψ_{HA} and Ψ_{HB} the respective atomic orbitals of the two hydrogen atoms A and B. The corresponding energy diagram, as illustrated in Fig.2.2-2 with the electron presence probability clouds, demonstrates that the $3a_1$ molecular orbital of water is largely delocalized outside the water molecule. This orbital is responsible for the Lewis base character of water. The molecule is therefore able to donate a pair of electrons to a ligand group and build a s bond with it. On the other hand, the $1b_1$ molecular orbital is strictly non binding, and hence has a very weak π donor character.

It is possible to estimate the partial electrical charge δq (also called δ) carried by each atom from the molecular orbital wave functions. Since the wave functions Ψ_M of a molecule AB is obtained by linear combination of Ψ_A and Ψ_B :

$$\Psi_M = a\Psi_A + b\Psi_B \qquad\qquad (2.2\text{-}6)$$

the electronic charge density is obtained by integrating on the whole volume

$$\Psi_M^2 = a^2\Psi_A^2 + b^2\Psi_B^2 + 2ab\Psi_A\Psi_B \qquad\qquad (2.2\text{-}7)$$

The a^2 coefficient gives the electron charge density contributed by the A atom only, the b^2 coefficient by the B atom only, and the overlap integral of $2ab\Psi_A\Psi_B$ is attributed equally to the A and B atoms. The difference between the total charge density of the A atom in the molecule AB and in the isolated state gives an estimation of the partial charge $\delta(A)$ carried by the A atom. For the water molecule, this estimation gives a negative partial charge for the oxygen atom of $\delta(O) = -0.4$ and a positive partial charge $\delta(H) = 0.2$ for each hydrogen atom.

Since its molecules are polarized, water is an excellent liquid medium in which to dissolve ionic solutes. This property is expressed by the coulomb electrostatic force F between two electrical charges q and q' :

$$F = \frac{1}{4\pi\varepsilon_r\varepsilon_0} \frac{qq'}{r^2} \qquad\qquad (2.2\text{-}8)$$

in which $\varepsilon_0 = 8.8542 . 10^{-12}$ F.m^{-1} is the dielectric permittivity of vacuum and ε_r the relative dielectric constant which has therefore no dimension. Water at 25°C, has a relative dielectric constant of $\varepsilon_r = 78.4$; it consequently largely attenuates the coulomb interaction between two electrical charges.

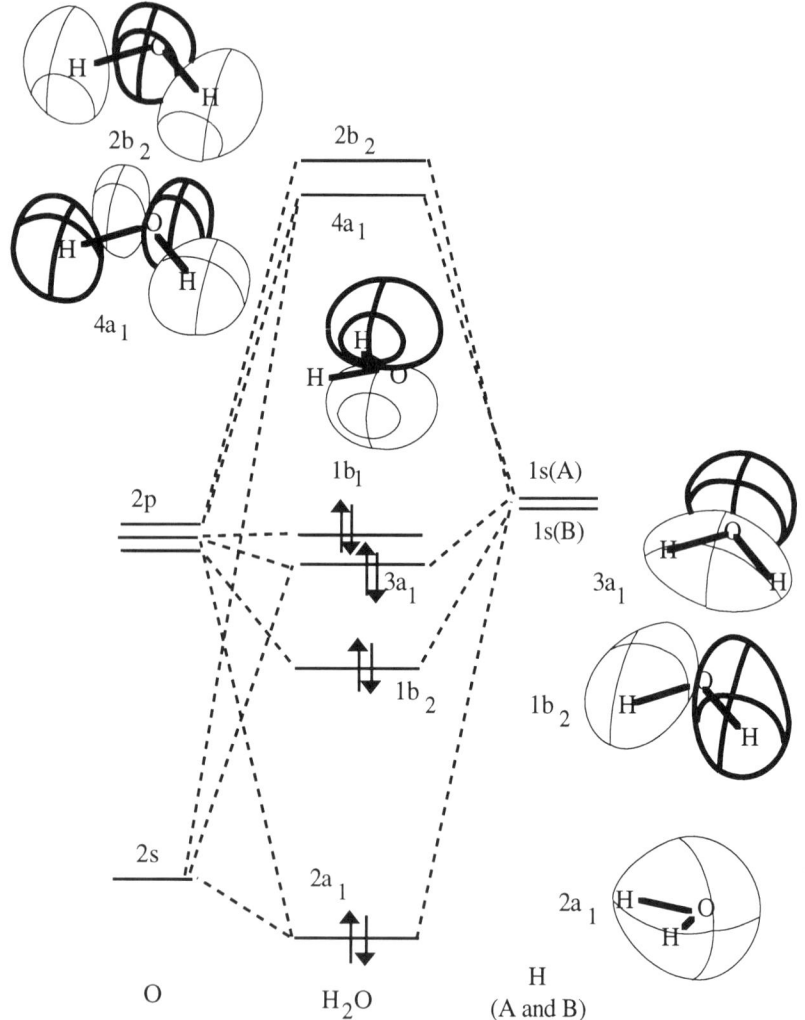

Figure 2.2-2 - Energy diagram and shape of the electron presence probability clouds for the molecular orbitals of water. After Jorgensen and Salem [3].

As illustrated in Fig.2.2-3, the polar structure of the water molecules leads to the formation of hydrogen bonds between oxygen and hydrogen atoms of different molecules. In this type of bond, a hydrogen atom fluctuates by tunnel effect between two minimum energy positions separated only by a small energy barrier. This barrier is of the order of 20 to 40 kJ mol^{-1}. Such bonds can also be formed with fluoride or nitrogen atoms instead of oxygen.

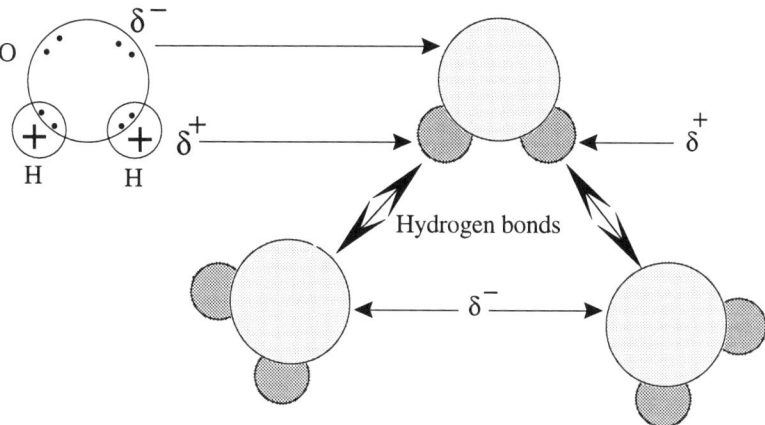

Figure 2.2-3 - Hydrogen bonding between two water molecules

A direct consequence of these bonds is that, according to molecular dynamic calculations [4,5], very few water molecules can be isolated in the pure liquid state. Most water molecules are bound by up to 4 hydrogen bonds to neighboring molecules. Those random groups have an average lifetime of the order of 10^{-10} s. Water molecules also auto-dissociate according to the reaction:

$$2H_2O \Leftrightarrow H_3O^+ + HO^- \qquad (2.2-9)$$

which equilibrium constant is $K_w = [H_3O^+][HO^-] = 10^{-14}$ at 25°C   (2.2-10)

Hence water is a protic solvent. Furthermore, the H_3O^+ and HO^- ions also associate themselves with other water molecules by hydrogen bonds, so that these two ions actually exist in liquid water as $[H_9O_4]^+$ and $[H_7O_4]^-$. The latter anion is the strongest possible base in aqueous solutions. In consequence, O^{2-} practically does not exist in water, and when a solid oxide is dissolved it immediately undergo an acid base protonation reaction.

NON AQUEOUS SOLVENTS

Apart for the molten salts, the polar non aqueous solvents have a molecular structure characterized both by a permanent dipole moment μ and by a relative dielectric constant ε_r. ε_r not only depends on m, but also on the polarizability α of the molecule, which is itself defined according to the relation

$$\mu* = \alpha E \qquad (2.2\text{-}11)$$

where $\mu*$ is the induced dipole moment when the molecule is submitted to an electric field E.

A high relative dielectric constant ($\varepsilon_r > 40$) is often due to the existence of a permanent dipole moment. Such molecules have good ionizing properties and can therefore dissolve other polar solute. On the other hand when the solvent's relative dielectric constant is low ($\varepsilon_r < 20$), it has a weak ionizing property and can only dissolve less polar solute. Table.2.2-1 gathers a list of frequently used solvents with their relative dielectric constant and dipole moment.

Table. 2.2-1 - List of some solvents with their dielectric properties.
After Lagowski [6].

Solvent	ε_r	μ (D)	Type
Acetone C_3H_6O	20.7	3.00	aprotic
Acetic acid $C_2H_4O_2$	6.2	0.99-1.51	protic
ammonia NH_3	16.9	0.90	protic
Benzene C_6H_6	2.3	0.00	aprotic
chloroform $CHCl_3$	4.8	1.11	aprotic
dimethylsulfoxide $(CH_3)_2SO$	45	3.90	aprotic
dioxanne 1,4 $C_4H_8O_2$	2.2	0.39	aprotic
water H_2O	78.5	1.85	protic
Methanol CH_3OH	32,6	1.70	protic
ethanol C_2H_5OH	24.3	1.71	protic
formamide CH_3ON	110.0	3.39	protic
Dimethylformamide C_3H_7NO	36.7	3.86	aprotic
nitrobenzene $C_6H_5NO_2$	34.8	3.99	aprotic
Tetrahydrofuran C_4H_8O	7.3	1.63	aprotic
Carbon tetrachloride CCl_4	2.2	0.00	aprotic
Diethyl ether $C_4H_{10}O$	4.3	1.15	aprotic
Pyridine C_5H_5N	14.2	2.19	aprotic

Solvents are classified as protic when they can exchange a proton, and as aprotic when they cannot do so. They can also be classified as acidic, in the Brønsted sense, when they are able to donate a proton or in the Lewis sense when they are able to accept a pair of electrons. Similarly, a base, according to Bronsted, is able to accept a proton, and according to Lewis to donate a pair of electrons. A solvent is amphoteric when it can behave both as a base and as an acid. Amphoteric solvents include:
- mineral acids (HCN, HX, HNO_3, H_2SO_4, H_2S)
- carboxylic acids R-COOH
- water, the first alcohols (CH_3OH, C_2H_5OH,....), and phenol C_6H_5OH
- ammonia NH_3 and amines (RNH_2, $RR'NH$)

- amides (R-CO-NH$_2$, R-CO-NHR').

Organic solvents are frequently used in sol-gel processing as they allow the control of the reaction of alkoxide precursors with water, and hence to direct with more flexibility the structure of sol-gel products.

2.3 - BASIS OF PRECURSORS TRANSFORMATIONS IN SOLUTION

THE PARTIAL CHARGE MODEL

Sol-gel precursors undergo chemical reactions both with water and with the other species present in the solution. One of the most efficient model used to predict those reactions is the Partial Charge Model (PCM) which has been recently elaborated by Henry and Livage [7,8] after a principle developed by Sanderson. It is based on the electrical interactions between the partial electric charges, δ, carried by each atom and molecule. Since the chemical potential of the electrons in an atom i or a molecule C depends on the partial electric charges $\delta(i)$ or $\delta(C)$ carried either by i or by C, and since the electronegativity $\chi(i)$ of i or $\chi(C)$ of C is directly related to this chemical potential, this model can also be expressed in terms of the particles' electronegativities [9].

The derivative of the function $E = f(n_e)$ which associates the total energy E of an isolated atom to the number of electrons n_e of this atom (Fig. 2.3-10) describes the influence of the electric charge carried by an atom on its electronegativity. The first ionization energy, I_1, of an isolated atom is also a function of the first and second partial derivatives of $f(n_e)$ with respect to n_e: [9]

$$I_1 = E(z-1) - E(z) \approx -\frac{\partial E}{\partial n_e} + \frac{1}{2}\frac{\partial^2 E}{\partial n_e^2} \tag{2.3-1}$$

The electron affinity of this atom is:

$$A = E(z) - E(z+1) \approx -\frac{\partial E}{\partial n_e} - \frac{1}{2}\frac{\partial^2 E}{\partial n_e^2} \tag{2.3-2}$$

The electronegativity of an isolated atom is therefore defined as:

$$\chi^a = \frac{1}{2}(I_1+A) = -\frac{\partial E}{\partial n_e} \tag{2.3-3}$$

Since its hardness is defined as

$$\eta^a = \frac{1}{2}(I_1-A) = +\frac{1}{2}\frac{\partial^2 E}{\partial n_e^2} \tag{2.3-4}$$

and the chemical potential of the electrons in this atom is

$$\mu_e^a = \frac{\partial E}{\partial n_e} \qquad (2.3\text{-}5)$$

we have

$$\chi^a = -\mu_e^a \qquad (2.3\text{-}6)$$

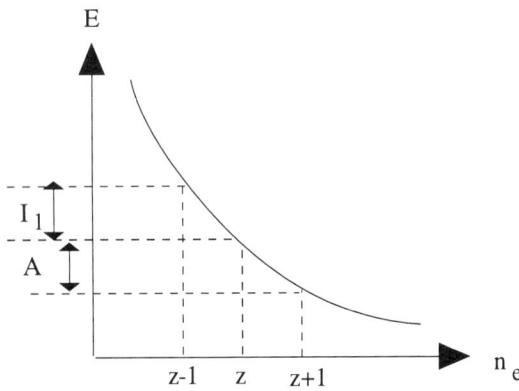

Figure 2.3-1 - First ionization and affinity energies of an atom. After Chermette and Lissilour [10]

For any chemical transformation, equilibrium is obtained when all phases have the same chemical potential. In the same manner, the electrons of a molecule transfer from atom to atom until they all reach the same chemical potential or electronegativity. Each atom thus gains an electronic charge (δn_e). The definition of their hardness then leads to the following equation:

$$\mu_e(i) \qquad = \qquad \mu_e^a(i) - \eta_i^a (\delta n_e) \qquad (2.3\text{-}7)$$

Since the electronegativity of an atom $\chi(i)$ is the opposite of the chemical potential of the electrons $\mu_e(i)$ this is equivalent to:

$$\chi(i) \qquad = \qquad \chi_i^a + \eta_i^a (\delta n_e) \qquad (2.3\text{-}8)$$

Electric charge transfer thus continues until all atoms in all species reach an equilibrium where they all have the same electronegativity. In order to reach such an equilibrium an ion such as H^+ may leave the complex if its partial charge reaches $\delta(H)=+1$, an anion X^- if its partial charge reaches $\delta(X)=-1$, or a molecule HX if $\delta(HX)=0$. In this manner, the partial charge model provides the thermodynamical basis needed to understand the possible evolution of a complex in solution.

Several electronegativity scales exist. The absolute electronegativity χ_i^a and hardness η_i^a concern an isolated atom i. The Mulliken electronegativity χ_i^M and hardness η_i^M take into account the electronic state of an atom in its average valence structure. At last, the Pauling electronegativity χ_i^P and hardness η_i^P involves the average structure configuration in which the atoms are engaged. It is therefore related to the binding enthalpies ΔH_{ij} between atoms i and j, ΔH_{ii} between two atoms i, and ΔH_{jj} between two atoms j according to the relation:

$$\chi_i^P - \chi_j^P = [\Delta H_{ij} - \frac{1}{2}(\Delta H_{ii} + \Delta H_{jj})]^{1/2} \qquad (2.3\text{-}9)$$

The Pauling χ_i^P and Mulliken χ_i^M electronegativities are linked by the following equation:

$$\chi_i^M = \frac{\chi_i^P}{0.335} + 0.615 \qquad (2.3\text{-}10)$$

In the Partial Charge Model, it is the Allred and Rochow's electronegativity χ_i^o and hardness η_i^o which is used because it takes into consideration both the valence state and the shape of an atom X in its average polarization. They are reported in Table 2.3-1. Furthermore, in this scale we have the following relationship:

$$\eta_i^o = 1,36\sqrt{\chi_i^o} \qquad (2.3\text{-}11)$$

Let us consider in an aqueous medium and at a given pH a complex C such as $[M(OH)_y(H_2O)_{N-Y}]^{(z-y)+}$. According to the partial charge model we have at equilibrium

$$\chi(C) = \chi(H^+) = \chi(H_2O) \qquad (2.3\text{-}12)$$

We can therefore predict which complex of a precursor M will exist in a solution by simply calculating the electronegativities of all the possible ones. Furthermore, since for a simple atom i with a partial charge δ_i:

The Chemistry of Precursors Solutions

$$\chi(i) = \chi_i^{\,o} + \eta_i^{\,o}\,\delta_i \qquad (2.3\text{-}13)$$

Table 2.3-1 - Allred-Rochow electronegativities. After Jolivet [5].

H 2.10																	He 3.20
Li 0.97	Be 1.57											B 2.02	C 2.50	N 3.07	O 3.50	F 4.10	Ne 5.10
Na 1.01	Mg 1.29											Al 1.47	Si 1.74	P 2.11	S 2.48	Cl 2.83	Ar 5.10
K 0.91	Ca 1.04	Sc 1.23	Ti 1.32	V 1.56	Cr 1.59	Mn 1.63	Fe 1.72	Co 1.75	Ni 1.80	Cu 1.75	Zn 1.66	Ga 1.82	Ge 2.00	As 2.20	Se 2.50	Br 2.69	Kr 3.10
Rb 0.89	Sr 0.99	Y 1.19	Zr 1.29	Nb 1.45	Mo 1.56	Tc 1.67	Ru 1.78	Rh 1.84	Pd 1.85	Ag 1.68	Cd 1.60	In 1.49	Sn 1.89	Sb 1.98	Te 2.15	I 2.33	Xe 2.60
Cs 0.87	Ba 0.89	Lant	Hf 1.36	Ta 1.50	W 1.59	Re 1.88	Os 1.99	Ir 2.05	Pt 2.00	Au 2.02	Hg 1.80	Tl 1.60	Pb 1.92	Bi 2.03	Po 2.12	At 2.28	Rn 2.30
Fr 0.86	Ra 0.95	Uran															

Lant	La 1.18	Ce 1.17	Pr 1.18	Nd 1.19	Pm 1.20	Sm 1.20	Eu 1.13	Gd 1.27	Tb 1.24	Dy 1.26	Ho 1.28	Er 1.30	Tm 1.30	Yb 1.24	Lu 1.36
Uran	Ac 1.12	Th 1.24	Pa 1.22	U 1.24	Np 1.22	Pu 1.24	Am 1.25	Cm 1.20	Bk 1.20	Cf 1.20	Es 1.20	Fm 1.20	Md 1.20	No 1.20	Lw 1.20

where $\chi_i^{\,o}$ is the reference electronegativity and $\eta_i^{\,o}$ the reference hardness, we have:

$$\delta_i = \frac{\chi - \chi_i^{\,o}}{\eta_i^{\,o}} = \frac{\chi - \chi_i^{\,o}}{1.36\sqrt{\chi_i^{\,o}}} \qquad (2.3\text{-}14)$$

In the case of a complex molecule C^{z+} composed of several elements, we must consider its total, or formal charge which is defined as:

$$z = \sum_i \delta_i \qquad (2.3\text{-}15)$$

Since at equilibrium all atoms in C have the same electronegativity; $\chi(i) = \chi(C)$ and hence

$$\chi(C) = \frac{\sum_i \sqrt{\chi_i^o} + 1.36\, z}{\sum_i 1/\sqrt{\chi_i^o}} \qquad (2.3\text{-}16)$$

For water at pH = 7, $\delta(H) = +0.2$ and $\delta(O) = -0.4$ (see section 2.2)

$$\text{As} \qquad z = \sum_i \delta_i = \delta(O) + 2\,\delta(H) = 0 \qquad (2.3\text{-}17)$$

$$\chi(H_2O) = \frac{2\sqrt{\chi_H^o} + \sqrt{\chi_O^o}}{\dfrac{2}{\sqrt{\chi_H^o}} + \dfrac{1}{\sqrt{\chi_O^o}}} = 2.491 \qquad (2.3\text{-}18)$$

For an acidic or basic solution, pH $\neq$ 7 and the water molecules carry a partial charge $\delta(H_2O) \neq 0$ either positive or negative. This partial charge is the average of the H^+ and OH^- charges as if they were evenly shared between all the H_2O molecules. As mentioned in section 2.2, this partial charge is actually a result of the fast transformation of complex groups of water molecules bound to each other by hydrogen bonds. The Nernst equation then provides the expression of the chemical potential of a proton:

$$\mu(H^+) = \mu^o(H^+) - 0.06\ \text{pH} \qquad (2.3\text{-}19)$$

Finally, if you consider that $\mu(H^+)$ is proportional to $\chi(H^+)$ and that $\chi(H^+)$ takes the values 2.491 and 2.631 at pH = 7 and pH = 0 respectively where it is present as the species $[H_7O_3]^+$ and $[H_9O_4]^+$, you reach the following formula:

$$\chi(H^+) = \chi(H_2O) = 2.631 - 0.02\ \text{pH} \qquad (2.3\text{-}20)$$

TRANSFORMATION MECHANISMS OF COMPLEXES

Various atomic or molecular groups called ligands can bind to a complex C or a cation M either directly or by substituting another ligand. The mechanism of the transformation depends on the partial charge of the different atoms in the species. Those with a negative partial charge are nucleophilic, and those with a positive charge are electrophilic. Similarly, in a substitution reaction, the new ligand with the highest partial negative charge, Y, is the nucleophile while the group in the metal complex with the highest positive charge, X, is the leaving group.

The direct addition of a new ligand occurs when the coordination number of the cation in the complex is not fully satisfied. The mechanism then involved is a nucleophilic addition symbolized by A_N, which is known and rather complex. There is substitution of a ligand by another when the coordination number of the metal is already full; this can be done by one of three different mechanisms. In this case the reaction is expressed as an exchange of a Lewis base by another in order to form a Lewis acid. In the following example

$$Cl^- + [Co(OH_2)_6]^{2+} \Rightarrow [CoCl(OH_2)_5]^+ + H_2O \qquad (2.3\text{-}21)$$

Cl^- substitutes H_2O.

Three different substitution mechanisms exist: the dissociative, the associative and the interchange mechanisms. In the first step of the dissociative substitution mechanism, enough thermal energy must be available in order to break the bond between the leaving group, X, and the complex. This X ligand then becomes labile, not very stable. Consequently, an intermediary complex in which the metal M has a reduced coordination number is formed; it can therefore be observed by analytical techniques such as NMR. (Fig. 2.3-2a) In the second step, the entering group, Y, completes again the normal coordination number of the cation M.

Since the rate constant of such a reaction does not depend on the concentration of the entering Y ligand, this dissociative mechanism is a unimolecular nucleophilic substitution, S_N1. This is for instance the case of the substitution of H_2O by a ligand L such as ammonia (NH_3) or pyridine in $[Ni(OH_2)_6]^{2+}$.

Water is formed in the first step

$$[Ni(OH_2)_6]^{2+} \Rightarrow [Ni(OH_2)_5]^{2+} + H_2O \qquad (2.3\text{-}22)$$

while L enters in the second step:

$$[Ni(OH_2)_5]^{2+} + L \Rightarrow [Ni(OH_2)_5L]^{2+} \qquad (2.3\text{-}23)$$

In the associative substitution mechanism, the entering group Y binds to the complex before departure of the other ligand. There is therefore in this first step an intermediary complex in which the cation M has an increased coordination number (Fig.2.3-2b) This intermediate is also observable by analytical techniques. The X leaving group separates from the complex only in the second step. As the rate constant of this mechanism depends on the concentration of both the entering and the leaving group, it is a bimolecular nucleophilic substitution, S_N2. Some examples are explained further on.

Finally, in the interchange substitution mechanism there is no intermediary complex in which the metal M has either an increased or decreased coordination number. The reaction thus proceeds in only one step and this is a S_N2 substitution. (Fig.2.3-2c)

Various factors can influence the reaction mechanisms leading to the exchange of ligands. Steric strain on the reaction center, for instance, inhibits associative reactions in favor of the dissociative ones. Chelating ligands also have an importanteffect as they are polydentate and can therefore bind to a metal atom by several bonds. The acetylacetonato ion $[CH_3COCHCOCH_3]^-$ is, for example, bidentate; if L is the chelating ligand, it can form a ring (LML) with the metal. The distance between the two binding points is the "bite distance" (Fig.2.3-3). Table 2.3-2 gives a list of the most important chelating ligands.

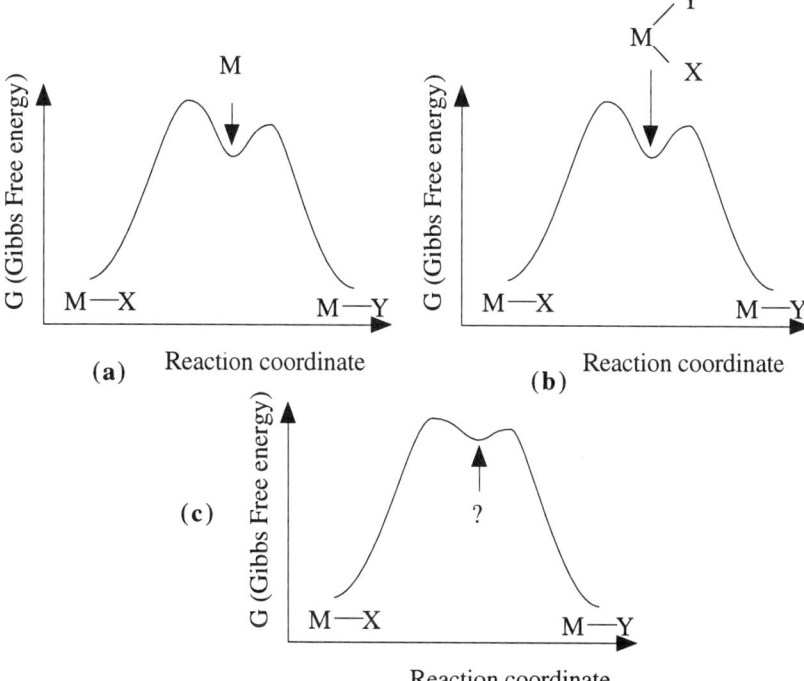

Figure 2.3-2 - Substitution mechanisms of a ligand Y for A ligand X. (a) dissociative S_N1 mechanism; (b) associative S_N2 mechanism; (c) Interchange S_N2 mechanism. After McMurry [11]

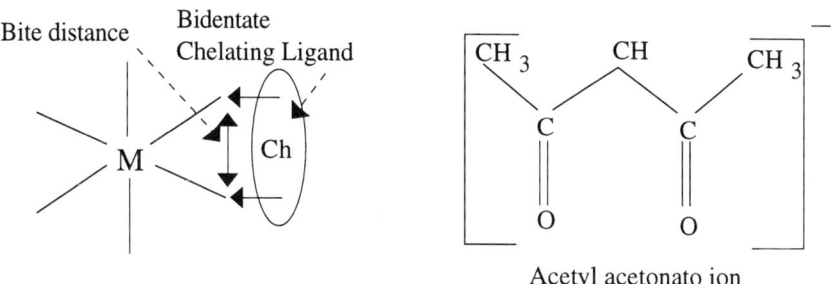

Acetyl acetonato ion

Figure 2.3-3 - Binding of a chelating ligand with a metal complex. An example of a bidentate chelating ligand: the acetylacetonato ion. After Shriver et al. [12].

Table 2.3-2 - Typical chelating ligands (2: bidentate, 3: tridentate, 4: tetradentate, 6: hexadentate). The coordinating atoms are in parenthesis. Adapted from Shriver and al. [12].

Name	Formula	abbreviation	Chelating type
Acetylacetonato	$[CH_3COCHCOCH_3]^-$	acac	2(O)
2,2-Bipyridine		bipy	2(N)
Diethylenetriamine	$NH(C_2H_4NH_2)_2$	dien	3(N)
Ethylenediamine	$H_2NCH_2CH_2NH_2$	en	2(N)
Ethylacetoacetate	$CH_3COCH_2COCH_2CH_3$	etac	2(O)
Diethylenediamine-tetraacetato		EDTA	6(N,O)
Glycinato	$N[NH_2CH_2CO_2]^-$	gly	2(N,O)
Maleonitriledithiolato		mnt	2(S)
Nitrilotriaceto	$N(CH_2CO_2^-)_3$	nta	4(N,O)
Oxalato	$C_2O_4^{2-}$	ox	2(O)
Tetraazacylclotetradecane		cyclam	4(N)
2,2'2"-Triaminotriethylamine	$N(C_2H_4NH_2)_3$	trien	4(N)

2.4 - METAL SALTS SOLUTIONS

In sol-gel processing, when metal salts are used they are often dissolved in an aqueous medium. The metal salt MX dissociates into ions which disperses in the solution, and the anions' negative charge X^{z-} balances the positive charge of the metal atom M^{z+}. The cation and the anion then have the same absolute formal charge z. These anions are sometimes considered as impurities; they must, for example, be eliminated in order to produce pure oxide ceramics. However, they can also be invaluable in channeling the chemical transformations within the solution. In any case, the ions firsts solvate with the water molecules, a reaction due to the polar nature of the solvent.

IONS SOLVATATION

Since water has a dipolar moment, the positive charge z+ of a cation attracts the partial negative charge, that is the oxygen atom, of H_2O molecules ($\delta(O) < 0$). As a consequence, the cation is entrapped by a number N of water molecules which constitutes, since they are the first neighbors, the first solvatation shell. (Fig 2.4-1) This shell is tightly bonded to the metal cation M^{z+} so that the chemical formula of the complex formed by the solvated ion is $[M(H_2O)_N]^{z+}$. The value of N is fixed for a given type of metal; its value often ranges from 4 to 8 and is frequently equal to 6, such as in $[Al(H_2O)_6]^{3+}$. Water also solvates the proton and section 2.2 reports that the most frequent value of N for H^+ is 4, hence the complex $[H(H_2O)_4]^+$ also written $[H_9O_4]^+$.

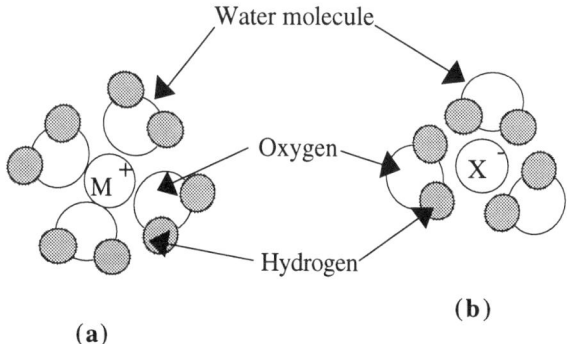

Water molecule

Oxygen

M $^+$

X

Hydrogen

(a)

(b)

Figure 2.4-1 - Solvatation of: (a) a cation and (b) an anion.

There is also a second shell of water molecules beyond this first solvatation shell. In it the oxygen atoms are turned toward the hydrogen atoms of the first shell. However this second shell is much less rigid than the first one and, therefore, does not have to be taken into consideration in the chemical evolution of the precursor. Water also solvates anions. In this case the solvent's molecules are turned the other way around with the hydrogen atoms oriented toward the anion (Fig 2.4-1). Nevertheless, the solvatation of anions is not as important as the one of cation that is, in fact, responsible for the hydrolysis of precursors, a complex chemical transformation leading to the formation of oxides.

HYDROLYSIS

Hydrolysis is the deprotonation of a solvated metal cation; it consists in the loss of a proton by one or more of the water molecules that surrounds the metal M in the first solvatation shell [13]. As a consequence, the aquo ligand molecule, H_2O that is bonded to the metal is either transformed into an hydroxo ligand, OH^-, if only one proton leaves, or into an oxo ligand, O^{2-}, if two protons detaches. [14]

The formation of hydroxo ligands

There is formation of an hydroxo ligand when the solvated metal is an acid and when water therefore acts as a Lewis base [13]. This corresponds to the following reaction:

$$[M(OH_2)_N]^{z+} + H_2O \Leftrightarrow [M(OH)(OH_2)_{N-1}]^{(z-1)+} + H_3O^+ \qquad K_{11} \qquad (2.4-1)$$
$$\text{acid} + \text{Lewis base} \Leftrightarrow \text{conjugated base} + \text{conjugated acid}$$

K_{11} is the equilibrium constant of the first deprotonation reaction of a complex involving only one metal atom. This complex can undergo other successive deprotonations; hence the following global reaction for h consecutive loss of protons.

$$[M(OH_2)_N]^{z+} + h\,H_2O \Leftrightarrow [M(OH)_h(OH_2)_{N-h}]^{(z-h)+} + h\,H_3O^+ \qquad (2.4-2)$$

Here, in $[M(OH_2)_N]^{z+}$ all the ligands are water molecules, it is therefore the most acidic form of the metal complex. On the contrary, $[M(OH)_h(OH2)_{N-h}]^{(z-h)+}$ is the most basic form of M; it is also an "aquo-hydroxo" complex since it contains both aquo (H_2O) and hydroxo (OH^-) ligands.

If the metal has an acidic oxide, the following equivalent form of the deprotonation reaction (2.4-1) explains the formation of hydroxo ligands by addition of a base to the solution.

$$[M(OH_2)_N]^{z+} + \quad OH^- \quad \Leftrightarrow [M(OH)(OH_2)_{N-1}]^{(z-1)+} + H_2O \qquad (2.4-3)$$

This reaction has the following mechanism [5, 7] :

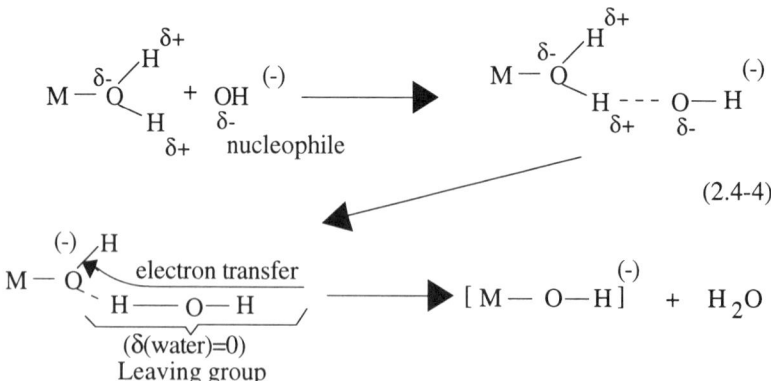

$$(2.4-4)$$

In this mechanism, a free OH^- nucleophilic ion attacks one of the hydrogen atom of one of the water molecule bounded to the metal M in the first solvatation shell. Since this hydrogen carries a positive partial charge ($\delta(H) > 0$), an electron charge transfer then occurs between the incoming OH^- ion and the original metal complex. Consequently, the partial charge, $\delta(H_2O)$, of the H_2O group composed of the incoming OH⁻ ion and the attacked H atom increases until it becomes null, $\delta(H_2O)$ = 0. At that moment, it leaves the metal complex as a water molecule. Such deprotonation reactions occur as long as, for an H_2O group, $\delta(O)_{\text{free water}}$ < $\delta(O)_{\text{complex}}$ < 0 , in which case they can be written as [5]:

$$[M(OH)_z(OH_2)_{N-z}] + H_2O \Leftrightarrow [M(OH)_{z-1}(OH_2)_{N-z-1}]^- + H_3O^+ \quad (2.4\text{-}5)$$

As indicated, the H^+ ion is solvated by water.

If the metals that have acidic oxides form hydroxo ligands, those which have basic oxides are characterized by the formation of oxo ligands, O^{2-}. For such metal, an hydroxo ligand can still be produced with an acid when a free H^+ ion attacks the nucleophilic oxygen of an oxo ligand [5,13]. In the case of water, the reaction is the following one:

$$[MO(OH_2)_{N-1}]^{(z-2)+} + H_2O \Leftrightarrow [M(OH)(OH_2)_{N-1}]^{(z-1)+} + OH^- \quad (2.4\text{-}6)$$
$$H^+ + OH^-$$

The complex reacts with an acid in a similar reaction:

$$[MO(OH_2)_{N-1}]^{(z-2)+} + H_3O^+ \Leftrightarrow [M(OH)(OH_2)_{N-1}]^{(z-1)+} + H_2O \quad (2.4\text{-}7)$$

The corresponding mechanism is then:

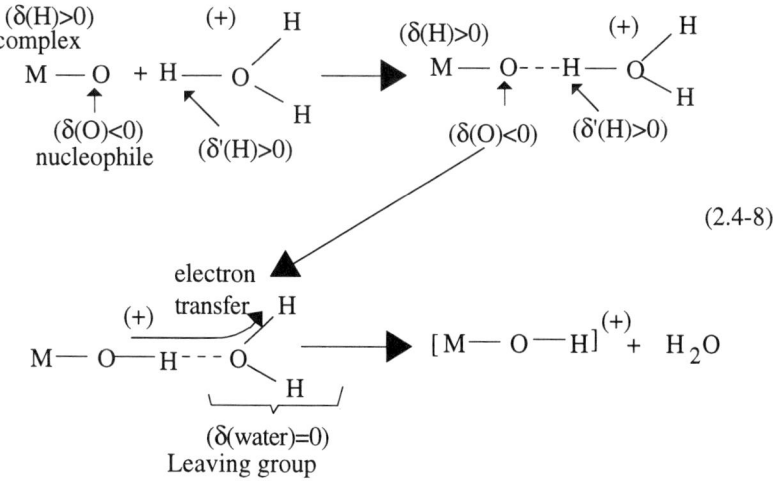

$$(2.4\text{-}8)$$

As before, this reaction continues as long as $0 < \delta(H)_{\text{complex}} < \delta'(H)_{\text{water}}$, [5].

Formation of oxo ligands
As mentioned previously, an oxo ligand is an O^{2-} anion bonded to a metal M within a complex. It is formed by the deprotonation of an hydroxo ligand according to the acid-base reaction [7]:

$$[M(OH)(OH_2)_{N-1}]^{(z-1)+} + H_2O \Leftrightarrow [MO(OH_2)_{N-1}]^{(z-2)+} + H_3O^+ \qquad (2.4-9)$$

| Acid | + Lewis base | $\Leftrightarrow$ conjugated base | + conjugated acid |

The product obtained, $[MO(OH_2)_{N-1}]^{(z-2)+}$, is an "aquo-oxo" complex since it contains both water and oxo ligands. Nevertheless, "oxo-hydroxo," and "oxo-hydroxo-aquo" complexes, which chemical formula is $[MO_x(OH)_y(OH_2)_{N-x-y}]^{(z-y-2x)+}$, also exist. This is, for example, the case of the vanadium complex $[VO(OH)_2(OH_2)_3]^+$ [5].

Application of the partial charge model to the hydrolysis of cations
According to the partial charge model, the electronegativity of any complex can be calculated from its formal charge and from the Allred and Rochow electronegativity of each atom present in the complex. The electronegativity of an aquo-hydroxo complex C= $[M(OH)_h(OH_2)_{N-h}]^{(z-h)+}$ is, for instance [5]:

$$\chi(C) \quad = \quad \frac{\sqrt{\chi_M^o}+N\sqrt{\chi_O^o}+(2N-h)\sqrt{\chi_H^o}+1.36(z-h)}{\dfrac{1}{\sqrt{\chi_M^o}} + \dfrac{N}{\sqrt{\chi_O^o}} + \dfrac{2N-h}{\sqrt{\chi_H^o}}} \qquad (2.4-10)$$

If the electronegativity of this complex, $\chi(C)$, is equal to the one of water then

$$\chi(H^+) \quad = \quad \chi(H_2O) = 2.631 - 0.02 \text{ pH} = \chi(C) \qquad (2.3-19)$$

and we find the relation [5]

$$h = \left(\frac{1}{1+0.014\text{pH}}\right)\left(1.36z - N(0.236-0.038\text{pH}) - \frac{2.621-0.02\text{pH}-\chi_M^o}{\sqrt{\chi_M^o}}\right) \qquad (2.4\text{-}11)$$

For instance:

$$\text{at pH= 0};\ \ h = 1.36z - 0.24N - \frac{2.621-\chi_M^o}{\sqrt{\chi_M^o}} \qquad (2.4-12)$$

$$\text{at pH = 14};\ \ h = 1.14z + 0.25N - \frac{0.836\,(2.341-\chi_M^o)}{\sqrt{\chi_M^o}} \qquad (2.4-13)$$

From this relation, we can estimate the number h of hydroxo ligands present in an aquo-hydroxo complex formed by a metal M. Likewise, other corresponding expressions exist for all types of complexes involving oxo ligands. Table 2.4-1 reports those most acidic and most basic forms of some metal cations as calculated by the partial charge model.

The partial charge model also gives the basis for the experimental determination of the domains in which pure aquo complexes, $[M(H_2O)_N]^{z+}$, and pure oxo anions, $[MO_m]^{(2m-z)-}$, are respectively formed. Those domains, which are represented in Figure 2.4-2, are function both of the formal charge z of the cation M and of the pH of the solution. This diagram reveals two extreme types of cations. Those with a

Table 2.4-1 - Most acidic and most basic complexes for some metal with formula $[MO_NH_{2N-h}]^{(z-h)+}$, together with the experimental and calculated value of h. The number h of hydroxo ligands is estimated by the partial charge model. Adapted from Jolivet [5].

M	z	Observed complexes	(h value)	N	calculated h
Mn	7	most acidic complex			
		$MnO_3(OH)$	(h=7)	4	7.8
		most basic complex			
		MnO_4^-	(h=8)	4	8.5
Si	4	most acidic complex			
		$Si(OH)_3(OH_2)^+$	(h=3)	4	3.8
		most basic complex			
		$SiO_2(OH)_2^{2-}$	(h=6)	4	5.1
Fe	3	most acidic complex			
		$Fe(OH)(OH_2)_5^{2+}$	(h=2)	6	1.9
		most basic complex			
		$Fe(OH)_4^-$	(h=4)	4	4
Li	1	most acidic complex			
		$Li(OH_2)_4^+$	(h=0)	4	-1.2
		most basic complex			
		$Li(OH)(OH_2)_3$	(h=1)	4	0.9

low formal charge such as z=1 form, in solution, very basic oxides and pure aquo cations, this is the case of Li^+ in $[Li(OH_2)_4]^+$. On the other hand, those with a high formal charge such as z=6 produce very acidic oxides and pure oxo anions. Sulfur, for instance, forms the acidic oxide SO_3 and the oxo anion SO_4^{2-}.

It is the electrostatic character of the hydrolysis mechanism as proposed by the partial charge model that justifies the correlation between the first deprotonation constant K_{11} of a cation M, its formal charge z, and the interatomic distance d between the metal and the oxygen atoms, M-O. This relationship, illustrated in Figure 2.4-3, obey to the following equation:[13]

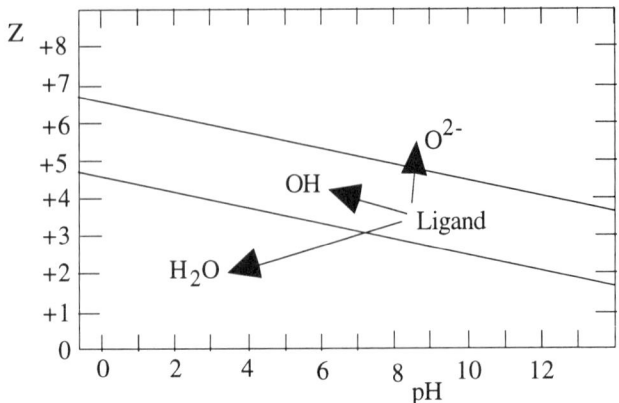

Figure 2.4-2 - Domains of formation of pure aquo or oxo ligands in function of the formal charge z of the cation and of the pH of the solution. Adapted from Jorgensen and Salem [3]

$$\log K_{11} = A + 11.0 \, \frac{z}{d} \qquad (2.4\text{-}14)$$

A, the constant of this relation, varies in function of the type of metal. In column I and II, the metals are the most resistant to deprotonation and hydrolysis, the constant has therefore a very low value; A=-22. The transition metals such as Cu^{2+} are slightly less resistant to hydrolysis and A=-20. As for the post transition cations (e.g., Bi^{2+}), since they undergo extensive hydrolysis, A has a higher value.

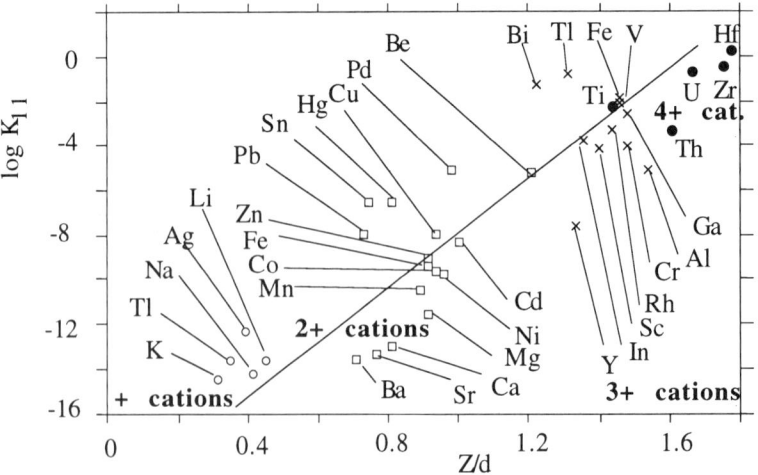

Figure 2.4-3 - Relation between $\log K_{11}$ and the ratio charge to M-O distance, $\frac{z}{d}$ for various cations. After Baes and Mesmer [13].

CONDENSATION   OF   CATIONS   IN   SOLUTION. POLYMERIZATION

Two mononuclear complexes of M, each comprising only one metal atom M, can react with one another in a polymerization reaction in order to form a polynuclear complex consisting of two metal atoms. Such a reaction, also called condensation, can, depending on the metal and if the conditions are right, keep on going so as to produce bigger polynuclear species.

Polymerization generally occurs if at least one hydroxo ligand (OH) is bounded to the cation M. This hydroxo ligand belongs either to an aquo-hydroxo complex of the type $[M\text{-}(OH)(H_2O)_{N-1}]^{(z-1)+}$, or to an oxo-hydroxo complex, $[M\text{-}(OH)(O)_{N-1}]^{(z-2N+1)+}$. We often simply write it as M-OH. As indicated in equations (2.4-1) and (2.4-6), those OH ligands are obtained by addition to the solution either of a base, in the case of metals forming acidic oxides, or of an acid, for metals forming basic oxides. The condensation occurring afterwards is responsible for the formation of one of two types of bridges between the two metal atoms.

Condensation by olation

The first step of any condensation reaction always include the construction of an "ol" bridge in which an hydroxo ligand is caught between the two metal atoms [5]. For the low charge cations, this is done through a dissociative S_N1 mechanism:

$$H_2O\text{-}M\text{-} \Leftrightarrow \text{-}M\text{-} + H_2O \qquad (2.4\text{-}15a)$$
$$\text{then}\quad \text{-}M\text{-}OH + \text{-}M\text{-} \Leftrightarrow \text{-}M\text{-}OH\text{-}M\text{-} \qquad (2.4\text{-}15b)$$

A nucleophilic addition reaction (A_N) is also possible when the coordination number of the metal can be increased as for $[Al(OH)_4]^-$. In this case

$$\text{-}M\text{-}OH + \text{-}M\text{-} \ OH \Leftrightarrow \text{-}M\text{-}OH\text{-}M\text{-}OH \qquad (2.4\text{-}16)$$

As for the transition elements, the mechanism is, as detailed in (2.4-17), an associative S_N2.

$$(2.4\text{-}17)$$

The condensation of the solution continues until a complex $[M(OH)_h(OH_2)_{N-h}]^{(z-h)+}$ in which $h<z$ cannot undertake anymore condensation, or until, as a result of unlimited polymerization of $[M(OH)_h(OH_2)_{N-h}]^{(z-h)+}$, a precipitate appears. Table 2.4-2 gathers the different types of "ol" bridges that exist.

Table 2.4-2 - Different Types of "ol" bridges. After Livage and al. [7].

Type of "ol" bridges	General formula	examples
μ_2-OH	M-(OH)-M	$[M_2(OH)(OH_2)_x]^{3+}$ with M^{2+} = Be, Mn, Co, Ni, Zn, Cd or Pb
$2\mu_2$-OH		$[Zr_4(OH)_8(OH_2)_{16}]^{8+}$ $[Al_{13}O_4(OH)_{24}(OH_2)_{12}]^{7+}$ $[M_4(OH)_6(NH_3)_{12}]^{6+}$ with M^{3+} = Cr, Co
μ_3-OH		$[M_4(OH)_4(OH_2)_4]^{4+}$ with M^{2+}=Pb,Co,Ni, Cd $[Pb_6O(OH)_6(OH_2)_4]^{4+}$ $[M_3(OH)_4(OH_2)_N]^{2+}$ with M^{2+} = Sn, Pb $[M_3(OH)_5(OH_2)_N]^{4+}$ with M^{3+} = Sc,Y,La,Ce
$3\mu_2$-OH		$[Co_2(OH)_3(NH_3)_6]^{3+}$.

Condensation by oxolation [5]

If, as shown in 2.4-18, there is condensation by oxolation, an "ol" bridge is first established between the two metal atoms before transforming into an "oxo" bridge in an S_N2 mechanism. In the intermediate complex of this mechanism, the maximum coordination number of the metal M is satisfied. As a consequence, both acids and bases can catalyze this reaction.

During the basic catalysis of the formation of oxo bridges, reaction (2.4-19), an OH$^-$ anion attacks the partially charged hydrogen atom, $H^{\delta+}$, of an hydroxo ligand belonging to the metal complex. This increases the negative partial charge of the oxygen atom of this ligand that consequently, becomes more nucleophilic and binds to another solvated hydroxyl group. Two water molecules, together with an OH$^-$ anion, then separates from the complex which is now composed of an "oxo" bridge. Similarly, during the acidic catalysis of condensation, reaction (2.4-20), an H$^+$ cation attacks the oxygen atom of an hydroxo ligand belonging to the metal complex.

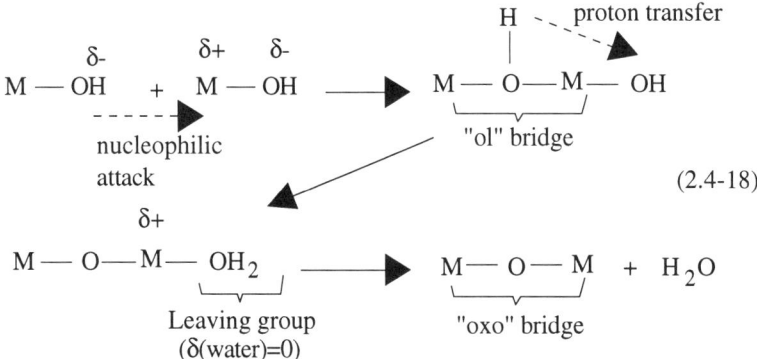

$$(2.4\text{-}18)$$

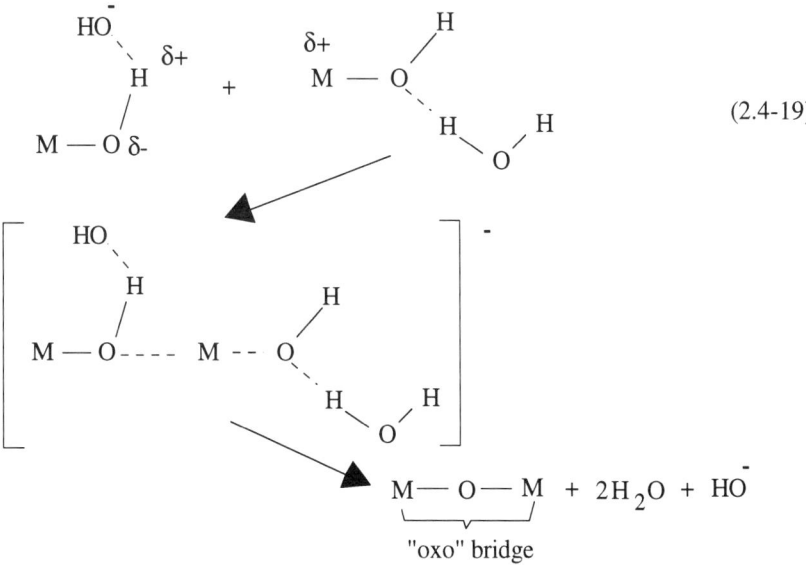

$$(2.4\text{-}19)$$

Hence, the newly formed corresponding H_2O ligand gains a positive partial charge, δ^+, and moves toward another OH ligand, thus forming an intermediate complex . The H_3O^+ ion then leaves while the complex is now formed of an "oxo" bridge.

A direct nucleophilic addition as in reaction (2.4-21) is also possible when the coordination number of a metal can be increased as in $[VO_4]^{3-}$.

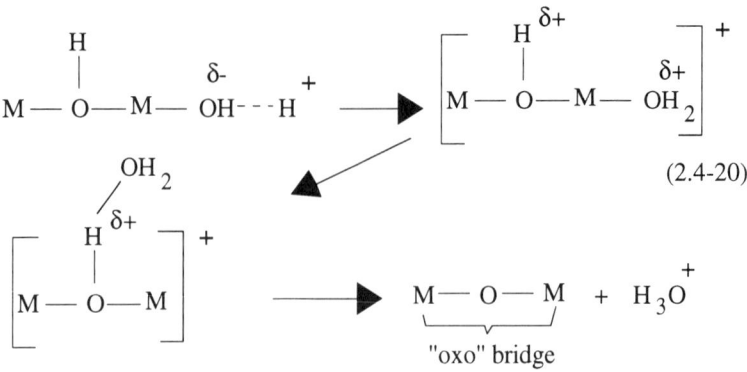

$$[-M-O]^- + -M-OH \Rightarrow [-M-O-M-OH]^- \qquad (2.4-21)$$

Condensation and the partial charge model

A nucleophilic substitution reaction requires three parts:
- An electron charge donor called the nucleophile. It is the atom or molecular group with the highest negative partial charge δ^- in absolute value
-An electron acceptor, called the electrophile. It is the metal cation M with a positive partial charge. It has been determined experimentally that we must have $\delta(M) > 0.3$.
-An easily separable leaving group; it is the ligand of the metal M with the highest partial charge δ^+.
Consequently, pure aquo species, that is those with formula $[M(OH_2)_N]^{z+}$, lack a nucleophile and hence cannot undergo any condensation reaction in water. Some examples are given in Table 2.4-3.
Similarly, in pure oxo species, that is those with formula $[MO_N]^{(2N-z)-}$, a strong π bond links the oxygen atom to the metal. Since no oxygen can leave, condensation cannot take place. Table 2.4-4 gives some examples of those pure oxo species.
For those complexes which contain hydroxo ligands, such as aquo-hydroxo, $[M(OH)_h(OH_2)_{N-h}]^{(z-h)+}$ and oxo-hydroxo, $[M(O)_{N-h}(OH)_h]^{(2N-h-z)-}$ complexes; since $\delta(OH) < 0$ the hydroxo ligand acts as a nucleophile. Consequently, since the aquo-hydroxo complexes also comprise a leaving group, a water molecule with $\delta(H_2O) > 0$, an "ol" bridge can form directly through an S_N1 mechanism (reaction 2.4-15) or, if the metal is a transition element, an S_N2 mechanism. In this case, the H_2O leaving group is very labile (i.e., not very stable.) For the oxo-hydroxo complexes, however, the H_2O leaving group is involved, as in reaction 2.4-17, in a proton transfer that transforms through an S_N2 mechanism the "ol" bridge linking the two metal atoms into an "oxo" one. These complexes cannot therefore undergo condensation.

Table 2.4-3 - Electronegativity and partial charges of some aquo species.
After Jolivet and al. [5]

$[Mn(OH_2)_6]^{2+}$	$\chi=2.657$	$\delta(Mn)=+0.59$	$\delta(H_2O)=+0.23$
$[Cr(OH_2)_6]^{3+}$	$\chi=2.756$	$\delta(Cr)=+0.68$	$\delta(H_2O)=+0.39$

Table 2.4-4 - Electronegativity and partial charges of some oxo species.
After Jolivet and al. [5].

$[MnO_4]^-$	$\chi=2.533$	$\delta(Mn)=+0.52$	$\delta(O)=-0.38$
$[CrO_4]^-$	$\chi=2.055$	$\delta(Cr)=+0.27$	$\delta(O)=-0.,57$

It has been determined experimentally that we must have $\delta(M) > 0.3$ for condensation to occur. In fact, even though $\chi=2.533$ and $\delta(O)=-0.48$ in $[PO_3(OH)]^{2-}$, $\delta(P)=+0.03$, the partial charge of the metal is too low and there is no polymerization. This additional condition applies in particular to silicon which undergoes condensation by oxolation. Table 2.4-5 reports a few mononuclear complexes of Si.

Table 2.4-5 - Partial charge model characteristics of a few mononuclear complexes of Si of the type $[SiO_4H_{8-h}]^{(4-h)+}$. After Jolivet and al. [5].

	Complex	χ	$\delta(OH)$	$\delta(Si)$
h=6	$[SiO_2(OH)_2]^{2-}$	2.10	−0.55	+0.20
h=5	$[SiO(OH)_3]^-$	2.37	−0.30	+0.35
h=4	$[Si(OH)_4]^0$	2,.8	−0.12	+0.47
h=3	$[Si(OH)_3(OH_2)]^+$	2.74	+0.03	+0.56

The silicon complex for h=6 exists only in strongly basic medium (pH=12) in which condensation is limited to the formation of dimers.

$$[SiO(OH)_3]^- + [SiO_2(OH)_2]^{2-} \Rightarrow [Si_2O_4(OH)_3]^{3-} + H_2O \qquad (2.4-22)$$

In this dimer $\chi=2.183$, $\delta(OH)=-0.22$, and $\delta(Si)=+0.25<0.3$. The partial charge of the metal dropped too low for condensation to continue.
The h= 5 complex of Si can also make the following dimer by condensation:

$$2[SiO(OH)_3]^- \Rightarrow [Si_2O_3(OH)_4]^{2-} + H_2O \qquad (2.4-23)$$

In this dimer $\chi = 2.346$, $\delta(OH) = -0.33$, and $\delta(Si) = +0.34 > 0.3$. In this case, the partial charge of Si is high enough for condensation to keep on going.

COMPLEXATION BY ANIONS

The anions brought to the solution either by the metal salt or by other constituents are Lewis bases like water and the other organic additives. They are therefore in competition with the hydroxo, oxo, and aquo ligands in the formation of the metal complex. The stability and lifetime of the corresponding complex formed therefore depend on their nucleophilic strength. In some cases, such as for phosphate, chromate and sulfate, the complex is very stable and the anion stays a ligand to the metal. In the other most frequent cases, if the metal participates only temporarily to the complex, it does so for a long enough time to determine its structure.

Complexation by X^- anions and the partial charge model [5]

The complexation reaction of a solvated cation, $[M(OH_2)_N]^{z+}$, with a monodentate anionic group X is

$$[M-(H_2O)_N]^{z+} + X^- \Leftrightarrow [M-X(OH_2)_{N-1}]^{(z-1)+} + H_2O \qquad (2.4\text{-}24)$$
$$\text{equivalent to } [M-(OH_2)_{N-1}]^{z+}X^-$$

The metal complex containing the X anion forms only if the partial charge of X is such that algebraically

$$\delta(X) > -1 \qquad (2.4\text{-}25)$$

Otherwise the anion X^- separates from the complex. This condition can be rewritten using electronegativities:

$$\chi(X^-) < \chi([M-(OH_2)_{N-1}]^{z+}) \qquad (2.4\text{-}26)$$

For a molecular group HX, the complexation reaction is slightly different:

$$[M-(OH)(H_2O)_{N-2}]^{(z-1)+} + HX \Leftrightarrow [M-X(OH_2)_{N-1}]^{(z-1)+} \qquad (2.4\text{-}27)$$
$$\text{equivalent to } [M-(OH)(H_2O)_{N-2}]^{(z-1)+}HX$$

In this case the product will form only if

$$\delta(HX) < 0 \qquad (2.4\text{-}28)$$

Otherwise the HX molecular group leaves the complex. This is explained by rewriting the requirement using electronegativities. For HX to remain attached to the metal M, the electronegativities must be so that

$$\chi(HX) > \chi([M-(OH)(H_2O)_{N-2}]^{(z-1)+}) \qquad (2.4\text{-}29)$$

The molecular group leaves when it cannot keep its excess negative charge that enables it to bind to the metal M.

Example: complexation of $[Fe(OH)_2(OH_2)_4]^+$ by bidentate anions [5]
A bidentate anion such as acetate, CH_3COO^-, can replace two of the water molecules of the iron complex $[Fe(OH)_2(OH_2)_4]^+$; it is therefore a chelating agent of the metal cation. Such anions are key factors in reducing the hydrolysis of the metal to the point of sometimes constraining the complex to a specific coordination and therefore to a particular solid phase. If you consider an X^- bidentate anion, the complexation reaction equivalent to (2.4-24) is:

$$[Fe(OH)_2(H_2O)_4]^+ + X^- \Leftrightarrow [FeX(OH)_2(OH_2)_2]^0 \qquad + 2\ H_2O \qquad (2.4\text{-}30)$$
$$\text{equivalent to } [Fe\text{-}(OH)_2(OH_2)_2]^+ X^-$$

As mentioned previously, X^- separates from the metal complex when

$$\chi(X^-) > \chi([Fe\text{-}(OH)_2(OH_2)_2]^+) \equiv \chi_X \qquad (2.4\text{-}31)$$

At the exact moment of separation

$$\chi(X^-) = \chi([Fe\text{-}(OH)_2(OH_2)_2]^+) = 2.68 = \chi_X \qquad (2.4\text{-}32)$$

Similarly, if you consider a molecular group HX, the complexation reaction equivalent to (2.4-27) is :

$$[Fe\text{-}(OH)_3(H_2O)]^0 + HX \Leftrightarrow [FeX(OH)_2(OH_2)_2]^0 \qquad (2.4\text{-}33)$$
$$\text{equivalent to } [Fe\text{-}(OH)_3(H_2O)]^0 HX$$

and the molecular group HX leaves the complex when

$$\chi(HX) < \chi([Fe\text{-}(OH)_3(OH_2)]^0) \equiv \chi_{HX} \qquad (2.4\text{-}34)$$

As for the X^- anion, at the instant of separation the electronegativities of the molecular group and of the complex are equal.

$$\chi(HX) = \chi([Fe\text{-}(OH)_3(OH_2)]^0) = 2.53 = \chi_{HX} \qquad (2.4\text{-}35)$$

The complexation of $[Fe(OH)_h(OH_2)_{4-h}]^{(3-h)+}$ with h=2 by a bidentate anion X^- such as HCO_3^- implies some conditions. They are summarized in Figure 2.4-4 together with the electronegativities of HCO_3 (χ_X) and H_2CO_3 (χ_{HX}). As indicated, the requirements for X^- and HX to remain in the complex are fulfilled so that there is formation of $[Fe(HCO_3)(OH)_2(OH_2)_2]^0$.
The pH of a solution alters the electronegativities of its species. Thus, instead of comparing the electronegativities of $\chi(X^-)$ and $\chi(HX)$ with those of the metal complex, χ_X and χ_{HX}, it is easier to determine the pH range in which complexation occurs and if the electrolyte HX can in fact reach this pH. The pH range corresponding to a complexation with H_2CO_3 is, for instance,

$pH_X = 1.5 < pH < 5.8 = pH_{HX}$. Table 2.4-6 reports the partial charge and electronegativities of different X^- anionic groups.

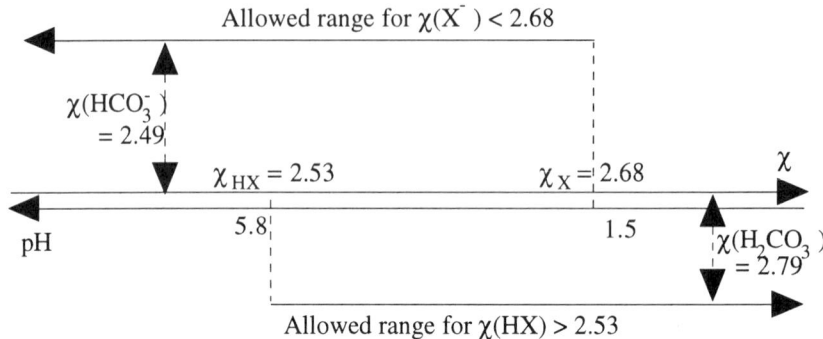

Figure 2.4-4 - Necessary conditions for the complexation of $[Fe(OH)_2(OH_2)_4]^+$ by $X^- = HCO_3^-$ as implied by the partial charge model.

Table 2.4-6 - X^- and HX electronegativities of different bidentate anionic groups together with the partial charge of X and HX in $[FeX(OH)_2(OH_2)_2]^0$. After Jolivet and al. [5].

X^-	CH_3COO^-	$H_2PO_4^-$	HCO_3^-	HSO_4^-	NO_3^-	ClO_4^-
$\chi(X^-)$	2.24	2.49	2.49	2.64	2.76	2.86
$\chi(HX)$	2.49	2.71	2.79	2.88	3.08	3.10
$\delta(X)$	-0.14	-0.64	-0.71	-0.94	-1.10	-1.26
$\delta(HX)$	+0.07	-0.38	-0.45	-0.65	-0.80	-0.94

According to this table, no complex of the type $[FeX(OH)_2(OH_2)_2]^0$ can form with the acetate ion ($X=CH_3COO^-$) and this because $\chi(HX) < 2.53 = \chi_{HX}$. Would there be a complex, $\delta(HX) > 0$, and the molecular group HX (CH_3COOH) would leave. Similarly, no such complex can form with the anions NO_3^- and ClO_4^- since $\chi(X) > 2.68 = \chi_X$. In this case, $\delta(X) < -1$ in the complex and it is the anion X^- which departs instead of the molecular group HX. Only the $H_2PO_4^-$, HCO_3^- and HSO_4^- anions can form stable complexes with Fe for h=2.
Similar calculations and reasoning can be made for all iron complexes of the type $[FeX(OH)_h(OH_2)_{4-h}]^{(3-h-1)+}$ thus determining, for the different value of h, the anions that can indeed stay a ligand to the metal.

Overall complexation of a metal M by anions.

By repeating the derivations previously done in order to determine the complexation of a bidentate monovalent anion we can predict the complex formed when only one X^{n-} anion replaces α water molecule in its reaction with a solvated metal cation of the type $[M(OH_2)_N]^{z+}$. In this overall complexation, the general formula of the final metal complex is $[MX(OH)_h(OH_2)_{N-\alpha-h}]^{(z-n-h)+}$. In order to determine it, it is necessary to compare the electronegativities both of $[XH_q]^{(n-q)-}$ with $[M(OH)_{h+q}(OH_2)_{N-h-\alpha-q}]^{(z-h-q)+}=\chi_{HqX}$ and of X^{n-} with $[M(OH)_h(OH_2)_{N-h}]^{(z-h)+}=\chi_x$ so as to respectively establish if either $[XH_q]^{(n-q)-}$ or X^{n-} leaves the complex. Similar reasoning and calculations can be carried out for all metal complexes, even if the metal is a polynuclear specie.

For example, Figure 2.4-5 show the results obtained for Ti. Those results fully agree with experimental data. Matijevic [16], for instance, reported that some anions such as SO_4^{2-} and PO_4^{3-} favors the formation of big polymeric metal complexes and, hence, of amorphous materials. On the other hand, Cl^-, NO_3^-, and ClO_4^- (together with Al) favors small polymers with more crystalline structures. Since the complexes they form are the "building blocks" of the solid, the anions are, in fact, much responsible for the structure of a sol-gel.

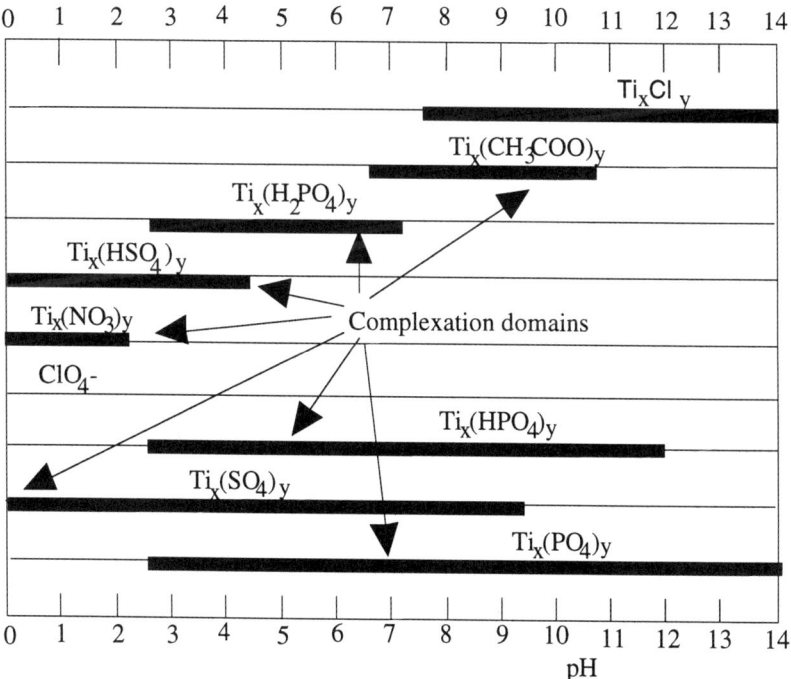

Fig. 2.4-5 - pH domains in which titanium complexes with various anions. Adapted from Livage and al. [15].

If those anions offer a large diversity in the structure of sol-gel materials, they can also present some inconveniences. Nitrate, for example, usually crystallizes during drying and consequently destroys the material's homogeneity achieved during sol-gel processing [17]. Furthermore, after hydrolysis, anions usually remain as impurities in the form of acids, such as HNO_3 or HCl [18]. The material must therefore be purified more or less efficiently by one of the following techniques.

(a) Addition of a water immiscible solution containing long-chain amines. The amines retain the hydrolytic acid while the anion to cation ratio is easily decreased down to 0.25.

(b) Extensive washing of the gelatinous precipitate followed by peptization. After re dispersion with nitric acid, the anion to cation ratio is close to 0.2. This method is specially used with Al and Ce [19].

(c) Thermal treatment, and especially for nitrates for which the solution is heated to 500°C [20]. This lowers the anion to cation ratio down to 0.05. A much smaller quantity of nitric acid is consequently necessary to re disperse the powder into a colloidal sol.

FORMATION OF A SOLID PHASE

Up to now, the transformations concerned complexes in solution. Yet, a limit of solubility exists between those soluble complexes and their corresponding solid phase. This limit corresponds to the equilibrium reached by a specific reaction. A simple example is the transformation of hydroxylated polynuclear complex into solid hydroxide.

$$xM(OH)_{z(s)} + (xz-y)H^+ \Leftrightarrow M_x(OH)_y^{(xz-y)+} + (xz-y)H_2O \qquad (2.4-36)$$

The corresponding equilibrium constant is:

$$K_{sxy} = \frac{[M_x(OH)_y^{(xz-y)+}]}{[H^+]^{xz-y}} \qquad (2.4-37)$$

When two or more polynuclear complexes with different x and y coefficients react to form a solid hydroxide, the total metal concentration at saturation is:

$$m_M^{satd} = \sum xQ_{sxy}[H^+]^{xz-y} \qquad (2.4-38)$$

This equation shows that, as was already described, the pH of the solution is an important factor in the formation of the solid. Yet it is not the only one as, as was seen previously, the nature of the anion together with the temperature and the nature and concentration of the metal cations are also significant parameters. In fact, the formation of metal complexes is often easier at higher temperatures.

Precipitation is mostly due to electrostatic interactions between molecules. This is particularly well illustrated by the good correlation between K_{11}, the constant of the first deprotonation reaction (2.4-1), and K_{s10}, the equilibrium constant of reaction (2.4-36) at saturation taken for x=1 and y=0. Those two reactions are the following:

$$M^{z+} + H_2O \Leftrightarrow [MOH]^{(z-1)+} + H^+ \qquad (2.4-39)$$

with equilibrium constant $K_{11} = \dfrac{[MOH]^{(z-1)+}[H]^+}{[M^{z+}]}$ (2.4-40)

and $M(OH)_{z(s)} + z\,H^+ \Leftrightarrow M^{z+} + z\,H_2O$ (2.4-41)

with the constant $K_{s10} = \dfrac{[M^{z+}]}{[H^+]^z}$ (2.4-42)

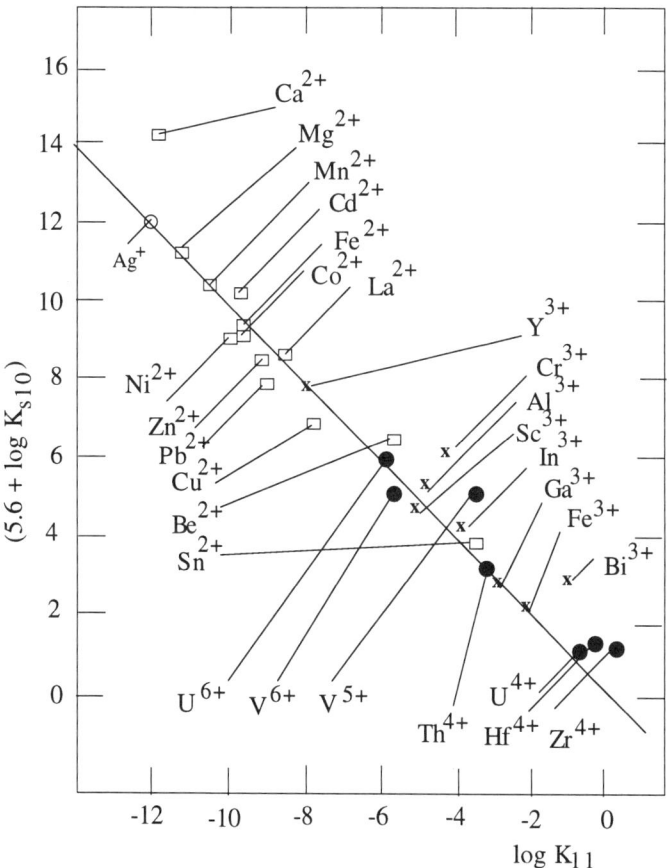

Figure 2.4-6 - Relation between the equilibrium constants K_{11} and K_{s10} of cations. The line corresponds to $(K_{11})^z K_{s10} = 10^{-5.6}$. Adapted from Baes and Mesmer [13].

The correlation between those two constants follows the relation

$$K = (K_{11})^z K_{s10} \approx 10^{-5.6}$$ (2.4-43)

So that the reaction corresponding to this new constant is:

$$M(OH)_{z(s)} + (z-1)M^{z+} \Leftrightarrow z[MOH]^{(z-1)+} \qquad (2.4-44)$$

From this reaction it appears necessary to decrease the concentration of $[M^{z+}]$ in order to increase the ratio of the concentrations of $[MOH^{(z-1)+}]$ to $[M^{z+}]$ while maintaining the solid phase in equilibrium. Thus, unless their concentration is already very low, the cations generally precipitate shortly after hydrolysis. Soluble hydrolyzed species form more favorably when the concentrations of the metals are low. According to equation (2.4-36), their total concentration at saturation, m_M^{satrd}, decreases as the pH increases until a minimum determined by the stability of each mononuclear complex is reached. However, when the pH becomes high enough, the anions participate in the formation of the complexes and consequently increase again their solubility. The ionic force of the solution also increases thus forming dimers. Those hydroxylated monomers and dimers favor crystal growth in specific structures so as to give crystalline ionic materials.

Since several soluble species generally react simultaneously to form a solid, it is often difficult to duplicate a given material with a specific structure. This difficulty often lies in the ability to add a base to the solution in an identical manner. As a matter of fact, the chemical reactions occurring around a basic drop changes as we go farther away from the drop. For instance, a drop of NaOH added to a solution of Fe(III) first deprotonates the aquo ligands of $[Fe(OH_2)_6]^{3+}$ into an hydroxo one, then, farther away, into an oxo one. This example, illustrated in Figure 2.4-7, corresponds to the following deprotonation reaction:

$$[Fe(OH)_i(OH_2)_{5-i}]^{(3-i)+} + OH^- \Leftrightarrow [Fe(OH)_{i+1}(OH_2)_{4-i}]^{(2-i)+} + H_2O \quad (2.4-45)$$

During polymerization, those mononuclear species bind to each other so as to form polynuclear complexes. Different types of polynuclear complexes can therefore be produced depending on the initial monomers. In order to predict which, or if any, complexes are formed, you must consider that deprotonation is a relatively fast reaction compared to the slower polymerization one. Furthermore, the deprotonation rate of the bigger polynuclear species is generally smaller than that of the monomers so the former often have a longer lifetime than the latter. Practically, depending on the manner in which mixing is carried out, and hence on the size of the basic drops and on the distance separating them, three extreme cases of polymerization are to be consider:

- (a) No polynuclear complexes are formed.
- (b) Polynuclear species are formed but in so small proportion that they remain in solution.
- (c) Polynuclear complexes are produced in large enough quantities to agglomerate into a solid.

Those generalizations justify why, in the previous example, the manner in which NaOH was added was much responsible for the formation of different kinds of iron products. Eventually, some of those will become polymeric gels while other will transform into more or less hydrated colloidal particles [22]. Schneider demonstrated that, for instance, if a base is added very slowly to a solution of Fe(III), sols that are stable for months are produced. If the base is added rapidly, however, an amorphous precipitate immediately forms [21]. The anions are, of course, as much important as the mixing technique; a precipitate is obtained more readily with SO_4^{2-} than with NO_3^- and ClO_4^- [23].

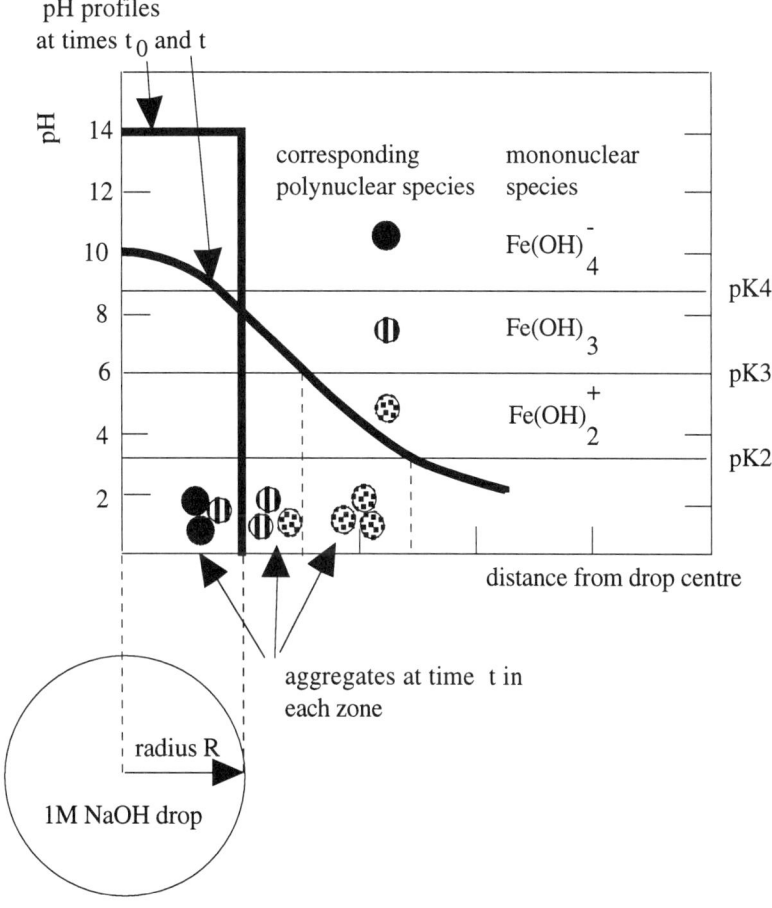

Figure 2.4-7 - Reactions occurring at increasing distance from an NaOH drop when added to an Fe(III) aqueous solution. Adapted from Schneider [21].

The limit of solubility of those polynuclear complexes also depends on the nature of the cation M and on the other complexing additives present in the solution. For example, when an aluminum nitrate salt is used, a buffer effect occurs during the titration of a base. The pH of the solution remains constant as the base is gradually added. (Figure 2.4-8) This buffer zone corresponds to the progressive formation of bigger aluminum polymers. When no more hydroxide can be formed, the pH rises steeply again.

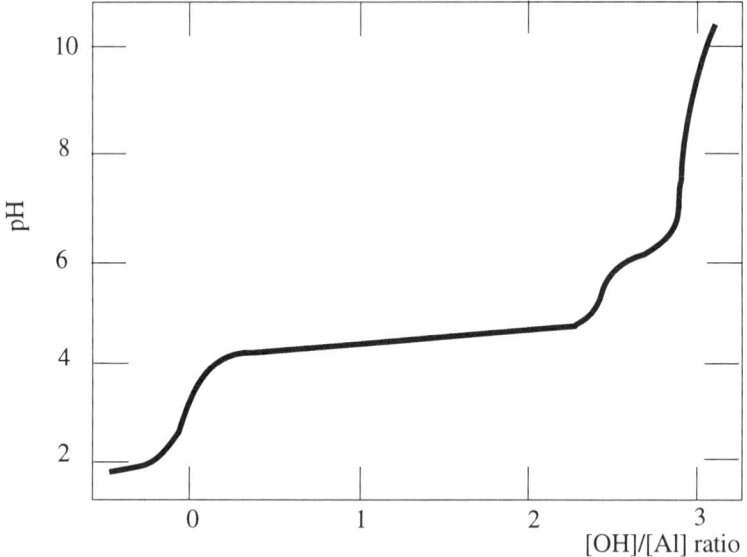

Figure 2.4-8 - NaOH titration of an Al(NO3)3 solution at a concentration of 0.07M . After Vermeulen et al. [24].

HYDROLYSIS OF CATIONS AS A FUNCTION OF THEIR NATURE

A study on the hydrolysis of cations was already published by Baes and Mesmer [13] and more recently by Livage et al. [7]. Hydrolysis is a complex technique that, depending on the conditions, gives rise to a great variety of colloidal structures ranging from metals to hydroxides and including oxides and oxihydroxides. Of those, the colloidal oxides are often so strongly solvated that the water molecules are tightly bonded to the complex and it is difficult to know the exact chemical formula of the particle. We therefore simply refer to them as "hydrous oxides" regardless of their actual structure [25]. The behavior of the cations in aqueous solutions can yet be summarized according to their nature and with respect of their final type of ligands as in Figure 2.4-9.

Cations with valence I

These cations have a low charge, a low electronegativity, and an oxidation number of I. They are not hydrolyzed in solution but remain as solvated cations of the form

$[M(H_2O)_N]^{z+}$. They form basic oxides which liberate OH^- anions when reacting with water. An example is Na.

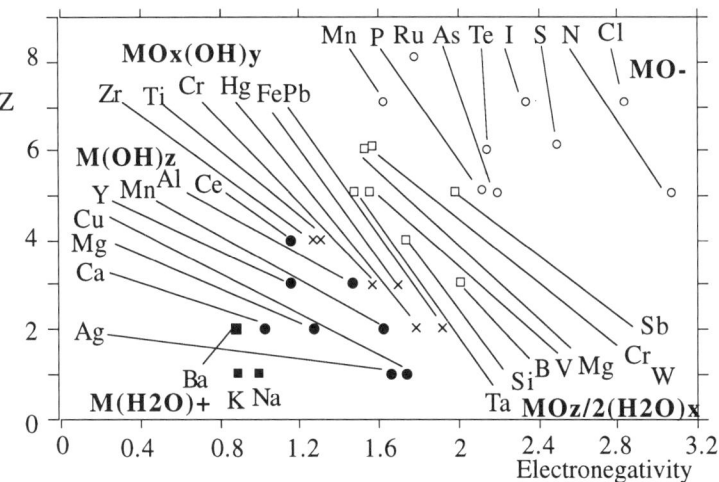

Figure 2.4-9 - Charge versus electronegativity. This relation shows the nature of the species formed by cations in aqueous solutions. Adapted from Henry and al. [8].

Cations with valence II
These cations have an oxidation number of II and a slightly higher charge and electronegativity than those with valence I. They therefore undergo condensation more extensively and precipitate in the form of hydroxides with formula $M(OH)_2$. In solution, most of those cations (Mn, Co and Ni) form compact tetramers that later transform into solid hydroxides with lamellar structures such as the one of Brucite, CdI_2. Cu creates linear polymers that crystallize as $Cu(OH)_2$ in a lamellar structure similar to that of boehmite, $AlO(OH)$.

Cations with valence III
Those cations with an oxidation number of III, such as Al, Fe, Cr, Sc, Y and the rare earth elements have a very rich aqueous chemistry. They easily polymerize in a variety of different polycations and constitute various solid phases, each more or less polymeric [25]. These phases are produced by a succession of olation and oxolation reactions which lead to the formation of several $[MO_x(OH)_h(H_2O)_{N-h-2x}]^{(z-h-2x)+}$ polynuclear complexes. Those complexes later form oxyhydroxide solids.
For example, we will discuss the considerably studied Al(III) salts and their hydroxides. In its corresponding solvated complex, $[Al(H_2O)_6]^{3+}$, the Al^{3+} cation binds to six water molecules [26]. This complex is such a strong acid that it even

dissolves metals [27]. Moreover, since its first deprotonation constant is low, $K_{11} = 1.12 \times 10^{-5}$, it is subject to extensive hydrolysis. Hence, the first polymers formed at pH>3 are dimers consisting of a double "ol" bridge. The mechanism of this dimerization condensation illustrated in (2.4-46) has for equilibrium constant $K_{22} = 1.12 \ 10^{-7}$.

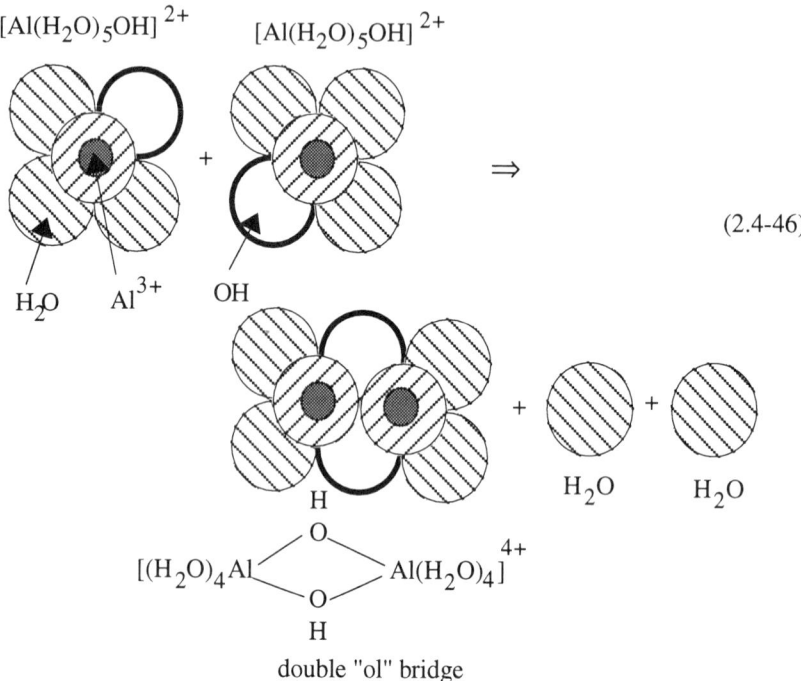

$$[Al(H_2O)_5OH]^{2+} \qquad [Al(H_2O)_5OH]^{2+}$$

(2.4-46)

$$H_2O \qquad Al^{3+} \qquad OH$$

$$+ \qquad H_2O \qquad H_2O$$

double "ol" bridge

This dimer is not detected by NMR technique while the corresponding trimer $[Al_3(OH)_4(H_2O)_9]^{5+}$ is. This trimer immediately loses one proton according to the following reaction :

$$[Al_3(OH)_4(H_2O)_9]^{5+} \Rightarrow [Al_3O(OH)_3(H_2O)_9]^{4+} + H^+ \qquad (2.4-47)$$

If it is in a basic environment, it then reacts to liberate three water molecules.

$$[Al_3O(OH)_3(H_2O)_9]^{4+} + 3 \ OH^- \Rightarrow$$
$$[Al_3O(OH)_3(O_2H_3)_3(H_2O)_3]^+ + 3 \ H_2O \qquad (2.4-48)$$

This reaction is necessary for the following polycondensation to occur. In this last step, a complex cation consisting of 13 Al atoms is formed; in it one Al occupies a tetrahedral site while all the others are in octahedral sites [26]. (Figure 2.4-10)

$$[Al(OH_2)_6]^{3+} + 4[Al_3O(OH)_3(O_2H_3)_3(H_2O)_3]^{+} \Rightarrow$$
$$[Al\{Al_3O(OH)_3(O_2H_3)_3(H_2O)_3\}_4]^{7+} + 6\,H_2O \qquad (2.4\text{-}49)$$

In this reaction, if a base is added to the solution or if the experimental conditions are such that the temperature is below 80°C, the tetrahedral site containing the Al disappears in order to form an octahedral arrangement of hexagonal rings. This new configuration is similar to that of trihydroxide bayerite, $Al(OH)_3$. On the other hand, if T > 80°C, a stable and big polycation composed of 13 Al atoms forms. It later transforms into a transparent layered gel with a structure similar to that of boehmite $AlO(OH)$ (figure 2.4-10).

In strongly basic medium, the aluminum is present in the solution as tetrahedral aluminate anions, $[Al(OH)_4]^{-}$. When the acid is added, it transforms into an aquo-hydroxo complex according to the following protonation reaction.

$$[Al(OH)_4]^{-} + H_3O^{+} \Rightarrow [Al(OH)_3(H_2O)]^{0} + H_2O \qquad (2.4\text{-}50)$$

Later on, as the hexameric intermediates rearrange into a trihydroxide $Al(OH)_3$ phase, this aquo-hydroxo complex transforms into the tetramer $[Al_4(OH)_{12}(H_2O)_5]^{0}$ thus founding the necessary nucleus for the growth of the boehmite gel.

Anions can, of course, modify significantly this intricate chemistry. Not only do acetate, like nitrate, chloride and sulfate, alter the chemical formula of the dimers [27, 28], but the rate of polymerization decreases from NO_3^{-} to Cl^{-} and SO_4^{2-}.

Contrary to Al, Cr and Fe do not produce polycations containing as much as 13 metal atoms. Their trimers only becomes tetramers. In the case of Cr(III), the $[Cr(H_2O)_6]^{3+}$ initially present in the aqueous solution slowly transform into first $[Cr_2(OH)_2(H_2O)_8]^{4+}$ then $[Cr_3(OH)_4(H_2O)_9]^{5+}$ and finally $[Cr_4(OH)_6(H_2O)_{10}]^{6+}$. When the base is then added, the blue hydroxide $Cr(OH)_3(H_2O)_3$ precipitates. In the last step, the oxo ligands substitutes the aqua ones in order to form the final tetramer, $[Cr_4O(OH)_5(H_2O)_{10}]^{5+}$. This last complex produces green gels with a global composition of $CrO(OH)$ [29]. With time, this chemical formula changes into Cr_2O_3.

In the case of Fe(III), the oxo ligands enter the complex in a similar manner as with Cr(III) and the precipitated gels obtained is called ferrihydrite. Its corresponding composition varies but is generally of the form $[Fe_pO_r(OH)_q]^{(3p-\,2r-q)+}$ [21]. During aging, these gels transform either into α-$FeO(OH)$, called goethite, or into α-Fe_2O_3, called hematite. When mixed Fe(II)/Fe(III) solutions are used, magnetite Fe_3O_4 gels with an inverse spinel type structure are produced.

In their metallic state with a valence of 0, the noble metals Ru, Rh, Pd, Pt, Ag and Au also precipitate as colloids [13].

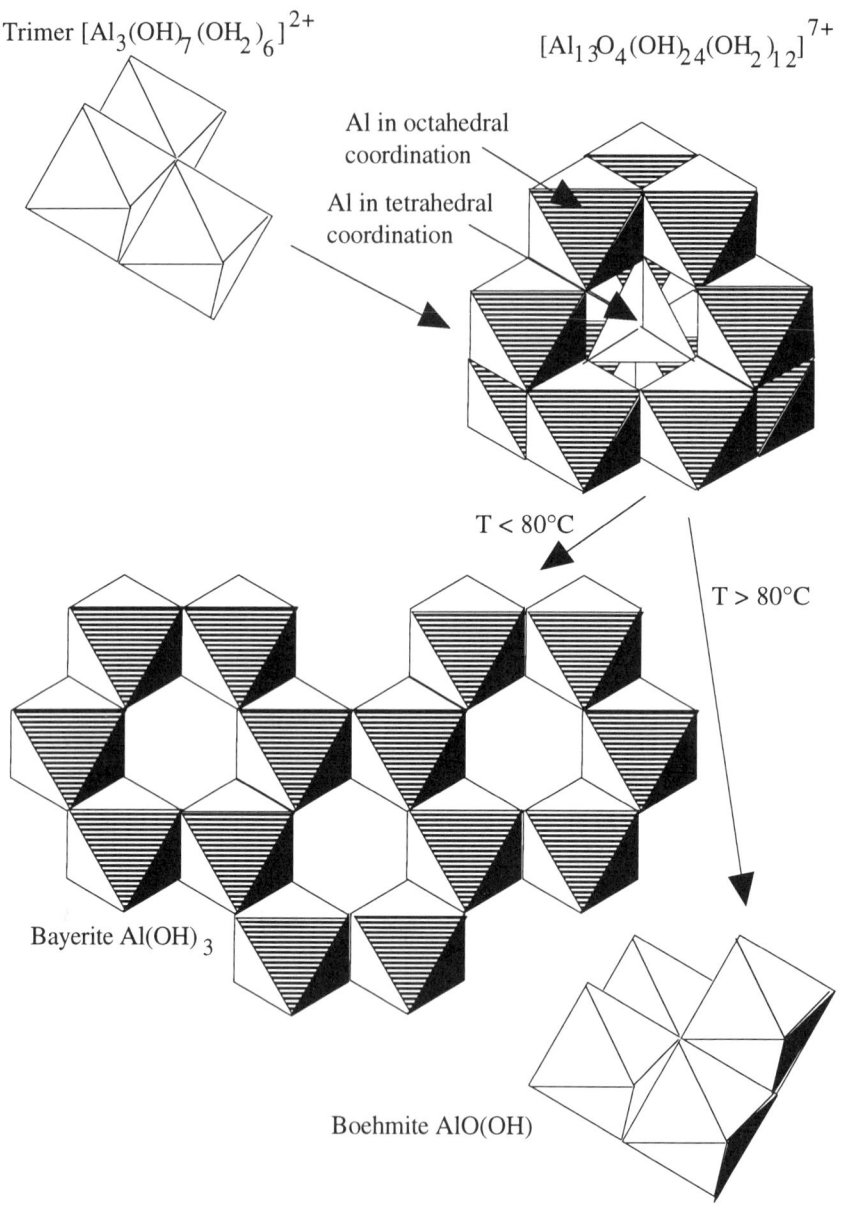

Trimer $[Al_3(OH)_7(OH_2)_6]^{2+}$

$[Al_{13}O_4(OH)_{24}(OH_2)_{12}]^{7+}$

Al in octahedral coordination

Al in tetrahedral coordination

$T < 80°C$

$T > 80°C$

Bayerite $Al(OH)_3$

Boehmite $AlO(OH)$

Figure 2.4-10 - Structure of complexes built from Al: trimer, polycation comprising 13 Al, bayerite nucleus, boehmite nucleus. Adapted from Henry and al. [8].

B(III), another trivalent element, behaves differently [5]. Contrary to the other ones, it has a high electronegativity and hence produces oxolated species of the type $[MO_x(H_2O)_{N-2x}]^{(z-2x)+}$ which later forms polynuclear molecules (Figure 2.4-9). Its final hydrated solid oxide has for general formula $MO_{z/2}.xH_2O$. Furthermore, boron presents two different monomers; the borate anion $[B(OH)_4]^-$ and the boric acid $B(OH)_3$ in which the metal respectively has a tetragonal and a trigonal coordination. Condensation in water can only occur if those two monomers are present and if the pH is such that 7<pH<11. The following reaction is a typical example of the corresponding condensation.

$$[B(OH)_4]^- + B(OH)_3 \Rightarrow [B_2O(OH)_5]^- + H_2O \qquad (2.4-51)$$

Dimers, trimers, as well as tetramers form only if those requisites are fulfilled. The polymers are then composed of rings consisting both of tetragonal and trigonal boron atoms. Note that at pH<7 only boric acid exists [5].

Cations with valence IV
Those elements, including Ti, Zr, and Hf, are soluble only in a very narrow pH range and build oxo-hydroxo complexes with general formula $MO_x(OH)_y$. (Figure 2.4-9) Their stability decreases as the size of the cation increases so that, for instance, Zr does not produce a hydroxide but an oxide [26]. Initially, a solution of Zr(IV) does form the stable tetramer $[Zr_4(OH)_8(H_2O)_{16}]^{8+}$. However, further condensation produces by slow oxolation reactions a white and gelatinous precipitate with formula $ZrO_{2-x}(OH)_{2x}.nH_2O$; n can of course vary. The experimental conditions then determine if either tetragonal or monoclinic ZrO_2 is finally obtained. As for Ti, its hydrolysis in a strongly acidic solution near the boiling point produces polymers consisting both of hydroxo and oxo bridges as in the octamer $[Ti_8O_8(OH)_{12}(H_2O)_x]^{4+}$ [30]. Complexes can also form with ligands other than OH. For example, sulfate produces with Ti the species $Ti(OH)_3(HSO_4)(H_2O)_2$ and $[Ti(OH)_2(HSO4)(H_2O)_2]^+$ (Figure 2.4-11). Similar bridges are formed with chlorine anions such as in $[Ti(OH)(H_2O)Cl]^{4+}$ and $[Ti(OH)_2Cl_4]^{2-}$ [31]. Further oxolation produces the rutile or anatase forms of TiO_2.

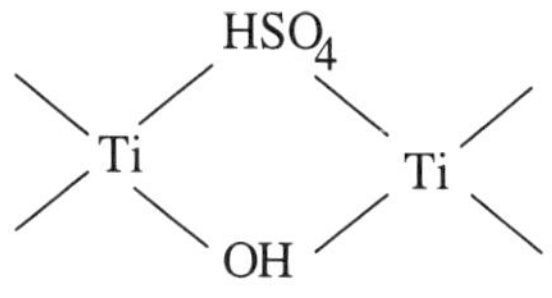

Figure 2.4-11 - Formation of mixed "ol-anion" bridges with Ti cations

Si, Ge, Sn and Pb react differently from the other cations with valence IV as they form more covalent bond with the oxo ligands [5,13]. They are soluble only in strongly basic solutions in which they build oxo or oxo-hydroxo anions. Furthermore, their oxides are acidic and their hydroxides $M(OH)_4$ unstable. They therefore spontaneously dehydrate in order to produce the hydrated oxide $MO_2.nH_2O$. Si, the most studied cation, is present in solution in the form of silicate obtained by dissolution of sodium metasilicate, Na_2SiO_3. This dissolution breaks the tetrahedron chains of SiO_4. Si presents many different monomers, of which the most important are reported in Table 2.4-5 [5]. Moreover, the other cations added to the solution also affects the nature of the polynuclear species obtained. For example, the polyanion $[Si_8O_{20}]^{8-}$ forms only if the tetramethylammonium ion, $[N(CH_3)_4]^+$, is in the solution. Similarly, Na^+ favors the formation of the complex $[Si_4O_{12}]^{4--}$. Lastly, if the pH<7, we can obtain more or less polymeric gels of $SiO_2.nH_2O$ with terminal hydroxo ligands.

Cations with valence V or higher [5]
Some of those cations, including V(V), Cr(VI), Mo(VI), W(VI) and Mn(VII) form, in basic solutions, monomers of the type $[MO_4]^{(8-z)-}$. V, Mo and W also build big polynuclear anions with as much as 10, 8 and 12 metal atoms respectively. In the case of V, in a weakly acidic solution, the decavanadic ion, $[H_2V_{10}O_{28}]^{4--}$, produces the fibrous gel of vanadic oxide, $V_2O_5.nH_2O$. Furthermore, polynanions composed of the two valence states of V, V(IV) and V(V), can form in water. Those mixed polynuclear complexes have a cage structure that can enclose an anion such as N_3^- hence forming the complex $[H_2V_8^{IV}V_{10}^{V}O_{44}(N_3)](NEt_4)_5$.
As for the cations with a high charge and electronegativity such as S and P, if they can also form oxo anions of the type $[MO_4]^-$, their electrophilic character is too low to condense into polynuclear compounds. They do, however, participate in the complexation of other cations and this often permanently. This is the case of Zr, Sb, V and W which precipitate into mixed [M,P] solids.

2.5 - ALKOXIDES SOLUTIONS

STRUCTURE AND PROPERTIES OF ALKOXIDES
An in-depth study of these particular chemicals has already been published by Bradley, Merhotra and Gaur [32]. Alkoxides, also called alcoholates, are compounds with chemical formula $M(OR)_z$ that are the result of direct or indirect reactions between a metal M and an alcohol ROH. Since many different alcohols exist, a great variety of alkoxides can be produced for each metal. Table 2.5-1 summarizes the nomenclature of these alcoholates.
The vast majority of the alcoholates are liquid or soluble in at least one organic solvent including, most of the time, their corresponding parent alcohol. This is to the exception of some alkoxides, including those of Cu, that are solid polymers insoluble both in water and organic solvents.

Table 2.5-1 - Nomenclature of alkoxides

alcohol R(OH)	alkoxy ligand to the metal M	alkoxide	abbreviation for OR
methanol CH_3OH		methoxide	OMe
ethanol C_2H_5OH		ethoxide	OEt
1, propanol (n-propanol) C_3H_7OH		1-propoxide (n-propoxide)	OPr^i
2, propanol C_3H_7OH (iso-propanol)		2-propoxide (iso-propoxide)	OPr^s
1, Butanol (n- butanol) C_4H_9OH		1 butoxide (n-butoxide)	OBu^n
2, Butanol C_4H_9OH		2 butoxide (sec- butoxide	OBu^s
2, methyl-propanol (iso butanol) C_4H_9OH		2, methyl propoxide (iso-butoxide)	OBu^i
2, methyl-prop,2,ol (tertio butanol) $C_4H_9(OH)$		tertio butoxide	OBu^t

It is often very hard to know the exact chemical formula of alkoxides, such as for aluminum butoxides, $Al(OC_4H_9)_3$. When the composition is know, as is the case of some aluminum alkoxides (see Figure 2.5-1), the compound frequently polymerizes. This is due to the oxygen of the alkoxy group (OR) which, when binding to a metal atom, can act as an electron pair donor for another metal. Contrary to the solvatation of a metal by water, the alkoxy groups can directly build bridges between different metal atoms even at the precursor level. This is especially the case if the solvent is the parent alcohol of the alcoholate.

The following diagram, Figure 2.5-1, shows some complex alkoxides in which several distinct alkoxy groups are bonded to the same central metal atom, M.

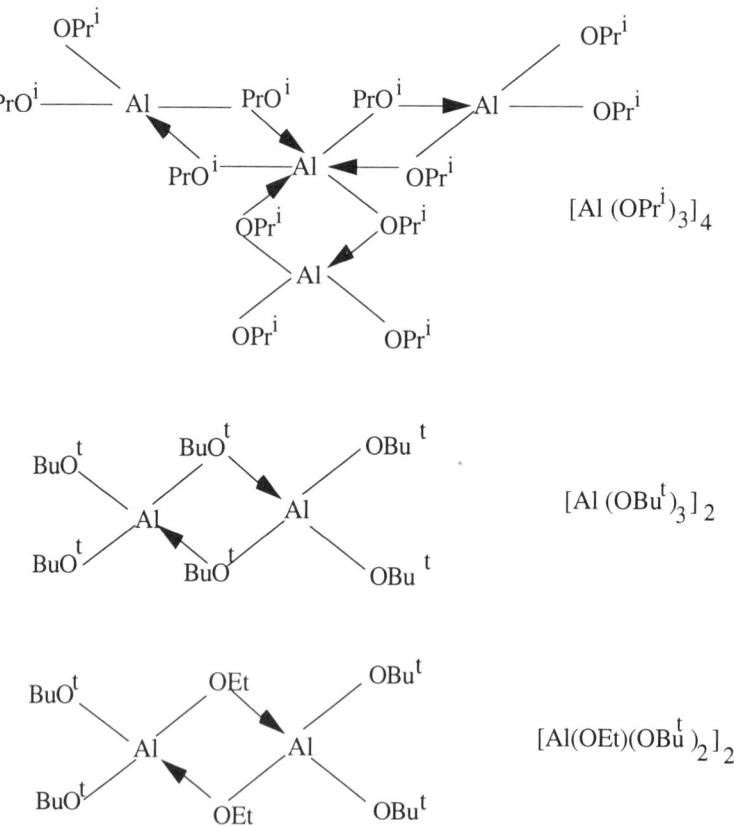

Figure 2.5-1 - A few known structures of aluminum alkoxides.
After Bradley et al. [32].

As is particularly well illustrated with Zr, alkoxides of a same metal present a great diversity of chemical and physical properties [33]. The primary alcoholates of Zr have similar volatility and all sublimate between 120°C and 160°C in a 10^{-4} mm of Hg vacuum. However, of those, $Zr(OC_5H_{11})_4$ is a viscous liquid while tetra-methoxide and tetra-ethoxide both form micro crystals. Similarly, among the secondary alkoxides, if $Zr(OBut^s)_4$ is a gum, $Zr(OPr^i)_4$ is a viscous solid that decomposes between 100°C and 125°C, and $Zr(OBut^t)_4$ is a stable liquid that vaporizes under atmospheric pressure between 190°C and 210°C [34]. Furthermore, if the secondary alcoholates are either monomeric, dimeric, or trimeric; the tertiary ones are all monomeric. In all cases, when the alkoxides are dissolved in their corresponding parent alcohol, they build stable coordination complexes with the solvent. In this complex, the oxygen of the alcohol gives an electron pair to the zirconium so as to form a strong bond (see below). Isopropanol, for instance, forms with $Zr(OPr^i)_4$ stable complex $Zr(OPr^i)_4.Pr^iOH$.

$$\begin{array}{c} R \\ \diagdown \\ O \colon\!\!\longrightarrow Zr \\ \diagup \\ H \end{array} \qquad (2.5\text{-}1)$$

Alkoxides are more reactive when in presence of moisture, heat, or light [35]. Unlike the metal salts, the impurities they create come mainly from the organic group. In fact, the alkyl group must stabilize sufficiently the alkoxide so that it is enough volatile for the M-OR and MO-R bonds to break cleanly hence liberating a free oxide impurity.

The ionic character of the M-O bond depends mostly on the size and on the electronegativity of the metal atom. If you consider Pauling's scale of electronegativity, a bond linking a metal and an oxygen is a least 50% covalent if the difference between the electronegativities is ≤ 1.7. This is the case of almost all alkoxides that, like Si, are sufficiently covalent to react differently from the metal salts in solutions. This contrast is further emphasized by the solvent since the alkoxides are often dissolved in an organic liquid alcohol that has much weaker ionizing properties than water. For those alkoxide precursors, water is not a solvent but a reactant that, when added in a controlled manner and at specific moment during the sol-gel process allows to regulate hydrolysis in a much greater extent than with metal salts. Even when water is added in large excess, there are never any anions coming form the salt. They can be added further on in order to modify the complexation reaction, but in any case they remain in much smaller concentrations than in metal salts solutions. Furthermore, those anions already build M-O bond at the precursor level. The aim is not to create those bonds, but to replace the OR groups either by hydroxo ligands so as to form M-OH, or by oxo ligands as in M-O.

The list of the elements that have been used to synthesize ceramics from such alkoxides is long. It includes many metals such as Si, Al, Ti, Zr, Pb (various references), La [36], Hf, Th, Y, Dy and Yb [34], Ba, Sn, Mg, and Sb [37], Ca [38]; but also some non metals such as B [39].

HYDROLYSIS

Alkoxides and metal salts have similar and as complex hydrolysis and polymerization chemistry [25,32]. We will therefore present the main steps of each key reaction in the same manner as was previously done with the metal salts. During hydrolysis, the alkoxy groups (OR) are replaced either by hydroxo ligands (OH) or oxo ligands (O). This reaction is influenced by various factors including

  (1) - the nature of the alkyl group
  (2) - the nature of the solvent
  (3) - the concentration of each specie in the solvent
  (4) - the water to alkoxide molar ratio $r_w = [H_2O]/[alk]$
  and (5) - the temperature [40,41].

Yet, practically, each oxide must be considered separately.

Formation of hydroxo ligands

In the first hydrolysis reaction, reaction (2.5-2), a first hydroxo ligand substitutes an alkoxy group.

$$M(OR)_z + H_2O \Rightarrow M(OH)(OR)_{z-1} + ROH \qquad (2.5-2)$$

This reaction is generally followed by a succession of other similar substitutions.

The mechanism of hydrolysis, illustrated in diagram (2.5-3), consist in a nucleophilic substitution in which a water molecule attacks the alkoxide. It is independent of the polymerization of the alcoholate.

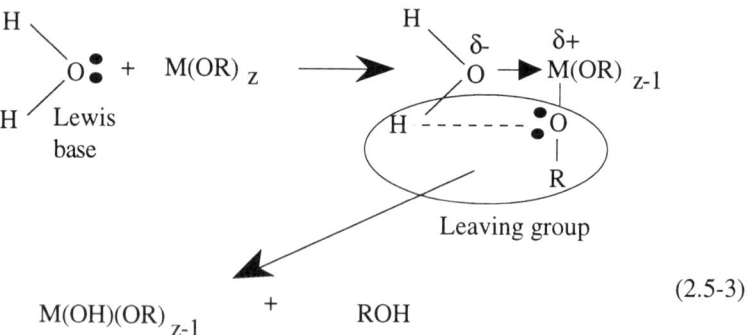

$$(2.5-3)$$

Moreover, experimental studies of its kinetics have determined that if hydrolysis is a very fast reaction for most cations, it is much slower for Si. This difference, due to the chemical characteristics of the corresponding alkoxides, is particularly well explained by the partial charge model.

In fact, according to the partial charge model, an alkoxide with formula $M(OR)_z$ undergoes hydrolysis if

(1) the metal M is an electrophile ($\delta(M) > 0$). (See table 2.5-2 for the comparison of the partial charge $\delta(M)$ of some metals in their corresponding alkoxide.)

(2) the oxygen is a nucleophile ($\delta(O) < 0$)

(3) a part of the complex, called the leaving group, can attain a positive partial charge thus enabling it to leave the intermediate structure. In most cases, this leaving group is either an alcohol (ROH with $\delta(ROH)>0$) or a water molecule (with $\delta(H_2O)>0$).

Table 2.5-2 - Partial charge $\delta(M)$ of the metal M in a few alkoxides

Alkoxide	$\delta(M)$
$Zr(OEt)_4$	+0.65
$Ti(OEt)_4$	+0.63
$VO(OEt)_3$	+0.46
$Si(OEt)_4$	+0.32

You will note from those data that the partial charge of Si in $Si(OEt)_4$ is smaller than that of the other metals in their corresponding alcoholates. It is therefore much more difficult kinetically to hydrolyze the silicium alkoxides than the other ones.

For tetravalent metal alkoxides $M(OR)_4$, a more simple reactivity criterium than the cation partial charge is its unsaturation (N-z) where N is the usual coordination number of the metal M in compounds and z the formal electric charge of the corresponding cation. The value of (N-z) increases from Si to Ce as shown in Table 2.5-3. Hence the hydrolysis reactivity increases in the order of cations Si << Sn<Ti<Zr<Ce [32]. Consequently, the hydrolysis of Si is usually catalyzed either by a strong electrophile such as H^+ or by a strong nucleophile such as OH^-.

Table 2.5-3 - Unsaturation of tetravalent metals in alkoxides.

Cation	N	N-z
Si	4	0
Sn	6	2
Ti	6	2
Zr	7	3
Ce	8	6

Moreover the alkoxides of alkaline and alkaline-earth elements react violently to the slightest humidity. It is therefore safer to dissolve them in appropriate solvents such as, for instance, in propanol for the propoxides [42].

Finally, the desired rate of a hydrolysis reaction is obtained by controlling various factors of which the most important are probably water and the alkyl group. For a slow reaction to occur, hydrolysis must be carried out in a dry environment such as, for instance, in a glove box. The reaction is also slowed down by adding a quantity of water smaller than that required stoechiometrically. As for the alkyl group, the longer it is, the harder it is to break the O-R bond and the slower the reaction. You must also consider that, beyond the first hydrolysis, it is harder for OH to substitute an alkoxy ligand.

Hydrolysis of boron alkoxides

Boron, in its hydrolysis, proceeds exactly as in reaction (2.5-2). As mentioned previously, its electron configuration differs from that of Si so that it has both a coplanar sp^2 and a tetrahedral sp^3 hybridization (Figure 2.5-2).

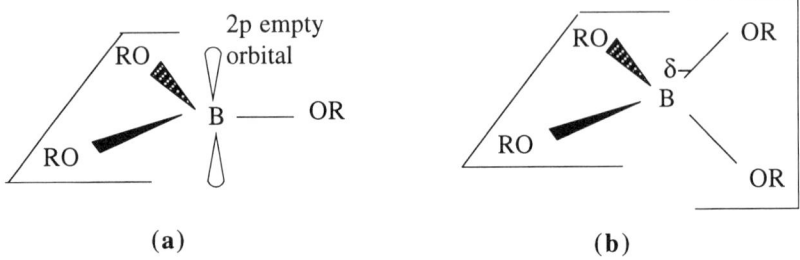

(a)   **(b)**

Figure 2.5-2 - (a) Tri-coordinated and (b) tetra-coordinated boron in alkoxides. Adapted from Brinker et al. [44].

When in its tetrahedral coordination, boron as neither a 2p nor a low d orbital available as would be necessary for an eventual sp^3d hybridization. It is therefore quite stable and prevails in solutions [43]. The trigonal configuration of boron has, however, an empty $2p_z$ orbital accessible to other electrons, hence its electrophilic character. If you further consider that the oxygen atoms of the water molecules carry a negative partial charge ($\delta(O) < 0$), then trigonal boron can undergo hydrolysis according to an S_N2 mechanism when attacked by H_2O. (mechanism 2.5-4).

possible leaving groups

(2.5-4)

In the intermediary higher coordination boron species of this mechanism, two molecular groups have acquired the necessary positive partial charge that enables them to leave the complex. One of them is the incoming H_2O molecule, its departure corresponds to the reverse reaction, and the other, the alcohol ROH, is the actual leaving group. Trigonal boron can also undergo alcoholysis according to a similar mechanism in which it is an alcohol molecule that attacks the alkoxide.

Hydrolysis of aluminum alkoxides

Aluminum alkoxides are sufficiently stable to vaporize without decomposing. During their hydrolysis, an hydroxo ligand (OH) replaces an alkoxy group (OR) according to the general reaction (2.5-2) which, in that case, is written as:

$$Al(OC_4H_9)_3 + H_2O \Rightarrow Al(OC_4H_9)_2(OH) + C_4H_9OH \qquad (2.5-5)$$

Hydrolysis of titanium alkoxides

In their hydrolysis reactions, the titanium alkoxides follow the general rule of equation (2.5-2). As for most other alcoholates, hydrolysis is not only quicker than condensation but also much more exothermic [44]. Titanium is particular in that three of its precursors: ethoxide, $Ti(OC_2H_5)_4$, isopropoxide, $Ti(OC_3H_7)_4$, and butoxide, $Ti(OC_4H_9)_4$, form peroxo complexes that, in solution, have a deep orange color. $Ti(O_2)(OH)_{aq}(OC_2H_5)$ (aq stands for the aqua group) is an example of those peroxo complexes. The mechanism leading to their formation is, however, yet unknown [45].

Hydrolysis of zirconium alkoxides

Unlike most other alcoholates, when zirconium undergoes hydrolysis, it is not hydroxo ligands that replace the alkyl groups but either oxo or aqua ones. The corresponding reaction is therefore a de-alcoholation of the following type:

$$Zr(OR)_x(OH)_y \Rightarrow ZrO_{x-x1}(OR)_{x1}(OH)_{y-x+x1} + (x-x_1)ROH \qquad (2.5-6)$$

The corresponding accepted mechanism proposed by Mazdiyasni [34] for the hydrolysis of $Zr(OBu^s)_4$ consist, as illustrated in figure (2.5-3) of four main steps.

In the first one, an OH ligand substitutes an OR molecular group as in any typical hydrolysis. Yet, in the following step, this new hydroxo ligand separates from the complex together with another alkoxy group, hence allowing the creation of the first Zr=O double bond. Those two first steps are repeated for the last two OR groups of the initial alkoxide so that the final product, ZrO_2, is a pure oxide. The first step being the slowest, the total reaction rate depends solely on the initial alkoxide concentration.

Hydrolysis of silicon alkoxides [46]

The result of the hydrolysis of silicon alkoxides is an Si-OH group called silanol in which an hydroxo ligand is bonded to a silicon atom. As was already mentioned, it is a very slow process. Still, silicon alkoxides, and especially tetraethoxysilane, are probably the most studied alcoholates.

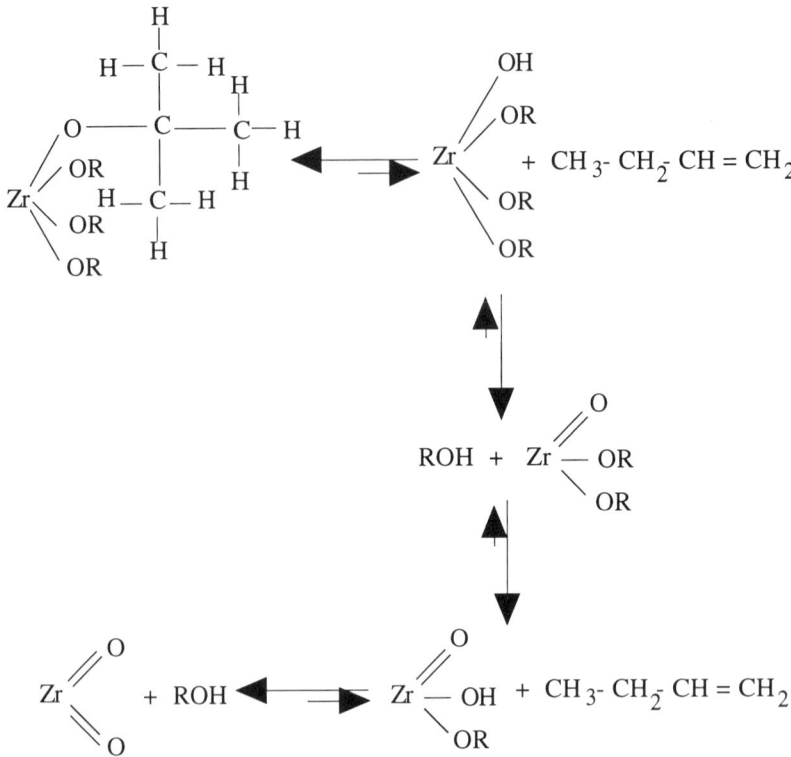

Figure 2.5-3 - Hydrolysis of Zr tetra tertio-butoxide. After Mazdiyasni et al. [34].

Tetraethoxysilane, also called tetraethoxysilicate and commonly abbreviated TEOS, has for formula $Si(OC_2H_5)_4$. Its chemistry, as will later be explained in the next chapter, varies in function of the pH. In fact, when the pH of an aqueous solution is ≈ 2.5, the silicate particles are not electrically charged. However, when the pH<2.5, the solution being quite acidic and the silicate particles being negatively charged, the relatively high concentration of protons catalyzes the hydrolysis reaction. The mechanism then corresponds to an electrophilic substitution in which an $[H_3O]^+$ hydronium ion attacks the oxygen of one of the alkyl group [47]. It is illustrated in the reaction diagram (2.5-7).

In the intermediary complex of this mechanism, the coordination of Si increases. The reaction rate depending as much on the concentration of H_3O^+ as on the one of the alkoxides, the mechanism is consequently an S_E2. Note that the steric strain is also an important factor as the rate of hydrolysis decreases as the length of the alkyl group increases.

$$
\begin{array}{c}
\text{RO} \\
\text{RO} - \overset{|}{\underset{|}{\text{Si}}} - \ddot{\underset{\delta-}{\text{O}}} - \text{R} + \text{H} - \overset{\text{H}}{\underset{\text{H}}{\text{O}}}{}^{(+)} \\
\text{RO} \qquad \underset{\delta+}{\longleftrightarrow}
\end{array}
\quad \longleftrightarrow \quad
\begin{array}{c}
\text{RO} \\
\text{RO} - \overset{|}{\underset{|}{\text{Si}}} - \text{O} - \text{R} \\
\text{RO} \quad \overset{\text{OH}}{\underset{\text{H}}{}} \text{H}
\end{array}
$$

$$
\longleftrightarrow \quad
\begin{array}{c}
\text{RO} \qquad \text{OR} \\
\text{HO} \cdots \overset{|}{\text{Si}} \cdots \text{OR} \\
\text{(H)} \qquad \text{(H)} \\
\text{RO} \\
\text{Leaving groups}
\end{array}
\quad \longleftrightarrow \quad (2.5\text{-}7)
$$

$$
\begin{array}{c}
\text{RO} \\
\text{RO} - \overset{|}{\underset{|}{\text{Si}}} - \text{OH} \quad + \quad \text{ROH} \quad + \quad \text{H}^{(+)} \\
\text{RO}
\end{array}
$$

On the other hand, the addition of a base increases the pH of the aqueous solution above 2.5 so that it is no longer an hydronium but an OH$^-$ anion that attacks the alkoxide trough an S_N2 mechanism in order to form the silanol group [46,48].

$$
\text{HO}^{(-)} + \begin{array}{c}
\text{RO} \qquad \text{OR} \\
\underset{\delta+}{} \overset{|}{\text{Si}} - \text{OR} \\
\text{RO}
\end{array}
\quad \longleftrightarrow \quad
\begin{array}{c}
\text{RO} \qquad \text{OR} \\
\text{HO} \cdots \overset{|}{\underset{\delta-}{\text{Si}}} \cdots \text{OR} \\
\text{RO} \qquad \delta- \\
\text{Leaving group}
\end{array}
$$

$$
\longleftrightarrow \quad
\begin{array}{c}
\text{RO} \qquad \text{OR} \\
\text{OH} - \overset{|}{\underset{|}{\text{Si}}} \\
\text{OR}
\end{array}
\quad + \quad \text{OR}^{(-)} \qquad (2.5\text{-}8)
$$

$$
\text{OR}^{(-)} + \text{H}^{(+)} \quad \longleftrightarrow \quad \text{ROH}
$$

Since $\delta(\text{OR})_{\text{complex}} < \delta(\text{OR})_{\text{alcohol}}$, at least one OR or OR$^-$ ligand must leave the intermediary complex formed by Si. The anion then recombines with a proton so as to form an alcohol molecule. Another, more complex mechanism involving two intermediary silicon complexes was also proposed for this reaction.

Formation of oxo ligands

Since Lewis bases are strong nucleophiles, they can deprotonate the OH ligands of cations which form acidic oxides, thus creating oxo ligands. This corresponds to the following general reaction [45]:

$$-M\text{-}OH + :B \quad \Rightarrow \quad [M\text{-}O]^- + BH^+ \qquad (2.5\text{-}9)$$

In this reaction :B is a Lewis base such as OH^- or NH_3. For instance, if you consider the silanol group ($\equiv$SiOH) resulting from the hydrolysis of silicon alkoxide, you would note that in this reaction (reaction 2.5-10) a base is a necessary catalyzer for the formation of the oxo ligand, $\equiv$SiO$^-$.

$$(RO)_3 \equiv Si - (OH) + OH^- \Rightarrow (RO)_3 \equiv SiO^- + HOH \qquad (2.5\text{-}10)$$

It was also indicated by Mazdiyasni et al. [34] that traces of water vapor can also hydrolyze metal alkoxides thus transforming them into oxi-alkoxides. Such a hydrolysis follows a reaction of the type (2.5-11).

$$M(OR)_4 + H_2O \Rightarrow MO(OR)_2 + 2\ ROH \qquad (2.5\text{-}11)$$

CONDENSATION [15,47,49]

Condensation, in aqueous solutions, is the result of either an olation or an oxolation reaction. In either case, one must be quite careful as the oxygen present sometimes speeds up the reaction to a point where it becomes necessary to work in the neutral atmosphere of argon in order to control it. The CO_2 coming from the air is also an important factor to control as, and especially with the alkali elements, it produces hydrogen carbonates that modify the products of the polymerization process. [50].

Condensation by olation

Condensation by olation follows an S_N2 nucleophilic substitution mechanism in two steps of which the first one always consist in a nucleophilic addition. It is only in the second step that the leaving group separates from the complex. This leaving group is either an alcohol molecule coming from a protonated alkoxy ligand, as in reaction (2.5-12), or a water molecule coming from a solvated alkoxide, as in reaction (2.5-13).

$$\Rightarrow \qquad M-\overset{\overset{\displaystyle H}{|}}{O}-M \quad + \quad ROH \qquad (2.5\text{-}12)$$

$$-M\text{-OH} \quad + \quad M\blacktriangleleft\!:\!O\big\langle{}^{H}_{H} \quad \Rightarrow \quad M-O:\!\blacktriangleright M\blacktriangleleft:O\big\langle{}^{H}_{H}$$

alkoxide solvated by water / leaving group

$$\Rightarrow \quad M-\overset{H}{\underset{|}{O}}-M \quad + \quad H_2O \qquad (2.5\text{-}13)$$

In the case of a departing alcohol, the protonation of the alkoxy ligand is first established by reaction (2.5-14) using an acid catalysis. Yet, in either case, before condensation occurs, the protonated alkoxy ligand and the solvated water molecule are both linked to the metal atom by a shared electron pair.

$$-M\text{-OR} \quad + \quad H_3O^+ \quad \Rightarrow \quad M\blacktriangleleft:O\Big\langle{}^{\delta+\ H}_{R} \quad + \quad H_2O \qquad (2.5\text{-}14)$$

Condensation by oxolation

As for metal salt solutions, in the condensation by oxolation of an alkoxide, an "ol" bridge must first be established before eventually transforming into an "oxo" one. Once this "ol" bridge has been built, its corresponding hydrogen needs only to transfer either to a terminal alkoxy ligand or to an hydroxo ligand to form the "oxo" one. In the first case, transfer of the H to an OR ligand according to reaction (2.5-15), condensation is more specifically referred to as alcoxolation.

$$-M\text{-OH} \quad + \quad -M\text{-OR} \quad \Rightarrow \quad M-\underset{H}{\overset{O:\blacktriangleright}{}}M-OR$$

leaving group / "ol" bridge

$$(2.5\text{-}15)$$

$$\Rightarrow \quad M-O-M\blacktriangleleft:O\big\langle{}^{R}_{H} \quad \Rightarrow \quad M-O-M \quad + \quad ROH$$

"oxo" bridge / "oxo" bridge

In the second case, transfer of the H to an OH group according to reaction (2.5-16), condensation is simply called oxolation.

Oxolation can also be the result of a de-etheration as in reaction (2.5-17).

$$-M\text{-}OH \ + \ -M\text{-}OH \ \Rightarrow \ M\text{---}O\overset{\underset{|}{}}{:}\text{---}M\text{---}OH$$
$$\underset{H}{}$$

leaving group   "ol" bridge

(2.5-16)

$$M\text{---}O\text{---}M \blacktriangleleft \ :Q \overset{H}{\underset{H}{}} \ \Rightarrow \ M\text{---}O\text{---}M \ + \ H_2O$$

"oxo" bridge   "oxo" bridge

$$-M\text{-}OR + RO\text{-}M\text{-} \Rightarrow -M\text{-}O\text{-}M\text{-} + ROR \qquad (2.5\text{-}17)$$

Condensation of aluminum alkoxides

When aluminum alkoxides condensate, a double "ol" bridge forms between two aluminum atoms. This reaction (2.5-18), catalyzed in acidic conditions, is similar to reaction (2.4-44) mentioned in the last section for the condensation of the Al metal salt solutions.

$$2 \ Al(OC_4H_9)_2(OH) \ \underset{H^+}{\Rightarrow} \ (C_4H_9O)_2Al \overset{\overset{H}{\underset{|}{O}}}{\underset{\underset{|}{\underset{H}{O}}}{}} Al \ (OC_4H_9)_2 \qquad (2.5\text{-}18)$$

Further condensation forms polynuclear compounds equivalent to those produced from an initial aluminum salt solution; including, when in acidic conditions, the Al_{13} cation.

Condensation of boron alkoxides

As was already explained, boron alkoxides can fix hydroxo ligands during hydrolysis only when in their trigonal configuration. As trigonal borates are always present in aqueous solutions together with tetrahedral ones, polymerization can follow. Yet, it is the presence of tetrahedral borate in the complex that guarantees the relative stability of each of those polymers in which each boron atom is bonded to other ones by oxo bridges. Figure 2.5-4 gives an example of a dimer, a trimer and a tetramer of boron.

The hydrolytic stability of each of those polymers decreases according to the following order:

$$[\equiv B\text{-}O\text{-}B\equiv] > [\equiv B\text{-}O\text{-}B=] > [=B\text{-}O\text{-}B=] \qquad (2.5\text{-}19)$$

Condensation of silicon alkoxides

In acidic conditions and at a pH<2.5, silicon alkoxides condense through a two-step S_N2 type mechanism. The first step consists in the protonation of a silanol group which increases the electrophilic character of the surrounding silicon atoms.

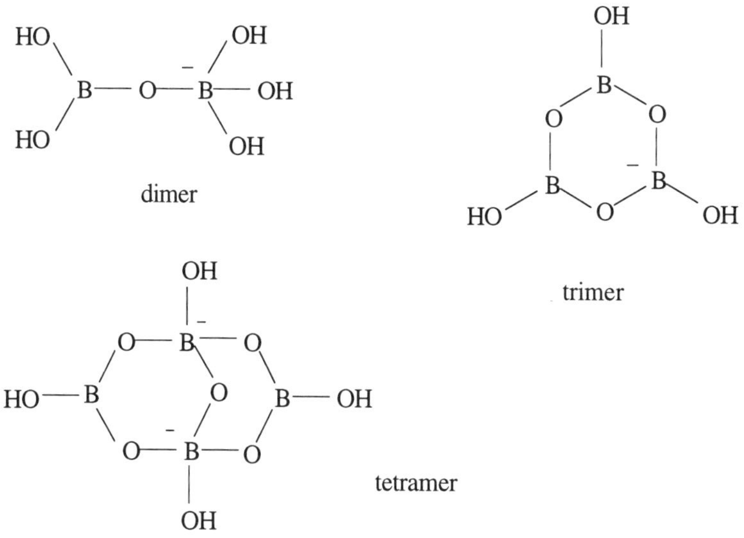

Figure 2.5-4 - Borate polymers : dimer, trimer and tetramer
according to Edwards and Ross [43].

As a consequence, this protonated silanol combines to another silanol group while
liberating an $[H_3O]^+$ ion. The two silicon atoms of the resulting polymer are then
linked through an oxo bridge called, in this specific case, a siloxane bond. Note that
the Si of the intermediary complex of this mechanism, as illustrated in (2.5-20) is
either tetra or penta coordinated [51].

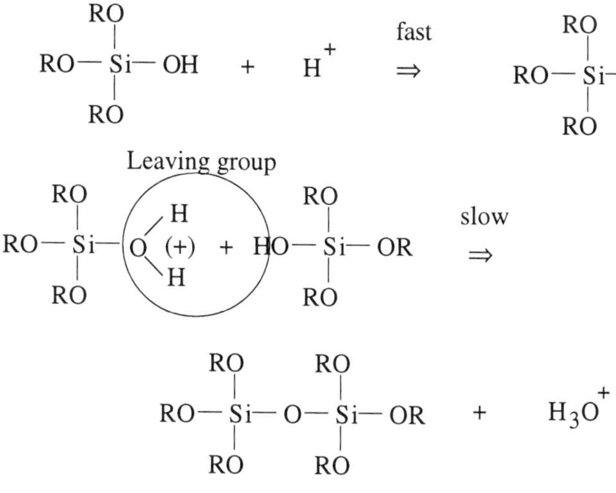

$$(2.5\text{-}20)$$

The rate of this condensation reaction depends on the second step of the mechanism as it is slower than the first one; it is, moreover, also proportional to the concentration of protons. Hence condensation is a slower transformation than hydrolysis. Furthermore, silanols protonate more easily when at the end of a polymer chain. The polymers obtained are therefore linear with scarcely any branching points.

On the other hand, when the conditions are such that the pH>2.5, the silanol groups are deprotonated according to reaction (2.5-10). They build siloxane bridges by another S_N2 mechanism that involves two intermediary complexes with penta-coordinated silicons. This corresponds to reaction (2.5-21) as proposed by Swain et al [52]. When the pH>4, the condensation rate is not only proportional to the concentration of OH⁻ anions but also superior to that of hydrolysis. Furthermore, since the reticulation inside the silicon polymers is more developed than when conditions for acidic catalysis are used, denser solids are obtained.

Transition state 1

$$\equiv SiO^- + Si(OR)_4 \Rightarrow \equiv SiO\cdots\overset{\delta^-}{Si}\cdots\delta^- \Rightarrow \equiv SiO\cdots Si^{(-)}$$

(2.5-21)

Transition state 2

$$\Rightarrow \equiv SiO\cdots Si \Rightarrow \equiv Si-O-Si\equiv + RO^-$$

Leaving group

Overall basic catalysts, including Lewis bases, accelerate condensation and alcohol molecules are better leaving groups than water. Efficient Lewis bases include for instance DMAP (dimethylaminopyridine) n-Bu₄NF and NaF [53, 54].

However, condensation is a reversible reaction and, hence, the siloxane bridges ≡Si-O-Si≡ built during this process can themselves be dissolved by alcohols between pH 3 and 8 in presence of OH⁻ catalyst. This corresponds to reaction (2.5-22). As a consequence, the monomers that have just been rebuilt can again condense with another dimer so as to form another trimer (reaction (2.5-23))

$$\overset{OH^-}{ROH + (RO)_3Si\text{-}O\text{-}Si(OR)_3 \Leftrightarrow Si(OR)_4 + Si(OR)_3(OH)}$$

(2.5-22)

$$Si(OR)_3(OH) + (RO)_3Si\text{-}O\text{-}Si(OR)_3 \Leftrightarrow (RO)_3Si\text{-}O\text{-}Si(OR)_2\text{-}Si(OR)_3 + ROH$$

(2.5-23)

In short, in the case of silicon alkoxides, three reactions: hydrolysis, condensation and redissolution are in eternal competition with one another so that the final composition of the solutions depends on the kinetics of those reactions which varies according to the pH (see figure 2.5-5).

The oxo bridges linking the different silicon atoms between them are, overall, constantly redistributed among monomers and polymers, thus explaining the formation of spherical particles in basic conditions.

$$\text{2 dimers} \underset{ROH}{\overset{OH^-}{\Longleftarrow \Longrightarrow}} \text{1 trimer + 1 monomer} \tag{2.5-24}$$

Formation of solid phases from alkoxides
Hydrolysis and condensation both keep on going thus gradually building up a tri-dimensional network that, at the end, often forms a solid phase. This process is accelerated by heat as the rate of both reactions increases together with the temperature [56]. Since the kinetics of hydrolysis and condensation, and hence the overall reaction and the type of polymers formed, depends on the pH, a great variety of materials with different structures can be obtained. These include linear polymers as well as dense colloidal particles and smaller ones with more or less weakly bonded cross-linked clusters of polymers.

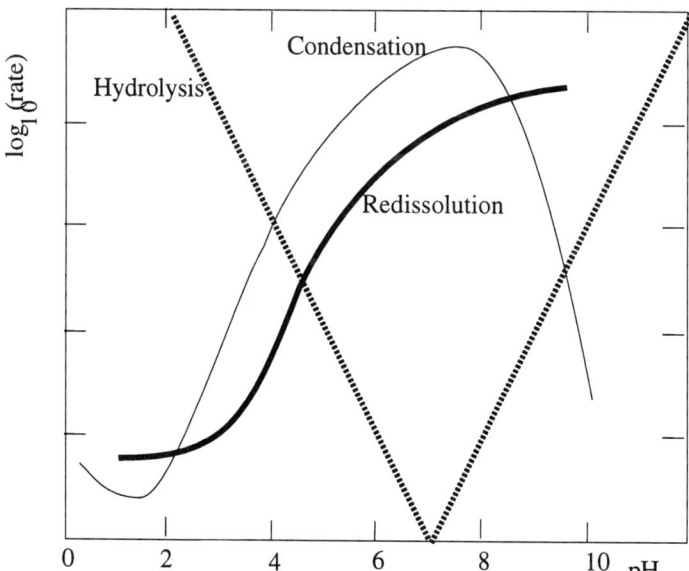

Figure 2.5-5 - Rates of hydrolysis, condensation and redissolution of TEOS. Adapted from Brinker [55].

Furthermore, except for Si, all alkoxides hydrolyze and condense so fast that the actual reaction rates are not yet determined. Still, although it is quite controversial, it is generally admitted that linear polymers form more favorably when hydrolysis is slower than polymerization. This would explain why, again to the exception of silicon, all hydrolysis must be slowed down with something like a chelating agent in order to form a gel; a gel having a much more linear polymeric structure than dense particles. When the rate of hydrolysis is not reduced, a solid with denser reticulation, a powder in most cases, precipitates. Moreover, the alkoxides of some elements, such as alkaline and alkaline-earth, do not condense. In those cases, no gel can possibly form. Instead, dense colloidal hydroxidic or oxidic particles are obtained according to the following two reactions [57]

$$M(OR)_z + zH_2O \Rightarrow M(OH)_z + zROH \qquad (2.5\text{-}25)$$
$$M(OH)_z \Rightarrow MO_{z/2} + z/2\ H_2O \qquad (2.5\text{-}26)$$

For those elements which alkoxides do indeed form gels, it is also easier to create large monoliths by using an excess of water in the first step of hydrolysis. In fact, the oxide content of the material increases when a greater proportion of water is used during hydrolysis so that, not only are they more extensive cross-linking, but the monoliths produced are stronger [58,59]

Silicon is the exception to all the other alkoxides and its chemistry has been therefore extensively studied. The results found are rather interesting as it is in acidic conditions when hydrolysis is faster and not slower than condensation that polymeric solids form. The type of alkoxide is also quite important since hydrolysis slows down drastically when the alkyl groups become bigger. Furthermore, the solids formed by TEOS when hydrolysis occurs in acidic condition, the catalyzer thus being H_3O^+, depend mostly on the proportion of water added. This is due to the fact that hydrolysis keep going on even after further condensation. Yet, the steric strain makes it increasingly difficult for an hydroxo ligand to substitute an alkoxy one. In fact the OR group initially belonging to a polymer is easier to substitute when located at the end, than in the middle of the polymeric chain. The overall reaction rate is proportional to the concentration of $[H_3O^+]$, the catalyzer, and is determined by the speed of hydrolysis rather than of condensation [60]. In those conditions, silicon alkoxides produce a great variety of materials including highly linear polymeric gels, randomly branched polymeric gels, and colloidal clusters of polymers inside which the cross-linking between the polymers is quite frequent [61].

When hydrolysis occurs in basic condition, either discrete colloidal silica particles or dense clusters form [61]. Those clusters are normally attached to one another by a few polymers; although, when the pH>7, they can no longer stay interconnected and, hence, a stable colloidal sol appears. It is the OH⁻ anions that are responsible for this difference as they have the ability to adsorb on some particles; thus, when they are present in high enough concentration in the solution they maintain the particles dispersed in the sol. Inversely, when the pH becomes low enough, the bigger particles continue to grow at the expense of the smaller ones that gradually depolymerize in a transformation similar to Ostwald ripening. Furthermore, the cross-linking of polymers at the surface of the silica particles changes constantly as they grow, so that during polymerization, as illustrated in Figure 2.5-6, both tetra and tri-siloxanes cyclic polymers form. The Si-O-Si bond angle of tri-siloxanes is smaller than of the tetra-siloxane cycles. As a consequence, the partial charge transfer between the Si and the O atoms make the Si-O bond of tri-siloxane more ionic and hence more susceptible to hydrolyze and depolymerize than tetra-siloxane. Experimentally, it has been determined that the rate of hydrolysis of tri-siloxane is 75 times greater that of tetra-siloxane. The tri-siloxanes are therefore present near the surface of the product where they feed the growth of other particles.

By comparison with silicon, aluminum alkoxides produce, depending on the pH, either tri-hydroxides, $Al(OH)_3$, or oxohydroxydes -- also called monohydroxydes, $AlO(OH)$, solids; each of which can exist under different structures. For example, if the reactions are achieved in acidic conditions than the solid obtained after evaporation is a gel with a structure similar to boehmite, $AlO(OH)$. Moreover, the polymerization reaction being fully reversible, the gel completely redissolves when replaced in water. In order to obtain an irreversible polymerization, the reactions need to occur in an organic solvent and the complexation of the alkoxide must be done by a chelating agent such as ethyl acetoacetate or acetic acid.

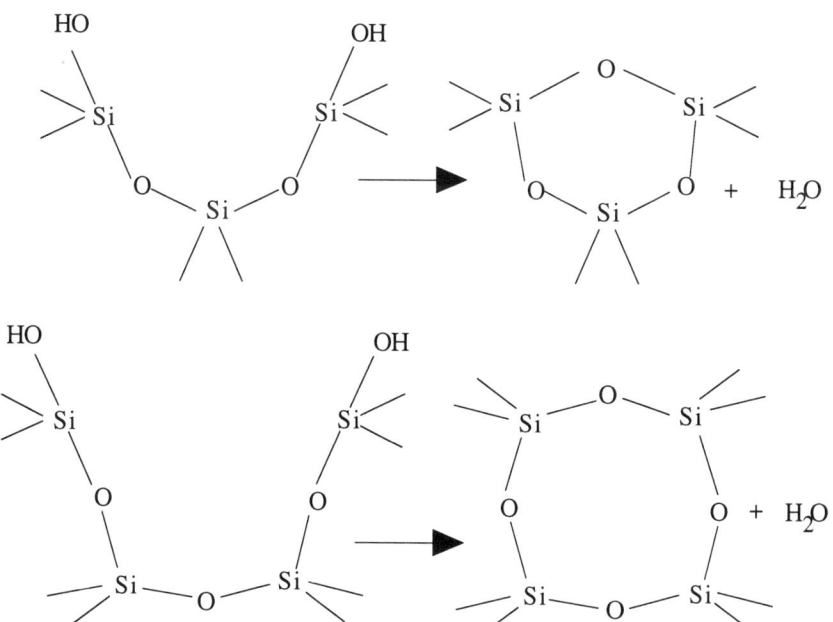

Figure 2.5-6 - Formation of tri-siloxane and tetra-siloxane units in the polymerization of silica. Adapted Brinker et al. [62].

As for zirconiumm alkoxides, hydrolysis being a much quicker reaction than condensation, the solid phase obtained consist of dense hydrated aggregates of ZrO_2 oxide. The final structure of the solid depends, like for most other alkoxides, on the proportion of water used during hydrolysis [63]. Some examples include the metazirconic acid, H_2ZrO_3, and the more polymeric solid oxide, $(ZrO_2)_n,mH_2O$. In order to produce soluble polymeric intermediates, it is obviously necessary to slow down drastically the hydrolysis reaction. This is achieved either by using a non-polar solvent such as cyclohexane or by undertaking hydrolysis in a gaseous environment with a controlled humidity. Yet, the most efficient technique still consists in complexing the zirconium alkoxide with a chelating additive such as acetylacetone in an organic solvent such as isopropanol [63]. The chelating complex then forms according to reaction (2.5-27):
The steric strain on those chelated compounds is much greater than on the corresponding non-chelated ones. They are therefore much less reactive and the OH ligands substitute the OR molecular group much more slowly. This slowing down has for consequence that more linear gel-type structures form in the solid phase.

$$(2.5\text{-}27)$$

Gels can also be produced from the tetra-isopropoxides of Zr, Ti an Nb by dissolving them in methanol and complexing them with tetramethylammonium hydroxide (TM) [64]. In the case of titanium, the final solid obtained is, according to Matsuda and Kato, either metatitanic acid, $TiO(OH)_2$, or orthotitanic acid, $Ti(OH)_4$ [65]. Yet, Barringer and Bowen demonstrated that some alkoxy groups always remain in the solid phase even if the ratio of hydrolysis is $r_w \geq 3$ so that the final product is never purely composed of those two acids [66]. Like for all other alkoxides, the hydrolysis and condensation of Ti alkoxides are controlled by two parameters: the concentration of the reactants and the hydrolysis ratio r_w [67]. Those two factors not only modify the rate of the reactions but also change the degree of hydrolysis [68]. For instance, if for the isopropoxide the ratio is $r_w < 1$ soluble linear polymers form while remaining bonded to some lateral alkoxy ligands. This corresponds to reaction (2.5-28).
If the hydrolysis ratio $r_w < 2$, then such solutions are suitable for the creation of fibers by the spinning technique.

$$n\ Ti(OR)_4 +\ (n\text{-}1)\ H_2O \Rightarrow$$

$$\text{(2.5-28)}$$

$$(RO)_3Ti\!-\!\!-O\!-\!\!\left[\,Ti\!-\!\!O\,\right]_n\!\!-\!Ti(OR)_3 +\ 2(n\text{-}1)ROH$$

Two other important factors also determine the structure of the polymer: the nature of the alkyl group and the catalyzer. The nature of the alkyl group regulates the length of the polymer chains; in fact the bigger groups slowdown the diffusion of the species in the solution and hence reduce the rate of hydrolysis so much that condensation only produces small polymers. As for the catalyzer, as was demonstrated by Yoldas, only two acids: HNO_3 and HCl allow clear solutions to form provided that the molar ratio [acid]/[alkoxide] < 0.30 [41]. At such concentrations, those acids catalyze the formation of linear chains [69].

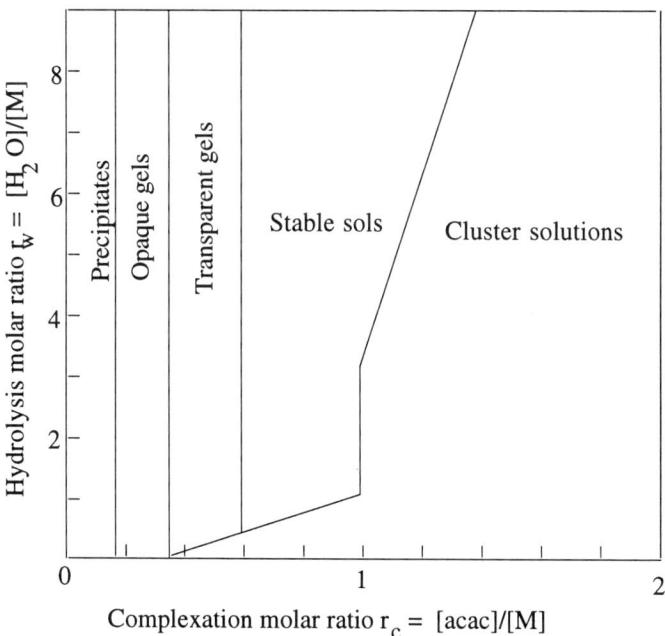

Figure 2.5-7 - Diagram showing the hydrolysis condensation behavior of complexed tetravalent metal alkoxides. Adapted from [70].

Good organic complexing ligands can also be used. They comprise β-diketones and their derivatives, polyhdroxylated ligands such as the polyols and α or β - hydroxyacids. For instance with acetylacetone [71] which can be written acacH,

$$M(OR)_z + x \; acacH \implies M(OR)_{z-x}(acac)_x + x \; ROH \qquad (2.5\text{-}29)$$

Hydrolysis-condensation reactions are governed by the hydrolysis ratio $r_w = \dfrac{[H_2O]}{[M]}$ and the complexation ratio $x = \dfrac{[acac]}{[M]}$. In the case of $Ti(OR)_4$ alkoxides, figure 2.5-7 shows that if $2 < r_w < 20$ and $0.15 < x < 1$, sols are obtained, which can be transformed to gels. On the other hand if $r_w \ll 1$, modified oxo-alkoxide clusters are formed. In the case of the titanium alkoxides with $x = 2$, $Ti(OR)_2(acac)_2$ hydrolyzes to $TiO(acac)_2$. For $x = 0.1$, the complex obtained is $Ti_{18}O_{22}(OBu^s)_{26}(acac)_2$ [72].

2.6 - OTHER PRECURSORS

A great number of precursors other than metal salts and alkoxides are also used for the production of oxidic sol-gels. Of those the more recent and probably the most promising are the organometallics which consist of a metal atom directly bonded to one of the carbons of an organic group [73]. Chelated complexes with organic anions can also be used. New functionalized macromolecules are also being studied. These other precursors expand the already quite large diversity of different materials conceivable with oxides. They especially allow a better control of the speed of hydrolysis that is often very, if not too fast with the alkoxides. Most of the time, they undertake either a simple hydrolysis or a thermal decomposition that produces a powder. Yet, their studies and those of the chemical reaction they undergo are far from completed and they still constitute a vast field of investigation that will hopefully greaten the potentialities of sol-gel processes.

ORGANOMETALLICS
Organometallic precursors differ from the metal alkoxides; in the first case the metal is bonded directly to the carbon atom while in the second it does so with the intermediate of an oxygen. The simplest organometallics are those in which the organic part simply consists of an alkyl group, the general formula thus being MR_n [74].
Organometallics are, like alkoxides, very pure compounds. Their metal to carbon bonds breaks readily during hydrolysis so they can easily produce oxides [75]. Organosilicon, organometallic in which Si is the metal, again behave differently from the other organometallics and they are still quite useful in making non-oxidic ceramics by the sol-gel technique. The organoalkoxysilanes are especially interesting.

Organoalkoxysilanes

They constitute the most useful family of organometallics, to synthesize oxide or hybrid oxide-organic materials. Trifunctional alkoxysilanes have the general formula $R'Si(OR)_3$. They mainly produce inorganic networks carrying organic groups. They hydrolyze much faster than $Si(OR)_4$ alkoxides in acidic conditions but they can easily be handled. Bifunctional alkoxysilanes $R'R''Si(OR)_2$ where R' and R" are methyl or phenyl groups do not normally produce three-dimensional networks. On the other hand, polyfunctional alkoxysilanes such as those in figure 2.6-1 are studied and they can be used to make sol-gel hybrid materials.

Other organometallics

Other M-C stable organometallic bonds can be made with Sn, Hg or Pt as the metal and used to synthesize hybrid organic-inorganic materials [76]. For instance, the hydrolysis of $butenylSn(OAm^t)_3$ yields clear solutions comprising complexes with a structure related to the cage-like cluster cation $[(BuSn)_{12}(\mu_3-O)_{14}(\mu_2-OH)_6]^{2+}$ which is surrounded by 12 butenyl chains. The butenyl groups can be polymerized with azobisisobutyronitrile (AIBN) .

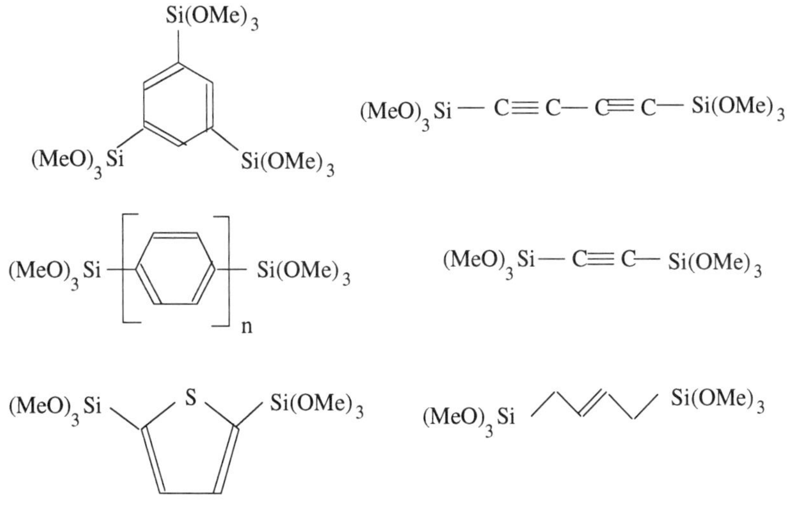

Figure 2.6-1 - Trimethoxysilylated organometallic precursors. Adapted from Shea et al. [77] and Corriu et al. [78].

On the other hand M-C bonds made with transition metals are energetically broken by hydrolysis and must be handled with care. It is possible to link such metals with organic groups by the intermediate of M-O-Si-C bonds, as this was done by organic polymerization of the anions $[SiW_{11}O_{40}(SiR)_2]^{4-}$ by the intermediate of an organic group R which can be a vinyl, allyl, methacryl or styril group [70].

METAL-ORGANIC CHELATES

Alkoxides belong to a larger group of compounds called metal-organic precursors which includes a great diversity of species, especially with the transition elements [79]. The ligands of those organic groups are often the anion of an organic acid, such as citric, oxalic or malonic acids. Citrates, for example, are easily obtained by dissolving in citric acid either a carbonate, such as $BaCO_3$, or an alkoxide, such as $Ti\text{-}(OPr^i)_4$ [80]. Sometimes, the acetyl acetonates of some metals, such as iron, can also be used. Most can only be dissolved in water, although for some, a few organic liquids can also be the solvent.

As seen previously, the anions of these precursors build chelated complexes around a coordination center which comprises a metal atom. Several bonds have to be broken for the ligand to completely separate from the coordination center, so that those compounds are quite stable and a catalyst is necessary to hydrolyze them. The catalyst can be either electrophilic, such as hydronium or any metal cation, or nucleophilic, like H_2O, OH^- or any halide anions. Those destructive catalysts proceed by binding either to the coordination center or to the free end of the chelating ligand, thus preventing the chelated compound to recombine through an inverse reaction. Moreover, the solvent may also play an important role in the catalytic hydrolysis. This is the case of $HClO_4$ or other strong acids in the cases of oxalic or malonic chelated complexes of Cr(III) or Co(III) formed with ethylene-diamine.

In any case, the nature of the cation used during the process is a major factor in determining the structure of the final material. For example, strongly cross-linked polymers used in the production of resins, paint and lacquers are feasible with aluminum, due to the nature of Al-O bonds [81].

NEW FUNCTIONALIZED MACROMOLECULES

Interesting precursors to synthesize hybrid organic-inorganic materials can be made either from organic macromonomers comprising some functionalities which are reactive towards inorganic precursors such as the alkoxides, or from inorganic oxopolymers comprising organic functionalities.

The first group includes the polysaccharides [82], cellulosic materials [83], vegetable oil derivatives [84] and macromonomers with reactive functionalities which can can be a hydroxy, carboxy or alkoxysilyl end, such as derivatives of alkylsilanes (PDMS, PDPS) [85-88], polystyrenes [89], polyoxazoline (POZO) [90], polyimides [91], polybutadiene (MPBP) [92], polyethylene oxides (MPEOU) [93], polytetramethylene oxide (PTMO) [94], polyether ketone (PEK) [93], polyoxypropylene (PPO) [92], polyorganophosphazenes (POP) [94], cyclophosphazenes (CP) [70] and polymethylmethacrylates (PMMA) [95]. For instance, the polyimide macromonomer in figure 2.6-2 with ethoxysilyl functionalities can react with TEOS, in solution in dimethylacetamide, to form hybrid materials.

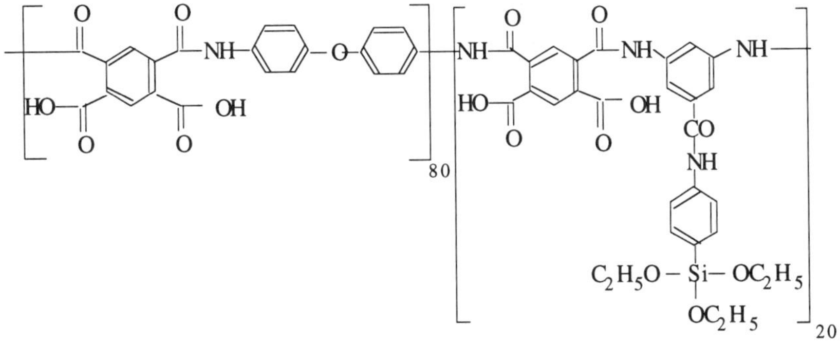

Figure 2.6-2 - Ethoxysilyl functionalized monomer polyamic acid used in the fabrication of polyimide-SiO$_2$ hybrids. Adapted from [91]

The second group is more limited. It comprises mostly siloxane clusters with organic functionalities, such as $[(CH_3)_2HSi]_8Si_8O_{20}$ and $[CH_2=CH(CH_3)_2Si]_8Si_8O_{20}$ [96]. Polymerizable ligands such as methacrylic acid and allylacetoacetate can also complex titanium ethoxide and zirconium propoxide to form big oxo clusters such as $Ti_6O_4(OEt)_8(OMc)$ and $[Zr_{10}(\mu_4\text{-}O)_2(\mu_3\text{-}O)_4(\mu_3\text{-}OH)_4(\mu_2\text{-}OPr^n)_8(OPr^n)_{10}$ where the complexing ligands are located at the periphery of the clusters [97,98]. These oxo clusters are very sensible to hydrolytic cleavage below a certain size and their polymerization must be performed in an organic solvent.

2.7 - PRECURSORS MIXING

Sol-gel processing offer the amazing possibility to synthesize complex oxides such as the titanate PbTiO$_3$ by mixing several oxide precursors. Those elaborate oxides can be established by mixing, after having previously and separately produced them by hydrolysis, the different solutions containing each a metal precursor. This method is therefore quite similar to the conventional technique that consist in mixing together the powders obtained for each oxide, though with this technique, the size of the particles or polymers present in the sol-gel stay within the colloidal range thus making the solution much more homogeneous. Lastly, in order to obtain a good interdiffusion of the metals within the product, it is necessary to heat the sol-gel at rather high temperatures.

Yet, another technique reduces considerably this thermal treatment temperature. It consists in the synthesis of another reactive complex from a direct reaction between all the different metal precursors present altogether in the same solution. For example, in order to create a binary oxide consisting of two metals M and M' with a good homogeneity even at the atomic scale, then the best method still consists in forming a new complex such as O-M-O-M'-O containing the two metals linked by metaloxane bonds. Since the two main types of precursors, alkoxides and metal salts, react completely differently, three different cases must be considered for this binary oxide:

    (1) mixing two alkoxides together
    (2) mixing two metal salts together
    (3) mixing an alkoxide together with a metal salt

MIXING TWO ALKOXIDES TOGETHER

Double alkoxides

Of all the different combinations of precursors possible, those concerning only mixed alkoxides are probably the most studied and the best known. Moreover, it has been determined that the different cations spread throughout the sol-gel with a greater homogeneity if they first form the so-called double-alkoxide complex which, as its name indicates, consist of two metal atoms. Some of those double-alkoxides are simply synthesized by mixing the two initial alkoxides in the same non-aqueous solvent and afterwards refluxing them together in a distillation bench. It is by this procedure, summarized in reaction 2.7-1, that $Al(OBu^s)_3$ and $Mg(OMe)_2$ produce the double alkoxide of Al and Mg [99]. Note that in this mechanism R is the butyl radical C_4H_9 and R' the methyl radical CH_3. Since the Al to Mg ratio of the resulting double-alkoxide is identical to that of the spinel $MgAl_2O_4$, this later can be directly synthesized by the hydrolysis and condensation of the new precursor and this without interdiffusion of the metal cations and at a temperature much lower than necessary with the conventional technique.

$$Mg(OR)_2 \ + \ 2\,Al(OR')_3 \ \Rightarrow \qquad\qquad\qquad\qquad (2.7\text{-}1)$$

Figure 2.7-1 shows the chemical formula of another double-alkoxide obtained this time with Al and Zr. A list of all the different double alkoxides produced up to now together with the different techniques leading to their synthesis has been published by Bradley et al [32]. It includes, for example, the complexes $Na_2[Zn(OEt)_4]$, $Mg[Al(OEt)_4]_2$, and $Al[Zn(OPr^i)_3]_3$ but also various precursors with formula $NaAl(OR)_4$. Some of them have a rather ionic nature and hence dissociate at least to some degree during hydrolysis. Yet, a non-negligible number of them are sufficiently covalent and stable to remain undissociated during the reactions. Those new products and the reactions leading to their formation still constitute a great research field that will inevitably lead to the creation of new and interesting materials.

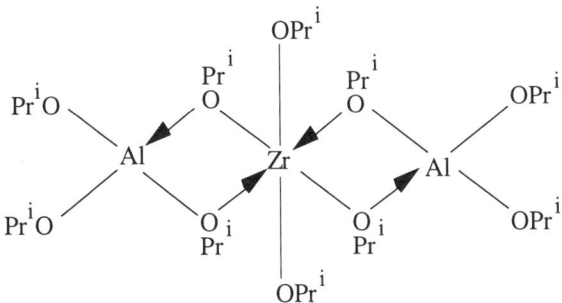

Figure 2.7-1 - Theoretical structure of the mixed alkoxide $ZrAl_2(OPr^i)_{10}$.
After Bradley et al. [32].

Simultaneous hydrolysis of simple alkoxides
In most cases, when two or more simple alkoxides are mixed together in an organic solvent, they immediately attach to one another by metaloxane bonds thus forming either heteropolar or homopolar complexes already stabilized by alkoxy ligands [37]. Alumino-silicates, one of the many families of mixed oxides, include several different compounds with which particularly much research is already accomplished. It is often the simultaneous refluxing method that is applied, as was the case with the isopropoxides of aluminum and silicon that were refluxed together in isopropanol in the study of Mazdiyasni and Brown [100]. Yet, although the mixed metaloxane bond Al-O-Si was not proven to exist in the mixed-alkoxide, a hydroaluminosilicate of formula $Al_3Si(OH)_{13}$ still precipitates after hydrolysis of the refluxed solution in diluted ammoniac. This corresponds to reaction (2.7-2):

$$6Al(OPr^i)_3 + 2\ Si(OPr^i)_4 + xH_2O \Rightarrow 2Al_3Si(OH)_{13}.xH_2O + 26\ Pr^iOH \quad (2.7-2)$$

After heat treatment this mixed hydroxide transforms into mullite of formula $Al_6Si_2O_{13}$.

The simultaneous refluxing method was of course applied to other combinations of precursors. Mixed alkoxides are also produced with yttrium isopropoxide, Y_2O_3, and zirconium of hafnium amyloxides, ZrO_2 or HfO_2 [35]. Titanates such as $PbTiO_3$ and $BaTiO_3$ are synthesized from an initial solution of Ti isopropoxide and Pb or Ba amyloxide refluxed in either isopropanol or benzene [101]. Furthermore, as demonstrated by Mazdiyasni and Brown, the isopropoxide of lanthanides and scandium can dope those titanates alkoxides if mixed together [102]. Smith et al. also produced the first $SrTiO_3$ and $SrZrO_3$ precursors [103]. Finally, of all those mixed alkoxides, the complex titanate known as PLZT has the most interesting electro-optical properties so that its resulting applications are thus quite important. PLZT is a titanate of zirconium, lead, and lanthanum; the precursors of those four metals must therefore be necessarily mixed somewhere in the formation of this ceramic. This titanate was the subject of many research, such as those done by Brown and Mazdiyasni [36], Haertling and Land [104] or Colomban [105]. Its hydrolysis and condensation are summarized in reaction (2.7-3) in which R, R', and R" correspond to the different alkyl groups.

$$0.9Pb(OR'')_2 + 0.1La(OR)_3 + 0.63375Zr(OR')_4 + 0.34125Ti(OR')_4 + 3H_2O \Rightarrow$$

$$Pb_{0.9}La_{0.1}(Zr_{0.63375}Ti_{0.34125})(OH)_6 + 0.3ROH + 3R'OH + 1.8R''OH. \quad (2.7-3)$$

Similar procedures lead to the formation of not only binary and ternary silicate glasses [106-109] but also of other interesting materials such as those synthesized from an initial mixture of TiO_2 and SiO_2, Al_2O_3 and SiO_2 [110], or CaO and SiO_2 precursors [38]. In all cases the presence of the multiple alkoxides could not be demonstrated, although it is quite probable that they do in fact exist for without them the excellent homogeneity of the cations in the final products could not be obtained. Yet another problem remains for the full development of this new technique, it is the search of a common solvent in which all the precursors could be mixed.

Matching the rate of hydrolysis of the different alkoxides
Double metaloxane bonds can eventually link different alkoxides together during hydrolysis and condensation if the precursors are hydrolyzed simultaneously in the same solution, and this even if they are not refluxed together. It is possible, for example, to produce a soluble alumino-siloxane mixed alkoxide from a partially hydrolyzed $Si(OR)_4$ complex and an $Al(OR')_3$ alkoxide. This corresponds to reaction (2.7-4):

$$
\begin{array}{c}
\text{OR} \qquad\qquad\qquad\qquad \text{OR} \\
| \qquad\qquad\qquad\qquad\qquad | \\
\text{RO}-\text{Si}-\text{OH} + \text{Al(OR')}_3 \Rightarrow \text{RO}-\text{Si}-\text{O}-\text{Al}-\text{OR'} + \text{R'OH} \quad (2.7\text{-}4) \\
| \qquad\qquad\qquad\qquad\qquad | \qquad | \\
\text{OR} \qquad\qquad\qquad\qquad \text{OR} \quad \text{OR'}
\end{array}
$$

The problem often lies in the difficulty in matching the rate of hydrolysis of the different alkoxides that renders the succeeding condensation impossible. This is, for instance, the case of a solution of $Si(OEt)_4$ and $Ti(OEt)_4$ alkoxides mixed in ethanol [111]. As demonstrated by Yamane et al., even though the two alkoxides hydrolyze simultaneously when mixed together, $Ti(OEt)_4$ hydrolyzes faster and the final product consists only of distinct SiO_2 and TiO_2 solid oxides. Two distinct oxides still form even if you reduce the gap between the reactions rates by replacing $Ti(OEt)_4$ with the tertiary amyloxide $Ti(OAm^t)_4$. To synthesize a single homogeneous oxide, the rate of the two hydrolysis reactions must be identical; this is achieved by refluxing $Si(OEt)_4$ together with $Ti(OPr^i)_4$ instead of $Ti(OEt)_4$.
This kinetic problem can be overcome by different methods. You can, for instance:

1) slow down the hydrolysis rate of all alkoxides by selecting an appropriate solvent.

2) hydrolyze the solution with an understoichiometric proportion of water.

3) first partially hydrolyze the alkoxide with the slowest hydrolysis rate.

The choice of an appropriate solvent is, as demonstrated by the work of Schmidt et al., a determining factor in reducing to that necessary the speed of the hydrolysis reactions. In fact, the metaloxane bond $=B-O-Si\equiv$ forms only if the two initials precursors, TEOS and ethylborate, are dissolved in ethanol [112]. This is also the case of the complex $NaOR-B(OR)_3-Si(OR)_4$ [88]. The global rates of hydrolysis and condensation can then eventually be controlled either by the amount of condensed water [113] or by that of residual hydroxo ligands [114].

As for hydrolyzing the solution with an understoichiometric proportion of water, this was already explained with the system of TiO_2 and SiO_2 alkoxides [110]. In those conditions, the OR and OH ligands remaining after hydrolysis and condensation are eliminated by further heat treatment while the linear polymers formed can be spinned. Similarly, in order to synthesize borosilicate glasses, the proportion of water needs to be kept lower than a certain predetermined level else the boron alkoxide hydrolyzes faster than TEOS. This predetermined ratio of water depends naturally on the pH. When the aqueous conditions are right, the solution of alkoxides remains clear and gels homogeneously.

Finally, you can also match the rate of hydrolysis of the different alkoxides by first partially hydrolyzing the alkoxide with the slowest reaction rate, as was, for instance, done with the Si alcoholates in the synthesis of mullite by Yoldas[115]. Nogami and Moriya [116] as well as Woignies et al. [117] also successfully applied this technique to the binary systems $SiO_2-B_2O_3$ and $SiO_2-P_2O_5$ as well as to the ternary one $SiO_2-B_2O_3-P_2O_5$. In the first two systems up to 50% of the molar proportion of B_2O_3 and 20% of P_2O_5 by mole were respectively incorporated.

It has been determined experimentally by Yamane et al. that the mixed alkoxides obtained from those three techniques do not always noticeably differ from each other. This is the case of the binary system TiO_2-SiO_2 [111]. You can of course combine those different techniques. It is, for instance, by using all of them that Liu and Wang synthesized the mixed oxide $Al_2O_3-SiO_2-B_2O_3-Na_2O$ [118]. However, with this method, the extreme volatility of boron due to the presence of saturated water vapor reduces the final boron content of the material. This problem is simply solved by working in a low temperature environment between - 20 and - 40°C , and in a dry ice-acetone mixture as was done by Sane in his study of the $Al_2O_3-SiO_2-B_2O_3-Na_2O$ glasses [119].

Two other techniques regulate efficiently the hydrolysis rate of alkoxides. The first one consists in reacting the alkoxides with the water while in its vapor state and in a gaseous environment. It is advantageous in that the process is considerably slowed down. The gas phase can simply consist of air with a controlled hygrometry. This method is the only one that allow the production of transparent glasses with composition $CaO.9SiO_2$ and $CaO.4SiO_2$ [38]. The second and most promising technique consist in chelating the precursor with appropriate additives. Debsikdar, for instance, synthesized $MgAl_2O_4$ gel monoliths by first complexing an Al alkoxide with acetylacetone thus producing a soluble chelate of formula

$(CH_3COCHCOCH_3).Al(OC_4H_9)_2$ that hydrolyzes slowly. Hydrolysis is then sufficiently slowed down compared to condensation for the double metaloxane bond M-O-M' to form between the two metal atoms M and M' [120]. This method agrees with the previously mentioned theory that states that, except for Si, single component oxides form rather polymeric gels.

It is not easy to understand why such a great variety of different structures can be accomplished for a same material by simply slightly modifying the procedure of hydrolysis. Even the distribution of the cations within the product depends on the method followed and, unless the critical parameter is already known, it is quite difficult to reproduce even a given result. These critical parameters differ for each material. For the Ba-Ti-O system, for instance, it has been demonstrated by Ritter et al. that it is the pH and the moment at which hydrolysis is done that critically determine the crystalline phase that the mixed alkoxides first produce [80]. In fact, in this case both $BaTi_4O_9$, $BaTi_5O_{11}$ and $Ba_2Ti_9O_{20}$ can form depending on the different factors.

MIXING TWO METAL SALTS TOGETHER

A good homogeneity of up to the atomic scale as that previously established with the alkoxides cannot be as easily accomplished by mixing metal salts together [37]. Furthermore, when several metal salts are used, they synthesize the final polymetallic compounds according to one of several theoretical phenomena such as coprecipitation, codecomplexation or syncrystallization [57]. In the coprecipitation mechanism, reaction (2.7-5), the two metallic salts mixed in the same solution precipitate simultaneously so that both metal species are present in each of the different complexes.

$$MCl_2 + M'Cl_3 + nOH^- \Rightarrow MM'(OH)_5 . yH_2O \qquad (2.7\text{-}5)$$

For codecomplexation to occur, reaction (2.7-6), the two metal cations must already be linked by a commune ligand within the same complex. It is this ligand that disappears during hydrolysis so as to form a mixed hydroxide precipitate.

$$(MM')(\text{complex ligand})_n + nOH^- \Rightarrow MM'(OH)_n + yH_2O \qquad (2.7\text{-}6)$$

As for the syncrystallization mechanism described in reaction (2.7-7), the solid obtained is the result of the direct precipitation of all the mixed complexes present in the solution. This is often what happens when both a carboxylic acid and ammonia are added to a solution of mixed metal salts.

$$MCl_2 + M'Cl_3 + AOO\text{-}H^+ + NH_4^+ \Rightarrow M(NH_4)_x[M'(AOO)_z]xH_2O \qquad (2.7\text{-}7)$$

Experimentally, it is very difficult to synthesize a double metaloxane bond from metal salts precursors. Hence the products obtained are often a mixture of different colloidal hydroxides. Yet, although this remains the main reason of why alkoxides are more extensively used than metal salts, some exceptions still exist. Rao et al., for instance, were able to synthesize various mixed carbonate solid solutions by mixing aqueous nitrate together with either excess ammonium or sodium carbonate [121]. The mixed carbonates produced included those of formula

(1) $Mn_{1-x}M_xCO_3$ with M= Mg, Ca, Fe, Co, Cd.

(2) $Ca_{1-x}M'_xCO_3$ with M'= Fe, Co.

(3) $Ca_{1-x-y}M'_xM'_yCO_3$ with M' and M" = Mn, Fe, Co.

Other solid solutions of the type $Ln_{1-x}M_x(OH)_3$ with Ln= La, Y or Nd and M= Ni, Fe, Cr or Co, known as hydrotalcites, can also be produced from metal salt precursors with hydroxide, nitrate or cyanide solutions. Lastly, a solid solution with composition $Mo_{1-x}W_xO_3$ can be obtained by adding hot concentrated nitric acid to ammonia solutions of MoO_3 and CoO_3. Further research are however still necessary in order to extend the different applications of mixed metal salt solutions in the synthesis of mixed cation complexes.

MIXING AN ALKOXIDE TOGETHER WITH A METAL SALT

Materials can also be synthesized by mixing together precursors of different types. Hence, when the alkoxide of an element is either not available or insoluble, it is common to use a metal salt simultaneously with another alkoxide or, in some rare cases, with an organometallic. Furthermore, in most cases, the metal salts comprise only the elements necessary in minor proportions in the procedure. Boron, for example, can easily be incorporated in a solution in the form of boric acid H_3BO_3 [122]. The Sr cations necessary for the synthesis of $SrO-SiO_2$ glasses are simply taken from the nitrate $Sr(NO_3)_2$ precursor [123]. The NASICON, a material of formula $Na_9Zr_4Si_2PO_{12}$ synthesized by Colomban, is a good example of what can be achieved by mixing together different types of precursors [57]. In his procedure, the precursors of Zr (Zr iso-propoxide) and Si (TEOS) were both alkoxides. Na, on the other hand, was mixed to the solution either as a butoxide, a salt, NaCl, or a base, NaOH. The P, finally, came either from $(NH_4)H_2PO_4$ or from tributyl phosphate, an organometallic. Another example is the transparent electrical conductor Cd_2SnO_4. Dislich did indeed produced it from Cd acetate and Sn alkoxide [37]. Moreover, if, as mentioned previously, the spinel monoliths $MgAl_2O_4$ can be synthesized from a double alkoxide, it can also be produced by mixing magnesium acetate tetrahydrate and Al-butoxide [124]. In this second method the lack of free MgO or Al_2O_3 in the final product proves that a good homogeneity has been established. These are just a few examples of the many materials that can be synthesized by mixing different type of precursors.

It is generally admitted that the elements originating from metal salt precursors cannot spread within the material as homogeneously as those coming from other type of precursors. This has been particularly well observed by Brinker and Mukherjee in the $SiO_2-B_2O_3-Al_2O_3-Na_2O-BaO$ system in which the Na and Ba came from acetates [125]. Indeed, the conditions necessary for a good diffusion of cations even at the atomic level differs in each case. Silicon cations, for example, are rather easy to incorporate in a material and the silicate complexes like the alkali-borosilicates of Konijnendijk et al. [126] and the silicates doped with rare earth of Mukherjee [127] are relatively easy to synthesize. Ti, on the other hand, remains with some other cations very difficult to insert in complexes, even in the silicates. A good solution to this problem is still to use an appropriate solvent that can produce a chelated complex as done by Higuchi et al. in their synthesis of hexaferrites doped with lanthanum [128]. Those hexaferrites were synthesized by mixing two alkoxides together. The first, La(OPr^i)$_3$, was previously dissolved in a mixture of acetylacetonate and isopropanol, and the second, Ba(OEt)$_2$, was dissolved in ethanol together with an acetylacetonate, Fe(acac)$_3$.

If necessary, fine solid powders can also be mixed to alkoxides and metal salts precursors. BaTiO$_3$, for instance, can also be synthesized by adding a fine Ba(OH)$_2$ powder to Ti(OPr^i)$_4$ [129], or by mixing aqueous citrate solutions or neodecavanate solutions in xylene [80]. Finally, Haertling and Land produced PLZT powders by mixing a fine oxide powder of (PbO) to a metal salt acetate La(Ac)$_3$ and to the alkoxides Zr(OBut^s)$_4$ and Ti(OBut^s)$_4$.

2.8 - NON-OXIDE SOLUTIONS

While the sol-gel processes are largely devoted to the synthesis of oxides, they are not limited to them. The field of materials science known as "inorganic-polymer chemistry" can be described within the sol-gel processes, in so far as reactions occur inside a liquid medium to produce a gel. Inorganic-polymer chemistry is not limited to sol-gel processes, however. For instance, a precursor solution can first be dried without gelation, that is to say without any formation of a 3-dimensional network inside the liquid. Then crosslinking can be achieved in the dry state by ultra-violet radiation [130-132]. No gel point can be observed at any stage, and the process cannot be considered as a sol-gel process. A precursor can also be directly pyrolyzed [133], or used in a vapor state for chemical vapor deposition [82,109], and these processes are at all sol-gel ones. However, the same can be said for the metal salts and the alkoxides.

SYNTHESES OF NITRIDES

Si(NH)$_2$ is a precursor of Si$_3$N$_4$ which was prepared as early as 1881 by Schützenberger and Colson [134]. It can be processed in a liquid medium by reaction between SiCl$_4$ and liquid ammonia at -50°C, as this was done in the early work of Vigouroux and Hugit [135] in 1903. A review of SiC and Si$_3$N$_4$ ceramics synthesized from siloxanes, silanes and silazanes was made by Wills et al. [107]. For instance, Verbeek and Winter [136] studied polysilanes and polysilazanes, Yajima et al. [137] the polymethylsilane [(CH$_3$)$_2$Si]$_n$, polycarbosilanes, and borodiphenylsiloxane [138]. Mazdiyasni et al. [139] studied the hexaphenylcyclotrisilazane and polymethylphenylsilane.

Some of these processes were carried out in a liquid medium, for instance Seyferth and Wiseman [140] synthesized various silazanes (H$_2$SiNH)$_x$ by ammonolysis of dichlorosilanes (e.g., H$_2$SiCl$_2$) in solution in benzene or in a more polar solvent (diethylene, dichloromethane). These silazanes can have a cyclic or a linear structure. A ring-opening polymerization occurs with trace amounts of a catalyst such as NH$_4$Cl [141], which produces a range of compounds including waxy or rubbery solids. At room temperature, these products transform to glassy solids by reaction with N_2. In a study by Cheronis [142] a solution of chlorosilane was prepared in diethylether by ammonolysis with liquid NH_3 at -40°C. After a thermal treatment at 60°C, or in the presence of air containing water vapor at room temperature, crosslinking occurred between the products of the reaction so that a clear material was formed. Hizawa and Najimoto [143], Zhinkin et al. [144] made similar syntheses in dry benzene solutions. In a patent of the Bayer company [145] chlorosilane solutions in methylene chlorine CH_2Cl_2 were refluxed in a NH_3 atmosphere. They transform according to the approximate reaction (2.8-1) [146]. The products of this reaction polymerized between 110°C and 130°C.

These inorganic polymeric materials are very sensitive to hydrolysis. It is therefore necessary to work in solvents very well dehydrated.

$$x \; Cl\!-\!\overset{\displaystyle Cl}{\underset{\displaystyle Cl}{Si}}\!-\!Cl \; + \; y \, NH_3 \; \underset{40°C}{\overset{CH_2Cl_2}{\Longrightarrow}} \; x \; \overset{\displaystyle CH_3}{Si}\!-\!\begin{bmatrix} H \\ N \end{bmatrix}_{1.5} + \; z \, NH_4Cl \quad (2.8\text{-}1)$$

SYNTHESES OF CARBIDES

At the limit between the chemistry of non-oxide ceramics and the chemistry alkoxides, White et al. [147] synthesized organo-silicon gels which themselves transform to SiC. The precursors leading to this carbide include alkoxysilanes $RSi(OR')_3$ comprising three alkoxy groups, and chlorosilanes $RSiCl_3$. The radical R could be saturated or unsaturated and comprised methyl, ethyl, hexyl, vinyl, allyl and phenyl. The chemical transformation was done by hydrolysis as for an alkoxide, according to a reaction which can be written:

$$RSi(OR')_3 + 3 \, H_2O \Rightarrow RSi(OH)_3 + 3 \, R'OH \quad (2.8\text{-}2)$$

The final products are the following "organosilsesquioxanes" which have the general formula:

$$\begin{bmatrix} R \\ | \\ Si\!-\!O \\ | \\ O_{0.5} \end{bmatrix}_n \quad (2.8\text{-}3)$$

Moreover, the hydrolysis of chlorosilane is a reaction which liberates some HCl. This acid is itself a catalyst of gelation which occurs according to an exothermic reaction. Also, Baney [74] showed that methyltrimethoxysilane can be hydrolyzed to produce methylsilsesquioxane $MeSiO_{3/2}$, a monomer which polymerizes according to reaction (2.8-5).

$$MeSi(OMe)_3 + 3H_2O \Rightarrow MeSi(OH)_3 + 3 \, MeOH \quad (2.8\text{-}4)$$

$$n \ MeSi(OH) \ \Rightarrow \ HO \left[\begin{array}{c} Me \\ | \\ Si \\ | \\ O \\ H \end{array} - O \right]_n H \ + \ (n-1) \ H_2O \qquad (2.8-5)$$

The polymerization reaction keeps proceeding until gelation occurs [148]. However, the gel formed is not stable, except when it is maintained in contact with a colloidal silica suspension. The final dry gel which can be made comprises therefore both $MeSiO_{3/2}$ and SiO_2. After heat treatment at 1200°C the ceramic which remains has an average composition close to $Si_{1.0}C_{0.5}O_{1.6}$.

SYNTHESES OF SULFIDES

The sulfides are another class of materials which can be directly synthesized in a liquid medium as a sol or as a gel. Hydrolysis by water which is used to produce oxides, must be replaced by thiolysis with H_2S. Thiolysis occurs for instance with germanium alkoxides, according to reaction (2.8-6) and a GeS_2 gel was first made by Melling [149].

$$2H_2S + Ge(OR)_4 \Rightarrow GeS_2 + 4 \ ROH \qquad (2.8-6)$$

By comparison with OH and OR ligands, SH ligands have an intermediate position. The electron donor ability and the ease to substitute a ligand, increases in the order [150]:

$$-OH < -SH < -OEt < OPr^i \qquad (2.8-7)$$

Hence, SH ligands can substitute for alkoxy OR ligands and the substitution is easier on isopropoxy than on ethoxy groups. However, they will be substituted by OH ligands. That is to say, to be efficient, thiolysis reactions must be carried out in an organic solvent in which all residual traces of water are eliminated. Seddon [151] actually repeated Melling's work and he actually proved by Infra Red spectroscopy that the gel was a mixture of GeS_2 and GeO_2 ; the oxyde was produced by the residual water in the solution.

Similar reactions can be made with other alkoxides, as Sriram and Kumta synthesized TiS_2 powders from $Ti(OPr^i)_4$ and H_2S [152] , while Kuta and Risbud Obtained La_2S_3 from La alkoxides [153]. More recently, Stanic el al. also made tungsten sulfide and Zinc sulfide gels from $W(OEt)_6$ and $Zn(OBu^t)_2$ [150].

Melling proposed that the thiolysis mechanism of $Ge(OEt)_4$ by H_2S is very similar to the attach of H_2O on TEOS. A hydrosufide ligand HS operates a nucleophilic substitution on the precursor by a S_N2 mechanism, while an alcohol molecule separates, as indicated below:

$$\equiv Ge\!-\!OR \;+\; H_2S \;\Rightarrow\; H\!-\!\underset{\substack{\begin{array}{c}H \quad OR\\ \delta+ \quad \delta-\end{array}\\ \text{Leaving group}}}{\overset{\substack{RO \;\; OR\\ \diagdown \diagup}}{\underset{\delta-}{S}\cdots Ge\cdots OR}} \;\overset{k_1}{\Rightarrow}\; \equiv Ge\!-\!SH \;+\; ROH \tag{2.8-8}$$

Once SH ligands are fixed on the Ge precursor, they can undertake condensation reactions by a S_N2 mechanism, so as to build a sulfo bridge Ge-S-Ge and liberate a H_2S or a ROH molecule, as shown in reaction (2.8-9) and (2.8-10).

$$\equiv Ge\!-\!SH \;+\; HS\!-\!Ge\equiv \;\Rightarrow\; RO\!-\!\underset{RO}{\overset{RO}{Ge}}\!-\!\underset{\delta-}{S}\cdots\overset{RO \;\; OR}{Ge}\!-\!OR$$

$$\overset{k_2}{\Rightarrow} \quad \equiv Ge\!-\!S\!-\!Ge\equiv \;+\; H_2S \tag{2.8-9}$$

$$\equiv Ge\!-\!SH \;+\; RO\!-\!Ge\equiv \;\Rightarrow\; RO\!-\!\underset{RO}{\overset{RO}{Ge}}\!-\!\underset{\delta-}{S}\cdots\overset{RO \;\; OR}{Ge}\!-\!OR$$

$$\overset{k_3}{\Rightarrow} \quad \equiv Ge\!-\!S\!-\!Ge\equiv \;+\; ROH \tag{2.8-10}$$

The intermediate $\equiv$Ge-SH compound is termed a mercaptide. Stanic et al. showed that a proportion remained in the final gel [125]. The thiolysis rate constant is k_1 while k_2 is the rate constant for condensation by formation of hydrogen sulfide and k_3 the rate constant for condensation of alcohol. Thiolysis is always the rate determining step, as it is much slower than condenstaion. As the thiolysis rate decreases, such as when the alkoxide concentration decreases, more and more polymeric gels (with smaller colloidal particles) are formed. This results support the opinion that slow hydrolysis and fast condensation tend to favor linear products .

Instead of alkoxides, it is possible to use organometallic compounds MR_n dissolved in toluene [74]. If the solvent is saturaed in H_2S, thiolysis of the precursors leads to the formation of powders. Powder of ZnS, Al_2S_3, $MgAl_2S_3$ and $ZnAl_2S_3$ [74] and a gel of ZnS [154] were produced by this technique. The reaction in the case of Zn for instance can be written:

$$Et_2Zn + H_2S \;\overset{\text{solvent}}{\Rightarrow}\; ZnS + 2EtH \tag{2.8-11}$$

The corresponding organometallic chemistry has some commune aspects with that of alkoxides. For instance, when the H_2S proportion is below the proportion required by stoichiometric, the reaction (2.8-11) does not proceed to completion, and some residual Zn-Et groups remain in the polymers formed. The type of the alkyl group also is important as the stability of the organometallic decreases in the order: Me_2Zn > Et_2Zn > -Bu^t_2Zn.

Many sulfides are very sensitive to decomposition by water. However the sulfide made with the divalent metals such as Zn, Pb, Cd, have strong covalent bonds. Hence they are not easily hydrolyzed by water [155]. Actually, they can be directly synthesized in an aqueous medium by reaction with H_2S [156]. The chemical transformation can be decomposed in deprotonation reactions such as (2.8-12) and polymerization reactions such as (2.8-13), exactly as for the oxides. Instead of hydroxo ligands, hydrosulfide ligands HS are linked to the metal. For CdS particles for instance, these reactions can be written:

$$Cd^{2+} + SH^- \Rightarrow CdS + H^+. \qquad (2.8\text{-}12)$$

$$CdS + Cd^{2+} + SH^- \Rightarrow (CdS)_2 + H^+. \qquad (2.8\text{-}13)$$

A convenient source for sulfur is thioacetamide (TAA). This compound can slowly release its anions at room temperature in an acidic medium, according to the reaction (2.8-14) with water:

$$CH_3-\underset{\underset{S}{\|}}{C}-NH_2 + H_2O \Rightarrow CH_3-\underset{\underset{O}{\|}}{C}-NH_2 + H_2S \qquad (2.8\text{-}14)$$

This source of sulfur was used by Matijevic [26] and Williams [157] to grow monodispersed CdS and ZnS powders. This CdS particles made by Matijevic were actually the first sulfide products synthesized in solution [158]. Other sulfur sources are hydrogen sulfide H_2S, ammonium sulfide $(NH_4)_2S$ [159] and dibenzyl disulfide $(C_6H_5CH_2S)_2$ [160].

OTHER NON-OXIDE SYNTHESIS

We already mentioned in 2.7 that mixed gels can be made by mixed condensation between a phosphorus precursor and transition metal precursors. These materials can also be considered as a product of reactions between metal precursors and phosphoric acid H_3PO_4, in which this acid simply replaces H_2S. The materials which can be obtained include unstable phosphate gels of zirconium [161], iron [162] and aluminum [159].

Moreover, phosphides and arsenides of cadmium and Zinc [163] can be made, as well as carbonates by reaction with CO_2.

The fabrication methods described up to this point can be considered as being "intrinsic" sol-gel process, in so far as a non-oxide compound is directly obtained in a liquid medium. Methods which could be termed "extrinsic" are also commonly used; they consist in making first an oxide gel which is transformed to a non-oxide in a further step. For the carbide SiC, a silica gel is first formed inside a liquid medium in which a source of carbon is dissolved. This carbon source can be a sucrose aqueous solution, or a phenolic or furfurylic resin in solution in acetone. Once the silica gel is formed, it is further calcined to get the carbide.

Borides of the elements in the group IVB (Ti, Zr, Hf), VB (V, Nb, Ta) and VIB (Cr, Mo, W), can be synthesized by the "extrinsic" method. For this purpose, it is necessary to mix a corresponding metal alkoxide with a precursor containing boron such as boric acid or the organometallic trimethyl borate [119]. Borides are formed after drying during a thermal treatment at a higher temperature.

As for the nitrides, they can be obtained by reaction of oxides with nitrogen, or NH_3, at high temperature [164,165]. However, the amount of nitrogen introduced in the material remains low, for instance 6% in silica gels according to Kamiya et al. [56].

These last examples show that, beyond the complex solution chemistry of oxide precursors, an almost virgin field awaits for investigation: the field of non-oxide syntheses in solution. It is possible to imagine a chemistry of metal complexes as rich as for the oxides [79], provided water is replaced by a non-aqueous liquid medium.

2.9 - REFERENCES

1 - Atkins P., "Physical Chemistry", 5th Edition, W.H. Freeman & Company, New-York (1994).

2 - Chabanel M., Gressier P., "Liaison chimique et Spectroscopie", Marketing Ed., Paris (1991)

3 - Jorgensen W.L., Salem L., "The Organic Chemistry Book of Orbitals", Academic Press, NewYork (1973).

4 - Stillinger F.H., Science 209 (1980) 4455.

5 - Jolivet J.P., Henry M., and Livage J., "De la solution à l'oxyde", InterEditions/ CNRS Editions, Paris (1994)

6 - Lagowski J.J., "The Chemistry of Non-Aqueous Systems", Vols.1-4, Academic Press, New-York (1965, 1967, 1970, 1976)

7 - Livage J., Henry M., Sanchez C., Progress in Solid State Chemistry, 18 (1988) 259.

8 - Henry M., Jolivet J-P., Livage J., Structure and Bonding, (1990) 1

9 - Parr R.G., Donnelly R.A., Levy M., Palke W.E., J. Chem. Phys. 68 (1978) 3801.

10 - Chermette H., Lissilour R., L'Actualité Chimique (France) , 4 (1985) 59.

11 - McMurry J., "Organic Chemistry", Brooks/Cole, International Thomson Publishing Company, 4th Edition, New-York (1995).

12 - Shriver D.F., Atkins P., Langford C.H., "Inorganic Chemistry", 2nd Edition, Freeman, New-York (1994).

13 - Baes C.F. Jr, Mesmer R.E., "The Hydrolysis of Cations", Wiley, New-York, (1976)

14 - Schmidt H., Kaiser A., Rudolph M., and Lentz A., in 'Science of Ceramic Chemical processing'", eds. L.L. Hench and D.R, Ulrich, Wiley, New-York (1986), pp 87.

15 - Livage J., Henry M., Jolivet J.P., Sanchez C., Mat. Res. Soc. Bulletin, January 1990, 18 .

16 - Matijevic E., Am. Rev. Mater. Sci., 15 (1985) 483.

17 - MacCarthy G.J., Roy R., J. Am. Ceram. Soc. 54 (1971) 639.

18 - Segal D.L., J. Non-Crystalline Solids 63 (1984) 183.

19 - Woodhead J.L.,   British Patent 1 342 893 , deposited on 10 Feb 1970 (German patent 2 105 912, 26 aug. 1971).

20 - Hardy C.J., in "Sol-Gel Processes for ceramic Nuclear Fuels", Proc. of a Panel, Vienna, IAEA, Vienne (1968) 71.

21 - Schneider W., Comments Inorg. Chem. 3 (1984) 205.

22 - Magini M., J. Inorg. Nucl. Chem. 39 (1977) 409.

23 - Segal D.L., "The preparation of magnetite from iron III ", AERE-R 9976 (1980) 25 pages.

24 - Vermeulen A.C., Geus J.W., Stol R.J., Debruyn P.L., J. of Colloid and Interface Science, 51 (1975) 449.

25 - Matijevic E., In "Ultrastructure Processing of ceramics, Glasses, and composites", L.L.Hench et D.R.Ulrich Eds., Wiley, New-York, (1984) 334.

26 - Cotton F.A., Wilkinson G., "Advanced Inorganic Chemsitry", John Wiley and Sons, New-York (1980).

27 - Akitt J.W., Milic N.B., J. Chem. Soc. Dalton Trans. (1981) 1624.

28 - Singh S.S., Can. J. Soil Sci., 62 (1982) 559 .

29 - Stünzi H., Marty W., Inorg. Chem. 22 (1983) 2145.

30 - Santacesaria E., Tonello M., Storti G., Pace R.C., Carra S., J. Colloid and Interf. Sci., 111 (1986) 44.

31 - Ciavatta L., Ferri D., G. Riccio G., Polyhedron 4 (1985) 15.

32 - Bradley D.C., Mehotra R.C., Gaur D.P., "Metal alkoxides", Academic Press, London (1978).

33 - Bradley D.C., Wardlaw W., "Zirconium alkoxides", J. Chem. Soc. London (1951) 280.

34 - Mazdiyasni K.S., Lynch C.T., Smith II J.S., J. Am. Ceram. Soc. 48 (1965) 372.

35 - Mazdiyasni K.S., Ceramics International 8 (1982) 42.

36 - Brown L.M., Mazdiyasni K.S.), J. of Am. Ceram. Soc. 55 (1972) 541.

37 - Dislich H., J. of Non-Crystalline Solids 57 (1983) 371.

38 - Hayashi T., Saito H., J. Mater. Sci. 15 (1980) 1971.

39 - Gossink R.G., Coenen H.A.M., Engelfriet A.R.C., Verheijke M.L., Verplanke J.C., Mat. Res. Bull. 10 (1975) 35.

40 - Bradley D.C., Carter D.G., Canadian J. Chem., 39 (1961), 1434-.

41 - Yoldas B.E., J. of Mat. Sciences 21 (1986) 1080.

42 - Mazdiyasni K.S., Lynch C.T., Smith II J.S., J. Amer. Ceram. Soc. 50 (1967) 532.

43 - Edwards J.O., Ross V., J. Inorg. Nucl. Chem., 15 (1960) 329.

44 - Karkamar B., Ganguli D., Trans. Indian Ceram. Soc., 44 (1985) 10.

45 - Muhlebach J., Muller K., Schwarzenback G., Inorg. Chem., 9 (1970) 2381.

46 - Brinker C.J., Scherer, G.W., "Sol-Gel Science- The Physics and Chemistry of Sol-Gel Processing", Academic Press, New-York, 1990

47 - Aelion R., Loebel A., Elrich F., J. Am. Chem. Soc. 72 (1950) 5705.

48 - Keefer K.D., in "Better Ceramics Through Chemistry", eds. C.J. Brinker, D.E. Clark and D.R. Ulrich, North Holland, New-York, 1984, pp 15.

49 - Stockmayer W.H., J. Chem. Phys., 11 (1943) 45.

50 - Prassas M., Phalippou J., Hench L.L., Zarzycki J.,   J. Non-crystalline Solids 48 (1982) 79.

51 - Pohl E.R., Osterholtz F.D., in "Molecular Characterization of Composite Interfaces", eds. H. Ishida and G. Kumar, Plenum, New-York (1985)   157.

52 - Swain C.G., Esteve M., and Jones R.H., " J. Chem. Soc., 11 (1949) 965.

53 - Corriu R.J.P., Leclercq D., Vioux A., Pauthe M., Phalippou J., in "Ultrastructure Processing of Advanced Ceramics", Mackenzie J.D., Ulrich D.R., Eds., Wiley, New-York, 1988, p. 113.

54 - Corriu R.J.P, Guerin C., Moreau J.E.E., in "Topics in Stereochemistry", 15, Eliel El., Wilen S.H., Allinger N.L., Eds., New-York, 1984, p. 43.
55 - Brinker C.J., J Non-Crystalline Solids, 99 (1988) 359
56 - Kamiya K., Ohya M., Yoko T., J. of Non-Cryst. Solids 83 (1986) 208.
57 - Colomban P., L'industrie Céramique 792-3 (1985) 186.
58 - Yoldas B.E., J. of non-Crystalline Solids 51 (1982) 105.
59 - Partlow B.P., Yoldas B.E., J. Non-crystalline Solids 46 (1981) 153.
60 - Zelinski B.J.J., Uhlmann D.R., J. Phys. Chem. Solids 45 (1984) 1069.
61 - Brinker C.J., Keefer D.W., Schaefer D.W., Ashley C.S., J. of Non-Crystalline Solids 48 (1982) 47.
62 - Brinker C.J., Bunker B.C., Tallant D.R., Ward K.J., J. de Chimie Physique, 83 (1986) 851.
63 - Debsikdar J.C., J. of non crystalline Solids 86 (1986) 231.
64 - Morosin B., Peercy P.S., Chemical Physics Letters, 40 (1976) 263.
65 - Matsuda S., Kato A., Appl. Catal., 8 (1983) 149.
66 - Barringer E.A., Bowen H.K., Comm. Amer. Soc. (1982) C199-.
67 - Komarneni S., Roy R., Breval E., J. Am. Ceram. Soc., 68 (1985) C41 .
68 - Boyd T., J. Polymer. Sci., 7 (1951) 591 .
69 - Sakka S, in "Treatise on Materials Science and Technology", Edited by M. Tomazawa and R.H. Doremus., Vol 22 (1982) 129.
70 - Sanchez C., Ribot F., New J. Chem., 18 (1994) 1007-1047.
71 - Mehotra R.C., Bohra R., Gaur D.P., "Metal β-diketonates and Allied Derivatives", Academic Press, London, 1978.
72 - Smith G.D., Gaughlan C.N., Campbell J.A., Inorg. Chem., 11 (1972) 2989.
73 - Baney R.H., in "Ultrastructure Processing of ceramics, glasses and composites", Edited by L.L. Hench and D.R.Ulrich. John Wiley, New-York (1984) 245.
74 - Johnson C.E., Hickey D.K., Harris D.C., Mat. Res. Soc. Symp., 73 (1986) 785.
75 - Rochow F.G., "The chemistry of Organometallic Compounds", Wiley, New-York, 1957.
76 - Bonhomme C., Henry M., Livage J., J. Non-Cryst. Solids, 159 (1993) 22
77 - Shea K., Loy D.A., Webster O., J. Am. Chem. Soc., 114 (1992) 6700.
78 - Corriu R.J.P., Moreau J.J.E., Thepot P., Wong Chi Man M., Chem. Mater., 4 (1992) 1217.
79 - Banerjea D., J. Indian Chem. Soc., LI (1974) 90.
80 - Ritter J.J., Roth R.S., Blendell J.E., J. Am. Ceram. Soc. 69 (1986) 155.
81 - Venezky D.L., in "Encyclopedia of Polymer Science and Technology", J. Wiley Ed., Interscience, New-York, 7 (1967) 664.
82 - Kramer J., Prud'homme R.K., J. Colloid and Interface Sci., 118 (1987) 294
83 - Monchâtre G., L'Actualité Chimique, January 1984, 11
84 - Kiselev V.S., Ermolaeva T.A., J. Appl. Chem. USSR 31 (1958) 100.
85 - Wilkes G.L., Orler B., Huang H.H., Polymer Prep., 26 (1985) 300
86 - Sur G.S., Mark J.E., Eur. polym. J., 21 (1985) 1051
87 - Burkhardt E.W., Burford R.R., Deatcher J.H., Chem. Mater. 1 (1989) 767.
88 - Parkhurst C.S., Doyle W.F., Silverman L.A., Singh S., Andersen M.P., McClug D, Wnek G.E., Uhlmann D.R., Mat. Res. Soc. Symp. Proc., 73 (1986) 769.
89 - Mourey T.H., Miller S.M., Wesson J.A., Long T.E., Kelts L.W., Macromolecules, 25 (1992) 45
90 - Chujo Y., Ihara E., Kure S., Suzuki K., Saegusa T., Makromol. Chem., Macromol. symp., 42/43 (1991) 303
91 - Morikawa A., Iyoku Y., kakimoto M., Imai Y., J. Mater. Chem. 2 (1992) 679
92 - Kohjiva S., Ochial K., Yamashita S., J. Non-Cryst. Solids, 119 (1990) 132.

93 - Boulton J.M., Thompson J., Fox H.H., Gorodisher I., Teowee G., Calvert P.D., Uhlmann D.R., Mat. Res. Soc. Symp. Proc., 180 (1990) 987.

94 - Glaser R.H., Wilkes G.L., Polymer Bull. 19 (1988) 51

95 - Coltrain B.K., Landry C.J.T., O'Reilly J.M., Chamberlain A.M., Rakes G.A., Sedita J.S.S., Kels L.W., Landry M.R., Long V.K., Chem. Mater. 5 (1993) 1445

96 - Hoebbel D., Pitsch L., Heidemann D., Jancke H., Hiller W., Z. Anorg. All. Chem., 583 (1990) 133

97 - Schubert U., Arpac E., Glaubitt W., Helmerich A., Chau C., Chem. mater., 3 (1992) 291

98 - Sanchez C., In M., Toledano P., Griesmar P., Mat. Res. Soc. Symp. Proc. 271 (1992) 669

99 - Roy R., J. Amer. Ceram. Soc., 52 (1969) 344.

100 - Mazdiyasni K.S., Brown L.M.,   J. Am. Ceram. Soc. 55 (1972) 548.

101 - Blum J.B., Gurkovich S.R., J. of Mater. Sciences 20 (1985) 4479.

102 - Mazdiyasni K.S., Brown L.M.,   J. Amer. Ceram. Soc. 11 (1971) 539 .

103 - Smith II J.S., Dolloff R.T., Mazdiyasni K.S., J. Amer. Ceram. Soc. 53 (1970) 91.

104 - Haertling G.H., Land C.E., J. Amer. Ceram. Soc. 54 (1971) 1.

105 - Colomban P., L'industrie Céramique 697 (1976) 531.

106 - Zarzycki J., Prassas M., Phalippou J., "Synthesis of glasses from gels. The problem of monolithic gels", J. of Mater. Sciences 17 (1982) 3371-3379.

107 - Wills R.R., Markle R.A., Mukherjee S.P.,   Am. ceram. Soc. Bull., 62 (1983) 904.

108 - Thomas I.M.,   U.S. Patent 3,709,833, January 09 (1973).

109 - Yoldas B.E., , J. Mater. Sci. 12 (1977) 1203.

110 - Sakka S., Kamiya K., in "Process. Kinet. Prop. Electron. Mag. Ceram. Proc. US-Jpn. Semin. Basic Sci. Ceram.", W.Komatsu, R.M. Fulrath, Y.Oishi, M. Koizumi, and S. Somiya, Eds. 1975, (Published in 1976) 135 .

111 - Yamane M., Inoue S., Nakazawa K., J. Non-crystalline Solids 48 (1982) 153.

112 - Schmidt H., Scholze H., Kaiser A., , J. Non-Cryst. Solids 48 (1982) 65.

113 - Schmidt H., Kaiser A., Glastech . Ber. 54 (1981) 338.

114 - Paoting Y., Hsiaoming L., Yuguang W., J. Non-crystalline Solids 52 (1982) 511.

115 - Yoldas B.E., J. Mater. Sci. 14 (1979) 1843.

116 - Nogami M., Moriya Y., J. Non-crystalline Solids 48 (1982) 359.

117 - Woignier T., Phalippou J., Zarzycki J., J. of Non-crystalline Solids 63 (1984) 117.

118 - Liu X., Wang Y., J. of Non-Crystalline Solids 80 (1986) 564.

119 - Sane A.Y., European Patent 0115745- Eltech Systems Corporation, 15. 08 (1984).

120 - Mukherjee S.P., J. Non- Cryst. solids, 42 (1980) 47 .

121 - Rao C.N.R., Gopalakrishnan J., Vidyasagar K., Ganguli A.K., J. Mater. Res. 1 (1986) 280.

122 - Dislich H.,   Angew. Chem. Internat. Edit. Engl. 10 (1971) 363.

123 - Yamane M., Kojima T., J. Non-Cryst. Solids 44 (1981) 181.

124 - Debsikdar J.C., J. of Mater. Sciences 20 (1985) 4454.

125 - Brinker C.J., Mukherjee S.P., Thin Solid Films 77 (1981) 141.

126 - Konijnendijk W.L., von Duuren M., Groenendijk H., Verres Réfractaires, 27 (1973) 11.

127 - Mukherjee S.P., Zarzycki J., Traverse J.P., J. Mater. Sci. 11 (1976) 341.

128 - Higuchi K., Naka S., Hirano S.S., Advanced Ceramic Materials 1 (1986) 104.

129 - Phule P.P., Raghavan S., Risbud S.H., J. Am. Ceram. Soc. 70 (1987) C108.

130 - West R., in "Ultrastructure Processing of ceramics, glasses, and composites", L.L. Hench and D.R.Ulrich. eds., J. Wiley, New-York (1984)235.

131 - Weber W.P., in " Ultrastructure Processing of ceramics, Glasses and Composites", L.L.Hench and D.R.Ulrich Eds. Wiley, New-York (1984) 292.

132 - Gilman H., Chapman D.R., J. Organometal. Chem. 5 (1966) 392.

133 - Rice R.W., Am. Ceram. Soc. Bull. 62 (1983) 889.

134 - Schutzenberger P., Colson A., C.R. Acad. Sci. Paris, 92 (1881) 1508.

135 - Vigouroux E., Hugot U., C.R. Acad. Sci. Paris 136 (1903) 1670.

136 - Verbeek W. Winter G., Ger. Offen. 2,218,960, Bayer, A.G., Nov.8 (1973)).

137 - Yajima S., Okamura K., Hayashi J., Omori M., J. Amer. Ceram. Soc. 59 (1976) 324.

138 - Yajima S., Shishido T., Hamano M., Nature, London, 266 (1977) 522.

139 - Mazdiyasni K.S.,West R., David L.D., J. Amer. Ceram., 61 (1978) 504.

140 - Seyferth D., Wiseman G.H., in "Ultrastructure Processing of ceramics, Glasses and Composites", L.L. Hench and D.R.Ulrich Eds., Wiley, New-York (1984) 265.

141 - Krüger C.R., Rochow E.G., J. Poly. Sci (A)1 (1964) 3179.

142 - Cheronis N.D., KowitZ A., Chamales C., Larson R., Newsbury E., Tripp G., Macartor F.L., Arvan P.G., Gustus E.L., Wildman P., Andler M., J. Am. Leather Chem. Assn. 44 (1949) 282.

143 - Hizawa K., Nojimoto E., Kyogyo Kagaku Zasshi 59 (1956) 1445.

144 - Zhinkin D.Y., Markova N.V., Sobolevskii M.V., Zh. obschch. khim. 33 (1963) 252.

145 - Winter G., Verbeek W., Mansmann M., Ger. Offen. 2,243,527, May 16, 1974 (U.S. Patent 3,892,583).

146 - Mazdiyasni K.S., Cook C.M., J. Am. Ceram. Soc. 56 (1973) 628.

147 - White D.A., Oleff S.M., Boyer R.D., Budinger P.A., Fox J.R., Advanced Ceramic Materials 2 (1987) 45.

148 - Wei G.C., Kennedy C.R., Harris L.A., Amer. Ceram. Soc. Bull. 63 (1984) 1054.

149 - Melling P.J., Am. Ceram. Soc. Bull., 63 (1984) 1427.

150 - Stanic V - "Synthesis of sulfides by the sol-gel process", PhD thesis, Dep. of Chem. Eng. , University of Alberta, Edmonton, AB, Canada.

151 - Seddon a.B., Hogdson S.N.B., J. Amer. Ceram. Soc., 26 (1991) 2599

152 - Sriram M.A., Kumta P.N., J. Amer. Ceram. Soc., 77 (1994) 1381

153 - Kumta P.N., Risbud, S.H., in "Ultrasrtucture Processing of Advanced Materials", edited by D.R. Uhlmann and D.R. Ulrich, John Wiley & Sons, Inc., New-York (1992) 555

154 - Johnson C.E., Hickey D.K., Harris D.C., in SPIE proceedings of "Infrared and Optical Transmitting Materials, 683 (1986) 112

155 - Wilhelmy D.B., Matijevic E., Colloids and Surface 16 (1985) 1-8.

156 - Fischer C-H., Weller H., Lume-Pereira C., Janata E., Henglein A., Ber. Bunsenges. Phys. Chem., 90 (1986) 46.

157 - Williams R., Yocom P.M., Stofko F.S.,   J. Colloid and Interface Sci., 106 (1985) 388.

158 - Matijevic E., Wilhelmy M.D., J. Colloid Interface Sci. 86 (1982) 476.

159 - Gorstein A.E., Baron, N.Y., Dubrovskaya N.Y., Chuprik V.F., USSR Patent 1,574,538 30, June 1990

160 - Czekaj C.L., PhD Thesis, Pennsylvania State University (1987)

161 - Vance E.R., Ahmad F.J., Mat. Res. Soc. Symp. Proc., 15 (1983) 105.

162 - Wilhelmy D.B., Matijevic E., J. Chem. Soc. Faraday Trans. 80 (1984) 563.

163 - Weller H., Fojtik A., Henglein A., Chem. Phys. Lett. 117, (1985) 485.

164 - Brinker C.J., J. Amer. Ceram. Soc. 65 (1982) C4.

165 - Pantano C.G., Glaser P.M., Armbrust D.J., in " Ultrastructure Processing of Ceramics, Glasses and Composites", L.L.Hench and D.R.Ulrich Eds., Wiley, New- York, (1984) 161.

COLLOIDAL PARTICLES
AND SOLS

3.1 - INTRODUCTION

Fine ceramic particles are in great demand to make parts which sinter well at a temperature as low as possible. In this chapter, we examine the direct synthesis of solid particles by nucleation and growth in a liquid medium. Such particles constitute one of the types of important intermediate solid products resulting from the chemical reactions reviewed in the preceding chapter. The second type of solids are gels which can form either directly from a solution, or after an intermediate step comprising colloidal solid particles. Hence, gels must be addressed in a further chapter. In some cases, all particles made in a given liquid medium process have the same shape, as well as a very narrow size dispersion; they are termed "monodispersed" or "monosized".

The synthesis of monodispersed microspheres presents a large theoretical interest to understand the optical, magnetic, electrokinetic, corrosion and catalytic properties of colloidal matter [1]. A recent technique uses organic solutes termed surfactants; a special chapter describes the related phenomenon.

When the particles synthesized in a liquid medium are the end product, they can simply be precipitated. On the contrary it is important to maintain them dispersed in a colloidal suspension, also termed a "sol", if the aim is to obtain a gel in a further step. The interactions involved in the precipitation or the stabilization of a sol are addressed in a separate section. Moreover, a diversity of curious phenomena which also depends on the interactions between particles can occur in sols, such as an ordered spacing of monosized particles. These aspects are discussed a separate section as well as the transformation of sols during aging.

In this chapter, a distinction is made between colloidal particles and conventional particles which have a bigger size. Very useful powders for practical applications can be obtained by agglomeration of colloidal particles, or by sol-gel after gelation. In the latter case, each particle can be viewed as a small spherical gel monolith, with a size of the order or larger than a μm. These non-colloidal particles constitute one practical application of the sol-gel process. They are examined in a further chapter.

3.2 - NUCLEATION AND GROWTH OF PARTICLES IN A LIQUID MEDIUM

RELATIONSHIP BETWEEN HYDROLYSIS, CONDENSATION AND THE FORMATION OF SOLID PARTICLES

As seen in chapter 2, the chemical reactions which transform a precursor solution, produce a large variety of complexes. This chapter addresses the case when these complexes build discrete and rather dense solid particles. In these conditions, the formation of solid particles can be described as a process of nucleation and growth, according to the sequence [1]:

$$\text{hydrolysis} \Rightarrow \text{condensation} \Rightarrow \text{nucleation} \Rightarrow \text{growth} \qquad (3.2\text{-}1)$$

A succession of events of this type is not always the rule: chemical reactions can also proceed so as to build at once a single monolith, by simultaneous merging of polymeric complexes. This last case corresponds to polymeric gelation and are examined in Chapter.4.

Solid particles can form as the result of heterogeneous nucleation on foreign inclusions, such as dusts or the products from uncontrolled hydrolysis. This does not make possible to obtain a well defined quality of powder, for instance easy to sinter. Rather, it is preferable to aim at [2]:
- a fine size (between 0.1 and 1 μm)
- a narrow size distribution
- an equiaxed shape (e.g. spheres)
- a non- agglomerated state.

To achieve these results it is necessary to avoid any heterogeneous nucleation. That is to say all initial foreign solid inclusions must be eliminated by ultrafiltration of the reagents. Moreover, the formation of the particles must be monitored by controlling the successive kinetic steps described further.

The formation of solid particles from a liquid is often described under the term "chemical precipitation", and it can occur with all the precursors described in chapter 2. The first investigations by Ewell and Insley date back from 1935 [3]. They show the existence of a solubility limit for each of the chemical complexes formed by hydrolysis and condensation, which depends on many parameters such as the pH and solute concentration. When the concentration of a complex exceeds its solubility limit, precipitation occurs.

PHASE TRANSFORMATION ACCORDING TO GIBBS

Two main kinetic processes of phase transformations were described by Gibbs [4] (figure 3.2-1). The first one, termed nucleation and growth, involves large change in composition inside a very small volume extent in a preliminary stage. The second one, termed spinodal decomposition, involves composition waves with small change in composition inside a large volume extent, in a preliminary stage.

The formation of solid particles from a precursor solution essentially involves the nucleation and growth process.

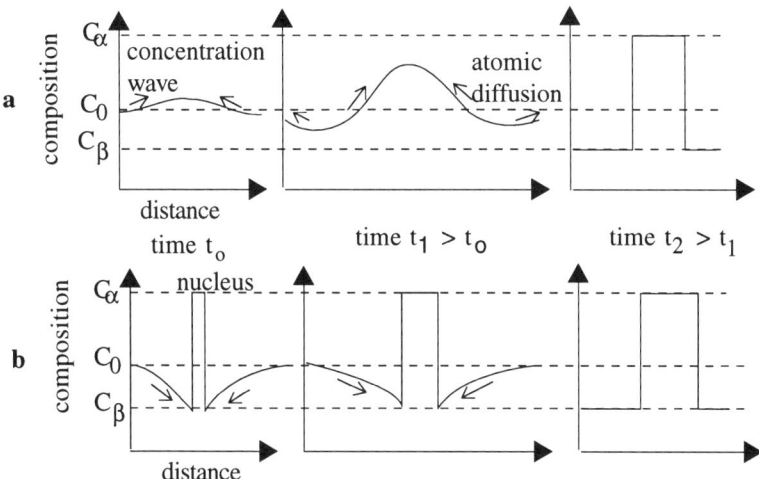

Figure 3.2-1 - Transformation of a phase with initial composition C_O in 2 phases with compositions C_α and C_β: (a) by spinodal decomposition; (b) by nucleation and growth. After Cahn [4].

Gibbs free energy of a spherical particle

In the process of phase transformation by nucleation and growth, the total free energy ΔG_r to form a spherical particle with a radius r comprises a contribution G_s to create the surface of the particle and of a contribution ΔG_{int} which is due to the phase transformation inside the particle

$$\Delta G_r = G_s + \Delta G_{int} \tag{3.2-2}$$

The contribution which is due to the surface is a positive one and it depends on the surface energy γ of the particle.

$$\Delta G_s = 4\pi r^2 \gamma > 0 \tag{3.2-3}$$

The contribution due to phase transformation inside a particle is negative and it depends on the difference ΔG_v between the free energies per unit volume of the new β phase and of the old one; α .

$$\Delta G_{int} = \frac{4}{3}\pi r^3 (G_\beta - G_\alpha) = \frac{4}{3}\pi r^3 \Delta G_v < 0 \tag{3.2-4}$$

The change in the free energy ΔG_r to form the particle, depends on r as illustrated in figure 3.2-2. At the beginning, when r is very small, the surface contribution dominates, so that the free energy change is $\Delta G_r > 0$. Therefore, this is not a stage which is thermodynamically favorable. However, after a critical value $r = r_c$ is passed, the internal contribution starts to dominate so that the evolution becomes thermodynamically favorable .

The value of r_c for which ΔG_r reaches a maximum is given by

$$\frac{\partial \Delta G_r}{\partial r} = 0 \qquad (3.2\text{-}5)$$

$$\text{That is to say } r_c = \frac{-2\gamma}{\Delta G_v} \qquad (3.2\text{-}6)$$

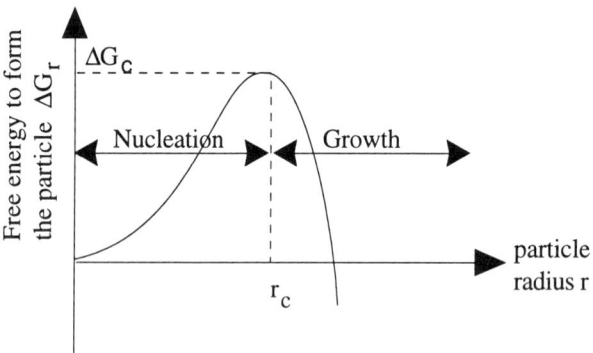

Figure 3.2-2 - Nucleation and growth of a spherical particle.
After Kingery et al. [5a].

When $r<r_c$, the new phase particle is termed an embryo and this stage of the process is termed nucleation. When $r>r_c$ the particle is termed a nucleus and this stage is termed growth. The free energy of a critical nucleus which corresponds to the maximum ΔG_c , is

$$\Delta G_c = \frac{4}{3} \pi r_c^2 \gamma = \frac{16\pi \, \gamma^3}{3(\Delta G_v)^2} \qquad (3.2\text{-}7)$$

Volume Gibbs free energies associated with nucleation and with final transformation

In sol-gel processing, solid particles are made from a solution. Hence, they have a composition which is different from the surrounding initial phase. The volume term ΔG_v, in the Gibbs free energy involved in nucleation, can be estimated with the help of a free energy diagram such as that shown in figure 3.2-3. However, because the solution keeps changing composition as nucleation and growth proceed, ΔG_v keeps also changing. In particular, two different ΔG_v are important, and they are illustrated in Figure 3.2-3 :

- A final $\Delta G_{v,f}$ which can be used when the particles are well developed.

- An initial $\Delta G_{v,n}$ which should be used at the beginning of the transformation, that is to say during nucleation

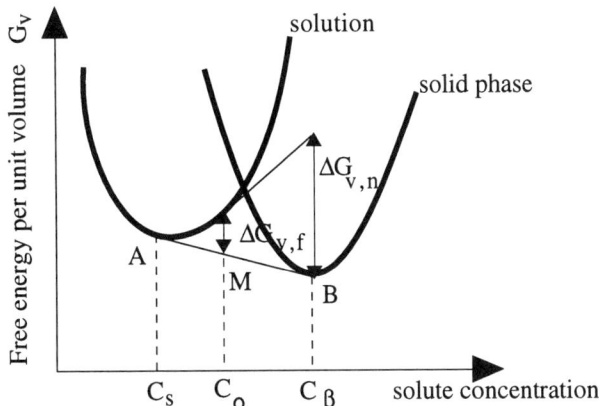

Figure 3.2-3 - $\Delta G_{v,n}$ for nucleation and average $\Delta G_{v,f}$ during a phase transformation with a change in composition. After Kingery et al. [5b].

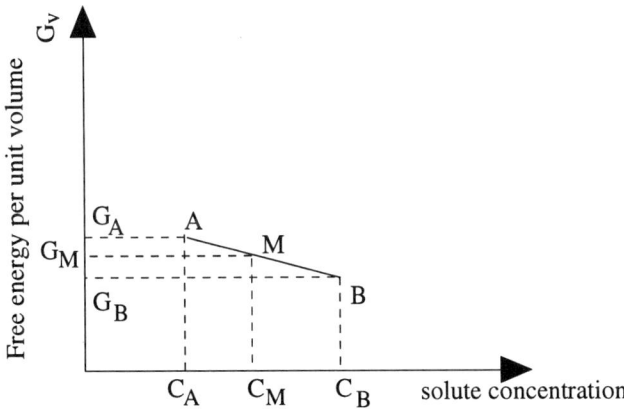

Figure 3.2-4 - linear relationship between an average free energy and the free energy of the two phases.

The final $\Delta G_{v,f}$ corresponds to the transformation

$$\text{Initial solution} \Rightarrow \text{Final solution + particles} \qquad (3.2-8)$$

In the initial solution, the solute concentration is C_0, while in the final solution, the solute concentration is C_s, and the solute concentration in the particles is C_B. Hence, $\Delta G_{v,f}$ can be deduced from the average final energy per unit volume $G_{v,f}$ as

$$\Delta G_{v,f} \quad = \quad G_{v,f} \quad - G(C_0) \qquad (3.2\text{-}9)$$

To derive $\Delta G_{v,f}$, we must consider that the average final composition C_0 and free energy $G_{v,f}$ correspond to point M in Figure 3.2-3, while the final solution and the solid particles correspond to points A and B respectively. These three points A, M, B are aligned and they are again represented in Figure 3.2-4 where the corresponding compositions and free energy are more simply termed C_A, C_M, C_B and G_A, G_B, G_M. The slope of the straight AMB line is

$$\text{slope} = \frac{G_M\text{-}G_A}{C_M\text{-}C_A} = \frac{G_B\text{-}G_A}{C_B\text{-}C_A} \qquad (3.2\text{-}10)$$

that is to say $G_M - G_A = (C_M\text{-}C_A)\dfrac{G_B\text{-}G_A}{C_B\text{-}C_A} \qquad (3.2\text{-}11)$

$$G_M = G_A + (C_M\text{-}C_A)\frac{G_B\text{-}G_A}{C_B\text{-}C_A} \qquad (3.2\text{-}12)$$

$$G_M = G_A\left(1 - \frac{C_M\text{-}C_A}{C_B\text{-}C_A}\right) + G_B \frac{C_M\text{-}C_A}{C_B\text{-}C_A} \qquad (3.2\text{-}13)$$

$$G_M = G_A \frac{C_B\text{-}C_M}{C_B\text{-}C_A} + G_B \frac{C_M\text{-}C_A}{C_B\text{-}C_A} \qquad (3.2\text{-}14)$$

The latter formula can be used to derive $\Delta G_{v,f}$.
It can also be transformed to the following form, which is useful to derive $\Delta G_{v,n}$ later on

$$G_B = G_M + (C_B\text{-}C_M) \frac{G_M\text{-}G_A}{C_M\text{-}C_A} \qquad (3.2\text{-}15)$$

If we now return to Figure 3.2-3 and replace C_A, C_M and C_B respectively by C_s, C_0 and C_β, we see that

$$G_{v,f} \quad = \quad G(C_s) \frac{C_\beta\text{-}C_0}{C_\beta\text{-}C_s} + G(C_\beta) \frac{C_0\text{-}C_s}{C_\beta\text{-}C_s} \qquad (3.2\text{-}16)$$

that is to say $\quad \Delta G_{v,f} = G(C_s) \dfrac{C_\beta\text{-}C_0}{C_\beta\text{-}C_s} + G(C_\beta) \dfrac{C_0\text{-}C_s}{C_\beta\text{-}C_s} - G(C_0) \qquad (3.2\text{-}17)$

$$\Delta G_{v,f} \quad = \quad G(C_\beta) - G(C_0) - (C_\beta\text{-}C_0) \frac{G(C_\beta)\text{-}G(C_s)}{C_\beta\text{-}C_s} \qquad (3.2\text{-}18)$$

To evaluate $\Delta G_{v,n}$ for nucleation, it is necessary to focus first on the equilibrium state of the solution. The solute saturation concentration C_s used in the previous equations and shown in Figure 3.2-3, is the saturation concentration in equilibrium with big solid particles. If we consider solid spherical particle with a very small radius r, as in a nucleus, the equilibrium solute concentration $C_s(r)$ increases drastically as r decreases in agreement with the Thompson Freudlich equation:

$$C_s(r) = C_s \exp \frac{2\gamma V_m}{RTr} \qquad (3.2\text{-}19)$$

where γ is the solid particle surface tension, V_m the molar volume of the complexes which constitute the solid, and R the universal gas constant.

Actually, nucleation is a problem of statistical fluctuations. An equilibrium state with solute concentration C_0 and Gibbs free energy $G(C_0)$ really is an average state. It corresponds to a point M which is located on the free energy curve of the solution, as shown in Figure 3.2-4.

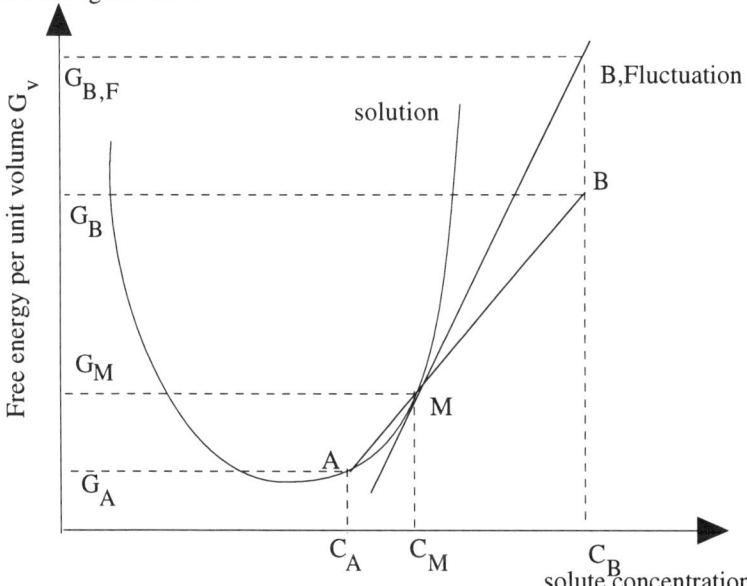

Figure 3.2-5 - Free energy of a fluctuation in a solution.

Within this equilibrium state, local fluctuations in composition and Gibbs free energy occur constantly around point M. One such fluctuation can be described as follows: inside a very small local solution volume the solute concentration is C_B, while the remaining solution has composition C_A. The corresponding fluctuation point B in Figure 3.2-5 is not on the free energy curve of the solution, because this is not an average state, and it can be very far from point M. On the other hand, the volume of the remaining solution is almost the same as the total solution, so that C_A and G_A also correspond to an average equilibrium state of the solution. That is to say the corresponding point A is located on the equilibrium free energy curve of the solution and it is very close to point M (Figure 3.2-5). The first problem is therefore to derive an expression for the free energy of the fluctuation, $G_{B,F}$ as a function of C_B, and this is provided by equation (3.2-16), leading to:

$$G_{B,F} = G_M + (C_B - C_M) \, \frac{G_M - G_A}{C_M - C_A} \tag{3.2-20}$$

As the volume of the fluctuation B comes closer zero, point A comes closer to point M, so that $\dfrac{G_M-G_A}{C_M-C_A}$ must be replaced by $\left(\dfrac{\partial G}{\partial C}\right)_A$ and (3.2-20) becomes

$$G_{B,F} = G_M + (C_B-C_M) \left(\frac{\partial G}{\partial C}\right)_A \qquad (3.2\text{-}21)$$

The straight line AMB becomes the tangent at point M to the free energy curve of the solution. If C_B and C_M are again respectively replaced by C_0 and C_β, we see that the fluctuation free energy is

$$G_{B,F} = G(C_0) + \left(\frac{\partial G}{\partial C}\right)_{C_0} (C_B - C_0) \qquad (3.2\text{-}22)$$

No overall free energy change is involved in the appearance of this fluctuation because this is part of the equilibrium state itself, in the solution. The local increase from G_M to in $G_{B,F}$ in the fluctuation is compensated by a slight decrease from G_M to G_A in the remaining liquid, so as to maintain constant the average solution free energy G_M.

To realize nucleation, it is now just necessary to transform the liquid fluctuation with solute concentration C_β,, to a solid particle with the same solute concentration C_β. The free energy change involved is

$$\Delta G_{v,n} = G_{B,\text{solid}} - G_{B,F} \qquad (3.2\text{-}23)$$

As per definition $G_{B,\text{solid}} = G(C_\beta)$ is the free energy of the solid phase

$$\Delta G_{v,n} = G(C_\beta) - G(C_0) - (C_\beta - C_0)\left(\frac{\partial G}{\partial C}\right)_{C_0} \qquad (3.2\text{-}24)$$

Overall, nucleation corresponds to the transformation

Initial solution $\Rightarrow$ (almost same) Initial solution + (very small)particles   (3.2-25).

As the solute supersaturation $S = \dfrac{C_0}{C_s}$, where C_s is the normal solute saturation shown in Figure 3.2-3, increases, the slope $(\dfrac{\partial G}{\partial C})_{C_0}$ at point M increases very fast, so that $\Delta G_{v,n}$ increases in absolute magnitude. Consequently, ΔG_c and r_c decrease. The free energy of formation of a particle, as a function of its radius r, is modified as shown in Figure 3.2-6. That is to say, nucleation becomes easier.

Nucleation rate
Nucleation is due to statistical thermodynamic fluctuations, in agreement with the Boltzmann statistics. Bimolecular collisions responsible for nucleation are enhanced

by aging, heating and the participation of chemical ligands. Aging gives more time for bigger fluctuations to occur. Heating accelerates all chemical steps. The ligands reviewed in chapter 2 modify the types of complexes from which nucleation can occur. For ionic precursors, the nucleation rate often depends on the metal cations concentration with a reaction order of 4 to 10, although the detailed kinetics depend largely on the nature of each cation [6].

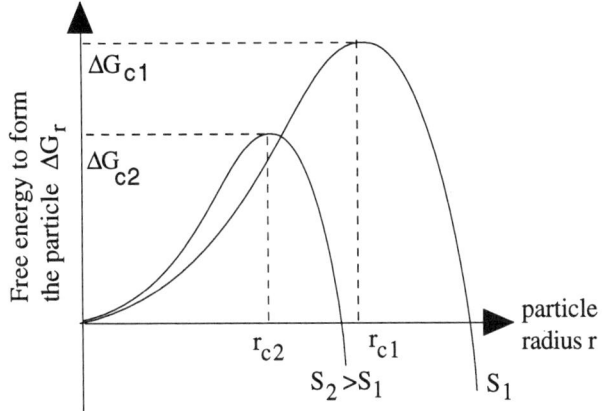

Figure 3.2-6 - Nucleation and growth of a spherical particle
for two values of the supersaturation $S = \dfrac{C_0}{C_s}$.

From the theoretical point of view, the rate of nucleation per unit volume and per second I_V is proportional to:
- the probability $P(\Delta G_c)$ that a thermodynamic fluctuation of Gibbs free energy amplitude ΔG_c can occur :

$$P(\Delta G_c) = \exp\left(-\frac{\Delta G_c}{kT}\right) \qquad (3.2\text{-}26)$$

- the number of molecules per unit volume n_0 which can be used as nucleation centers
- the successful atomic jump frequency Γ from one site to another one:

$$\Gamma = v_0 \exp\left(-\frac{\Delta G^*}{kT}\right) \qquad (3.2\text{-}27)$$

In this equation v_0 is the natural atomic vibration frequency and ΔG^* is the activation energy for diffusion per atom. This activation energy is related to the variation of the diffusion coefficient D of the complexes in solution, and to the temperature T, by

$$D = D_0 \ \exp\left(-\frac{\Delta G^*}{kT}\right) \qquad (3.2\text{-}28)$$

where D_0 is a constant

$$\text{So, } I_v = n_0 v_0 \exp\left(-\frac{\Delta G^*}{kT}\right) \exp\left(-\frac{\Delta G_c}{kT}\right) \qquad (3.2\text{-}29)$$

Equivalent relations can be obtained by replacing Γ with the viscosity of the solution η, according to relation (3.2-30), then the viscosity with the diffusion coefficient according to Stokes relation (3.2-31)

$$\Gamma = \frac{kT}{3 \ \pi \ \lambda^3 \ \eta} \qquad (3.2\text{-}30)$$

$$D = \frac{kT}{3\pi\lambda\eta} \qquad (3.2\text{-}31)$$

where λ is the diameter of a complex solute molecule.
The term ΔG_c in equation 3.2-29 is only due to a particle size effect. It does not take into account an activation energy term due to the olation or oxolation chemical reactions, such as the S_N2 mechanisms examined in chapter 2. If this polymerization activation energy is termed ΔG_P, then equation 3.2-29 must be replaced by:

$$I_v = n_0 v_0 \exp\left(-\frac{\Delta G^*}{kT}\right) \exp\left(-\frac{\Delta G_c + \Delta G_P}{kT}\right) \qquad (3.2\text{-}32)$$

Catalysts which are useful to accelerate the reactions of polymerization can decrease significantly ΔG_P and hence, accelerate nucleation.

Heterogeneous nucleation
Any foreign surface modifies the surface to volume ratio in a nucleus. Hence, in heterogeneous nucleation, ΔG_c must be replaced by:

$$\Delta G_{ch} = \Delta G_c \ f(\theta) \qquad (3.2\text{-}33)$$

where θ is the equilibrium contact angle between the nucleus and the foreign surface and $f(\theta)$ is a function of θ which depends on the geometry of the surface. For a planar foreign surface (Figure 3.2-7)

$$f(\theta) = \frac{(2+\cos\theta)(1-\cos\theta)^2}{4} \qquad (3.2\text{-}34)$$

In this case, the nucleation rate per unit foreign surface area, I_s is :

$$I_s \approx n_s^o \ n_0 \exp -\frac{\Delta G^*}{kT} \ \exp -\frac{\Delta G_{ch}}{kT} \qquad (3.2\text{-}35)$$

where n_s^o is the number of molecules which are in contact with a unit foreign surface area.

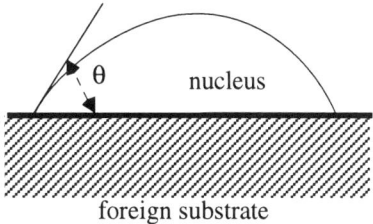

Figure 3.2-7 - Heterogeneous nucleation. After Kingery et al. [5].

LAMER MODEL, FOR HOMOGENEOUS NUCLEATION IN A SOLUTION.

If sufficient care is not taken, nucleation and growth occur simultaneously and they produce a polydispersed population of particles. In particular, when two liquids are suddenly mixed, strong local variations of the reagent concentrations occur and lead to the formation of an irregular distribution of particles. To obtain a monodispersed population, it is necessary to know at least qualitatively the growth and nucleation kinetics [7]. This knowledge can be used to nucleate first one single burst of simultaneous nuclei, then to stop nucleation while letting growth proceed. The most common variants of homogeneous nucleation used to synthesize monodispersed particles include [8]: forced hydrolysis, according to a term introduced by Matijevic; controlled release of anions; and controlled release of cations.

The control of homogeneous nucleation rests on the successive steps illustrated in Figure 3.2-8, according to a model by La Mer [9]. The growth rate G and the nucleation rate N are reported as a function of the concentration C in solute species. This solute can be any complex in the solution, resulting from hydrolysis and condensation of the precursors. The growth rate is non-zero above the solubility limit C_S, while nucleation requires a minimum supersaturation concentration C_{min}^{nu} > C_S to occur (Figure 3.2-8). Depending on the solution composition and on the nature of complexes which can reach more easily the minimum supersaturation necessary for growth, the growth and nucleation rates can be more or less fast with respect to each other. Nucleation occurs in region II, as well as growth. However, if the growth rate is relatively low by comparison with the nucleation rate, it is possible to maintain C above C_{min}^{nu} for an appreciable amount of time so that nuclei can form while they do not grow appreciably. If the relative magnitude of these rates is inverted, it is necessary to let the nucleation proceed only during a very short time. To stop nucleation, the solute concentration must be decreased below C_{min}^{nu} but maintained above C_S, so that only growth can keep proceeding.

Forced hydrolysis

The technique of forced hydrolysis, according to Matijevic's term, consists in letting enough time for nucleation to occur spontaneously in a solution containing a given initial concentration of precursors.

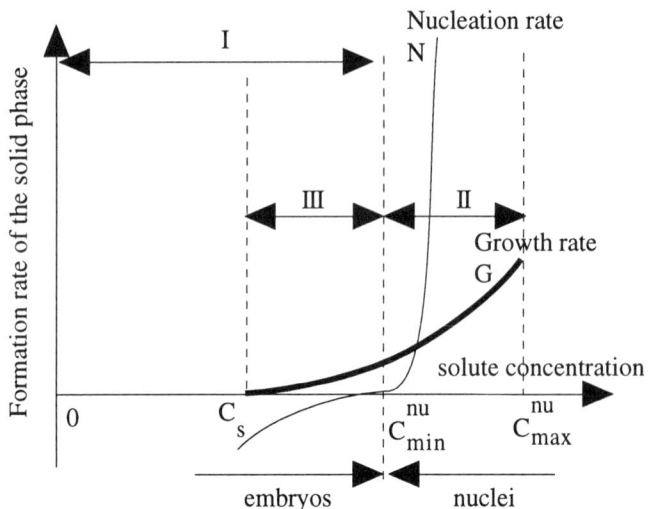

Figure 3.2-8 - Variation of the nucleation and growth rates with the solute concentration. After Haruta and Delmon [1].

When one type of complex, designated after as the solute, reaches its minimum concentration C_{min}^{nu} for nucleation, nuclei appear and grow (Figure 3.2-9). This makes the solute concentration to level out and to start decreasing. Hence, the concentration of solute falls again below the minimum level for nucleation and only growth can keep proceeding.

The initial precursor concentration must be selected so that nucleation only operates during a short time, so that all nuclei virtually grow simultaneously. Very often the precursors are metal salts and the results of hydrolysis, that is to say the complexes responsible for nucleation, depend on the anions. For instance with cobalt [10], only the acetate anions make it possible to obtain monodispersed particles.

Not all complexes enter in the composition of particles. As an example, for chromium hydroxide and basic ferric sulfate, it was shown that only selected complexes solutes, out of a large number, participate into nucleation and growth [11]. Depending on whether growth is carried out from polymeric metal complexes, or from smaller ionic complexes, amorphous or crystalline particles form. The final structure can be modified over a wide range by modifying the anions and the ionic strength of the solution.

Some examples of particles obtained by this process are: disk-like hematite (α-Fe_2O_3), and crystals of magnetite [12]; polyhedral copper oxide ; rodlike crystals of zinc oxide; vanadium pentoxide leaflets and elemental nickel spherulites [13]. One can also mention $Al(OH)_3$ particles with a color due to a chelating ligand [14], stabilized ZrO_2 particles [15] and antimony doped SnO_2 particles for gas sensors [16]. Here also the particle size, shape, and composition depend largely on the anion type and ionic strength. If a lower precursor concentration is used, more time is necessary so that one type of complex can reach supersaturation and nucleate particles. Another possibility is that different complexes, with a slower kinetics of formation but with a more stable thermodynamic structure may nucleate different particles. Hence, the types of particles obtained, for instance amorphous or crystalline, depend largely on the detail of the chemical process. All the precursors mentioned in chapter 2 can be used, including the alkoxides.

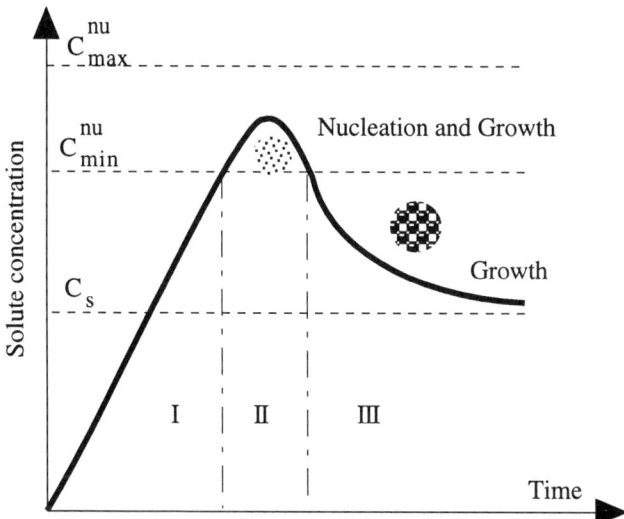

Figure 3.2-9 - Simplified evolution of the solute concentration with time, during the homogeneous nucleation of particles. The domains I, II and III are the same as in figure 3.2-7. After Haruta and Delmon [1].

Controlled release of anions or cations

A controlled release of anions makes it easier to tailor a precise size distribution. For instance, bimodal size distributions can be created, by monitoring the appearance of two bursts of nuclei at different times as illustrated in Figure 3.2-10. A similar result can be reached by controlling the release of cations. The appropriate ions can be released from chemical reagents playing the role of a reservoir, such as chelated complexes made with citrates or formaldehyde. Water is an OH^- reservoir, thioacetamide a S^{2-} reservoir. The release of desired species from a reservoir can be achieved by chemical reactions, such as when adding a reagent or modifying the pH or the temperature.

La Mer used a method of this type to synthesize sulfides by continuous decomposition of thioacetamide [9]. Referring to Figure 3.2-9, the release of the S^{2-} anions was monitored so as to be above the minimum concentration for nucleation C_{min}^{nu} in stage II, but below this minimum in stage III. In the synthesis of MoS_3 by Haruta and al. [17], the precursor was ammonium heptamolybdate, thioacetamide was the H_2S reservoir, acetic acid was used to maintain pH $\approx$ 4 so as to avoid polymerization of Mo complexes before they react with H_2S, and hydrazine N_2H_5 was used to accelerate the hydrolysis of thioacetamide when needed. The main soluble molybdenum species was MoS_4^{2-}. When a small population of nuclei was formed, bigger particles were grown, their diameter varied inversely with the degree supersaturation $S = \dfrac{C}{C_S}$. The standard deviation of the powders obtained in this example was about 15%.

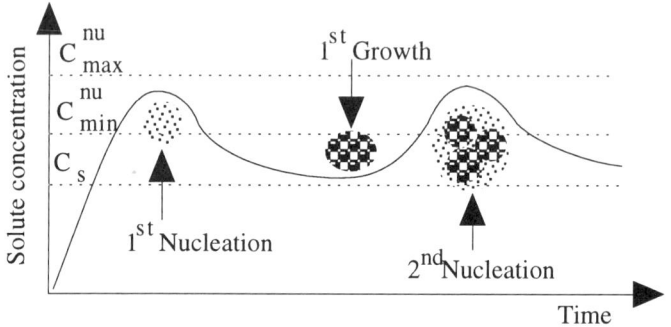

Figure 3.2-10 - Nucleation and growth sequence of a bimodal distribution of particles. After Matijevic [8].

Modification of the temperature
To synthesize CdS particles, Matijevic and Wilhelmy [18] modified the temperature T so as to change the saturation. The reactor was first placed at a temperature T_1 such that the solute concentration C was above C_{S1} but below C_{min1}^{nu} for nucleation at this temperature. In a next step, the temperature was lowered to a value $T_2<T_1$, so that C was above the new supersaturation concentration C_{min2}^{nu} for nucleation. Therefore, particles could nucleate. When the desired number of nuclei was reached, the temperature was again raised to T_1, which stopped nucleation and accelerated the growth process.

Use of separate reactors
Another method consists in using successively two separate reactors. In a first reactor, the solute concentration is $C>C_{min}^{nu}$ and in the second reactor the solute concentration is comprised between C_{min}^{nu} and C_S. The particles nucleate in the first reactor, from where they are transferred to the second reactor. This technique can be adapted to design a continuous flow process.

CRYSTALLINE GROWTH
Growth can be controlled by the diffusion of additional species towards the surface of a particle, or by the fixation onto the surface of this particle [18], as shown in Figure 3.2-11.

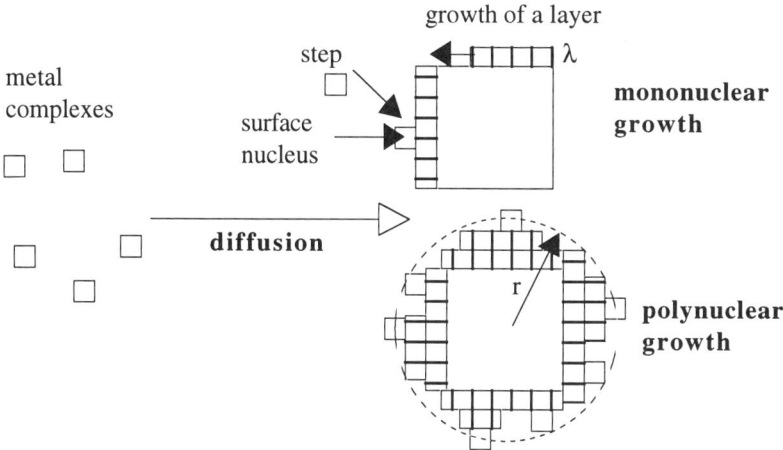

Figure 3.2-11 - The various growth regimes for particles of a new phase.

The effect of these three growth mechanisms on the surface aspect of the particles is summarized in Table 3.2-1.

Table 3.2-1 - Compared effects of the 3 growth mechanisms on the surface characteristics of particles.

Growth regime	mononuclear growth	polynuclear growth	Diffusion controlled
Atomic scale surface	smooth (crystallogra-phic planes)	rough	rough
Macroscopic scale surface	rough (faceted)	smooth	smooth

Growth controlled by the transport of new complexes
In new complexes which must be added to a particle move by diffusion, mathematical modeling gives an evolution of the radius r of particles in agreement with the following differential equation [19]:

$$U = \frac{dr}{dt} = \frac{D(C-C_S)V_m}{r} \qquad (3.2\text{-}36)$$

In this equation, D is the diffusion coefficient of the complexes and V_m is the molar volume of the material constituting the particle. By solving this differential equation the following growth law is obtained [8]:

$$r^2 = k_D t + r_0^2 \qquad (3.2\text{-}37)$$

where k_D is a constant which depends on the diffusion coefficient D.
For two particles with an initial radius difference δr_0, the relative radius difference decreases as time increases according to:

$$\frac{\delta r}{r} = \left(\frac{r_0}{r}\right)^2 \frac{\delta r_0}{r_0} \qquad (3.2\text{-}38)$$

Hence, this growth regime is favorable to the formation of monodispersed particles. Diffusion itself can be enhanced by convection.

Growth controlled by the fixation of new complexes - Mononuclear regime
In growth controlled by surface fixation, each layer needs to nucleate before it spreads onto the particle. The "nucleation" which is concerned in this paragraph is surface nucleation, as the particle already exists. Moreover, two regimes must be distinguished: a mononuclear regime and a polynuclear regime.
In the mononuclear growth regime, a layer has enough time to terminate its growth around a particle before the next layer nucleates. Therefore, growth proceeds layer by layer. This generally corresponds to well defined geometrical shapes where the particle faces are the most dense atomic planes. The growth U rate is proportional to the surface area of the particle. Hence [19]:

$$U = \frac{dr}{dt} = k_{m(}(C) \, r^2 \qquad (3.2\text{-}39)$$

where $k_{m(}(C)$ is a proportionality constant which depends on the solute concentration C. In this case, the relative size difference $\frac{\delta r}{r}$ between two particles which nucleated at a different instant, increases with time as

$$\frac{\delta r}{r} = \frac{r}{r_0} \frac{dr_0}{r_0} \qquad (3.2\text{-}40)$$

This growth regime is not favorable to the formation of monodispersed particles. It occurs when the entropy change which is associated with the phase transformation is [5] :

$$\Delta S_v > 4R \qquad (3.2\text{-}41)$$

where R is the universal gas constant.

Polynuclear growth regime

During polynuclear growth, surface nucleation is so fast that an atomic layer does not have the time to be completed before a second nucleation occurs. This growth regime is also termed normal growth as the growth rate U does not depend on the particle size, nor on the time [20]. The growth rate U depends on the temperature T and on ΔG_v according to the relation

$$U = \frac{dr}{dt} = k_p(C)\,\lambda\Gamma\left(1 - \exp -\frac{\Delta G_v}{kT}\right) \qquad (3.2\text{-}42)$$

where Γ is the successful atomic jump frequency which was previously defined for nucleation $\left(1 - \exp -\dfrac{\Delta G_v}{kT}\right)$ represents the probability to create a surface nucleus and $k_p(C)$ is a constant which depends on the solute concentration. This nucleus provides steps where new molecules can add up and easily fix to a layer. When $\dfrac{\Delta G_v}{kT}$ is small, the normal growth rate can be expressed by replacing Γ with an equivalent expression depending on the diffusion coefficient

$$U = \frac{k_p(C)D}{\lambda}\,\frac{\Delta G_v}{kT} \qquad (3.2\text{-}43)$$

The value ΔG_v for growth which must be used in these equations is different from ΔG_v for nucleation. The correct value to consider here corresponds to the average final $\Delta G_{v,f}$ value illustrated in Figure 3.2-3. This value can also be estimated from the solute supersatutation $S = \dfrac{C}{C_s}$ by

$$\Delta G_{v,f} = \frac{RT}{V_m}\,\mathrm{Ln}\,S \qquad (3.2\text{-}44)$$

The particles rather take on a spherical shape and their radius grow linearly with time [9], that is to say:

$$r = r_0 + k_p t \qquad (3.2\text{-}45)$$

where k_p is a constant. In this regime the relative initial radius difference $\dfrac{\delta r}{r}$ between two particles nucleated at a different instant, attenuates as time increases, as:

$$\frac{\delta r}{r} = \left(\frac{r_0}{r}\right)\frac{\delta r_0}{r_0} \qquad (3.2\text{-}46)$$

This relative size difference attenuation is slower than in the previous diffusion limited regime.

Transitions between growth regimes of particles.
From a comparison of these three mechanisms, it appears that the mononuclear growth regime dominates when the particles are very small. As the particles grow, the surface they offer to form surface nuclei, increases. Hence, a transition to the polynuclear regime is likely to occur. The diffusion of new complexes in the solution towards the particles becomes rate limiting when the particles are bigger, and the solute concentration which remains in the solution is low. The conditions according to which each mechanism predominates, as a function of the particle size and the solute concentration are illustrated in Figure 3.2-12 for ZnS particles synthesized by Williams et al. [20].

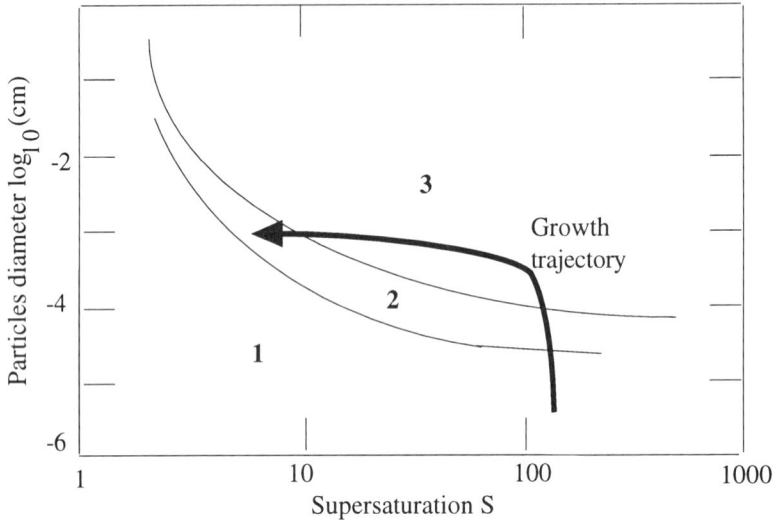

Figure 3.2-12 - Domains of occurrence of the three growth regimes of ZnS particles as a function of the particles diameter as a function of the supersaturation S: (1) mononuclear; (2) polynuclear; (3) diffusion limited. The dark line indicates the experimental growth trajectory of ZnS particles. After Williams et al. [20].

IMPORTANCE OF CRYSTAL DEFECTS. GROWTH OF AMORPHOUS PARTICLES
Crystalline packing defects provide permanent steps where new solute complex molecules can easily fix to a solid particle. An example of such defects is the screw dislocations illustrated in Figure 3.2-13, which helps to grow big monocrystals.
In this case, a distinction between mononuclear or polynuclear growth cannot be made. This is a growth regime known as normal growth. The growth rate U is constant, independent of the particle size. However, it can still be controlled by the diffusion of new complexes. In the case of amorphous particles, many steps are always available for growth. Hence the growth regime is always of the normal type.

Up to this point, no mention was made of the influence of thermal diffusion on the growth of particles. When a liquid material such as a metal solidifies, the difficulty in evacuating the heat of solidification is often responsible for the growth of dendrites. Since many hydrolysis and condensation reactions of oxide precursors are also exothermic, heat evacuation effects probably occur. As the oxide bonds are difficult to melt, it is possible that the local condensation heat accelerates the local condensation reactions, as well as the local diffusion of solutes. Hence, the formation of dendrites may also be possible.

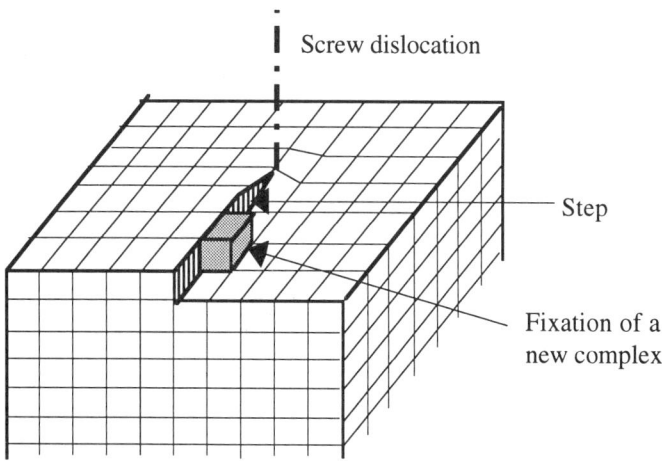

Figure 3.2-13 - Normal growth due to a screw dislocation

INFLUENCE OF THE CATION TYPE

Particles shape
In the present state of knowledge of the complex chemistry of cations in solution, it is not possible to predict which kind of particles shapes will be obtained as a function of the nature of the cations. However, an extensive amount of experimental results on hydrous oxides particles of uniform size and shape was gathered, in particular by Matijevic et al. [7]. A non exhaustive list is provided in Table 3.2-2.
To illustrate the variety of results, spindle shaped hematite particles by Matijevic and Cunas [24] are shown in Figure 3.2-14. Cubic PbS particles obtained by Wihlelmy and Matijevic [44] are shown in Figure 3.2-15. Figure 3.2-16 shows the variety in shape and composition of particles obtained by Ishikawa and Matijevic [29] from $CoSO_4$ solutions where sodium phosphate NaH_2PO_4 and urea were added.
May other particles were grown from alkoxides. Often, with these precursors, the emphasis was placed on the composition, purity and homogeneity rather than on the

size, such as in the mixed oxide powders made by Mazdiyasni [48]. However, it is also possible to obtain monodispersed particles, such as the particles listed in Table 3.2-3. In this table, examples of complex oxides powders made by homogeneous nucleation and growth are also reported. The technique is not limited to oxides; mixed PbS-xCdS and ZnS-xCdS particles could be grown by mixing a solution containing CdS nuclei in solutions of PbS or of ZnS precursors [44].

Table 3.2-2 - Particles made by nucleation and growth from metal salts solutions.

Particles	Shape	References
Hematite Fe_2O_3	ellipsoidal	[21]
	spherical	[22]
	cubic	[23]
	spindle	[24,25]
	disk	[12]
β FeOOH	rod	[26]
Fe_3O_4	spherical	[27]
Alunite $Fe_3(SO_4)_2(OH)_5.2H_2O$		[28]
Co_3O_4	cubic	[10,29]
Ni ferrites	spherical	[30]
Ni-Co ferrites		[31]
Amorphous AlOOH	spherical	[32]
AlOOH	layered spherical	[33]
CrOOH	spherical	[34]
CeO_2	spherical	[35]
Rutile TiO_2		[36,37,38]
ZrO_2		[39]
ZnO	rods	[13]
V_2O_5	leaflets	[13]
Cu(I)O	polyhedral	[13,40]
$Th(OH)_2SO_4$	spherical	[41]
Ni	spherical	[13]
SiO_2		[42]
ZnS	spherical	[43]
PbS	cubic	[44]
CdS	spherical	[45]
$CdCO_3$	cubic	[46]
PbS-xCdS		[44]
ZnS-xCdS		[44]
Al phosphate	spherical	[18]
Fe phosphate	spherical	[47]

Table 3.2-3 - Particles made by nucleation and growth from alkoxide solutions and complex particles.

Particles from alkoxides	ref.	Complex particles	ref.
TiO_2	[38]	$NiFe_2O_4$	[56]
ZrO_2	[49]	$(Mg,Mn)Fe_2O_4$	[57]
ZrO_2-Al_2O_3	[49]	Li_2TiO_3	[58]
SiO_2	[50]	$Pb_{1-x}La_x(Zr_yTi_x)_{1-x/4}O_3$	[59]
SiO_2-B_2O_3	[51]	$HfTiO_4$	[60]
ZnO	[52]	$BaTiO_3$	[61]
Ta_2O_5	[53]	$SrTiO_3$	[62]
HfO_2 (Y doped)	[54]	$SrZrO_3$	[62]
ZrO_2 (Y doped)	[55]	mullite $3Al_2O_3.2SiO_2$	[63]
		$Ba_{1-x}La_xFe_{12}O_{19}$	[64]

Figure 3.2-14 - spindle shaped hematite particles made at 100°C from $FeCl_3$. (a) seen under an electron transmission microscope; (b) seen under a scanning electron microscope. After Matijevic and Cimas [24] with permission.

Very often the particles shape transformation is extremely complex to describe. For instance, ZnS particles synthesized by Rowell et al. [65], or by Williams et al. [20] appear to be faceted at low magnification. However, at a higher magnification, these facets appear to be fibrous. Actually, the particles which can be observed in microscopy are often agglomerates of smaller ones. For instance the TiO_2 particles made by Santacesaria et al. [66] or by Barringer and Bowen [38] are well described by a double size distribution: a first one for crystallites which can be determined by X-rays (6 - 8 nm); a second for the particles observed under an electron microscope (0.2 to 0.5 µm).

A typical example of monodispersed particles made by homogeneous nucleation and growth, are ZnS particles made by Wilhelmy and Matijevic [44]. Their average size is 0.22 µm and their size distribution is indicated Figure 3.2-17. They show an effect known as the Higher Order Tyndall Effect (diffraction of visible monochromatic wavelengths). They are also quite stoichiometric and well crystallized. However, their B.E.T specific surface area, 66 m^2/g, indicates they have an internal structure composed of primary crystallites with a size of 3 to 10 nm [67].

Figure 3.2-15 - PbS particles made from thioacetamide and lead nitrate. After Wilhelmy and Matijevic [44], with permission.

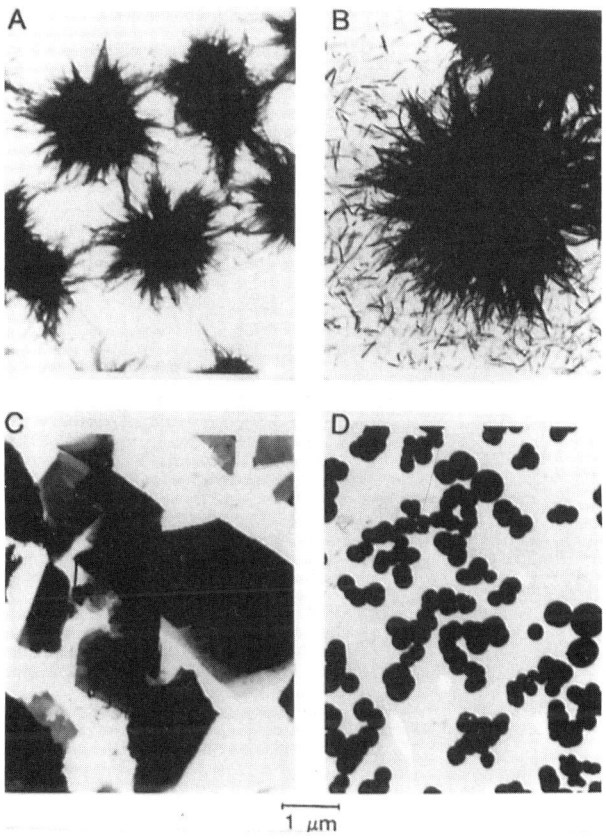

Figure 3.2-16 - Transmission electron microscope micrographs of $CoSO_4$ particles made at 80°C in the presence of urea and sodium phosphate NaH_2PO_4. Composition: (A) and (B) $CoCO_3.xH_2O$; (C) $Co(NH_4)PO_4.xH_2O$;   (D) $Co_3(PO_4)_2.xH_2O$. After Ishikawa and Matijevic [29] with permission.

Growth termination

Anoher interesting phenomenon is the termination of growth at a given particle size. Even the anions seem to have an effect not only on the structure and shape of the particles, but also on their size. As an example, the size of aluminum hydroxide spherical powders is a function of the proportion of SO_4^{2-} ions to Al in solution [68]. In the case of CdS particles [69] and MnO_2 [70] growth stops at a very precise final size which depends on the fabrication parameters (ion concentration, pH, solvent). With alkoxides, the amount of water for hydrolysis is also an important parameter which determines the final particles size. An excess of water usually results in finer powders.

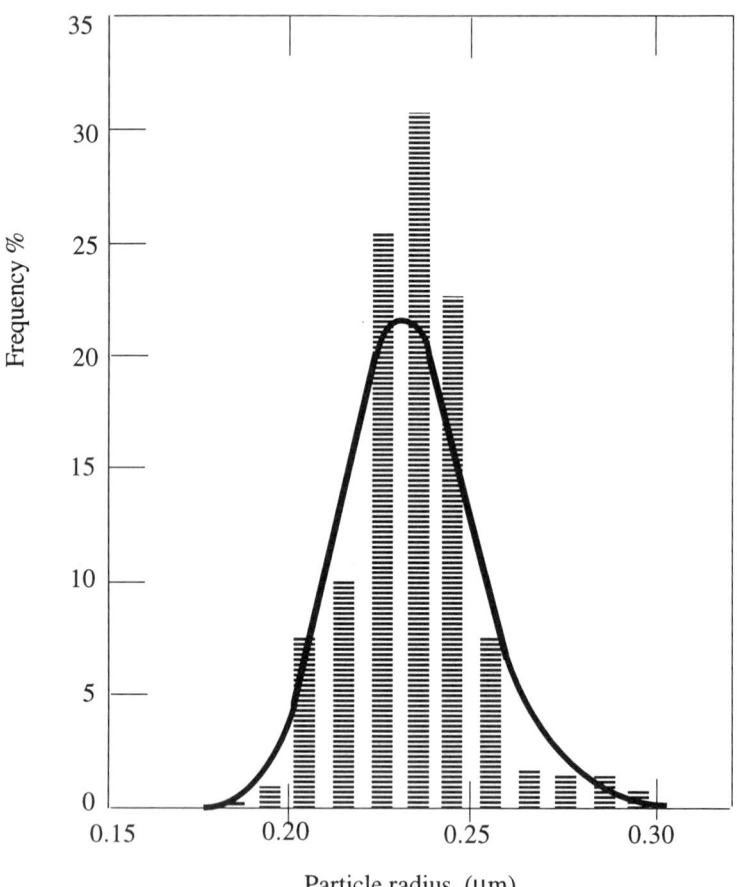

Figure 3.2-17 - Particle size distribution of a ZnS sol as determined from electron microscopy (histogram) and by light scattering (dashed lines). After Wilhelmy and Matijevic [43].

A possible explanation to these phenomena rests on different hydrolysis kinetics for different ligands. In particular, chelating ligands have a profound effect. With titanium alkoxides, the half time life of OR ligands is typically of the order of 20s., while it is of the order of 2.5 h for substituted diketonato groups such as acac [71].

Consequently, olation becomes less prominent than oxolation as the complexation ratio $x = \frac{[\text{acac}]}{[\text{M}]}$ increases. Kinetically, a system behaves as if $x = 0$ at the beginning of hydrolysis because the uncomplexed alkoxide groups hydrolyze first, so that solid TiO_2 particles can start to form. Then, more and more complexed alkoxides are mixed in the condensation product which fix on the particles. These particles end up being capped by acac groups, so that their growth terminates. However, the process does not stop there. During aging, some acac groups are slowly eliminated from the particles surface and condensation can again slowly operate between surface OH groups of neighbor particles: it produces colloidal gels. A behavior similar to that of Ti occurs with the tetravalent metals Zr, Ce, Ti, Sn and even with the trivalent Al. Zr tetragonal particles capped by acac have been made [72]. TiO_2 sols can be made stable at pH < 10 when hydrolyzing $Ti(OPr^i)_3(\text{acac})$; They absorb visible radiation and are stronger reducing agents than Other TiO_2 colloids.

The competition between particles growth and termination is quite general. It explains for instance how so called quantum CdS dots can be obtained by capping nano CdS particles with thiophenols as surface capping agents, illustrated in figure 3.2-18.

These organic additives constitute the first step towards the synthesis of mixed organic-inorganic materials

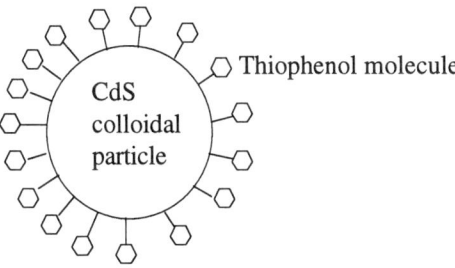

Figure 3.2-18 - Capping of CdS particles with thiophenol complexing molecules. Adapted from [71]

OTHER TECHNIQUES TO SYNTHESIZE SOLID PARTICLES FROM A SOLUTION

Hydrothermal process

It is possible to apply a nucleation and growth technique similar to forced hydrolysis above 100°C in an aqueous medium, in the so-called hydrothermal process. The liquid medium is very often strongly alkaline and comprises strong chelating agents. Here also, some complexes are formed which lead to the formation of hydrous metal oxide powders [68]. As the process is carried at a higher temperature than in usual nucleation and growth, the particles formed can be much bigger, and with a much better defined crystallographic shape.

Electrochemical precipitation
This process was developed by Beer and Planer [73] and applied to ZnO, CuO, SnO_2, TiO_2, NiO, MnO_2, Fe_2O_3 as well as to various ferrite compositions [74]. It consists in letting metal cations migrate from an anode to a cathode in an electrolyte. These cations are oxidized before they reach the cathode, so that they precipitates. The temperature makes it possible to control the particle size.

Particles nucleation and growth inside a gel
To avoid an agglomeration of these crystallites which often constitute a particle made by nucleation and growth, it is possible to make them nucleate and grow inside a gel network [75]. In this way, SbSI, SbSBr, SbSCl and SbSF crystals,   3 nm wide fibrillar iron oxide and 50 nm vanadium oxide ribbons, were made inside silica gels [76,77,78]. For this purpose, some precursors are mixed with the silica precursor before gelation, while the remaining reagents are introduced by diffusion after gelation. Other examples include .

Decompositon of gels
Some particles can only be obtained by phase transformation, from gels which decompose during aging. This is particularly the case with the magnetic oxides.

Use of emulsions
Microeemeulsions are very useful to synthesize oxide as well as non-oxide materials. A microemulsion is a dispersion of fine droplets ($\approx$ 10 nm) of an organic solution in an aqueous solution. They can be used to synthesize non-oxide particles. For this purpose a cation precursor can be dissolved in an organic liquid [79,80]. Monodispersed particles of Fe, Co and Ni borides were prepared by this method The chemical reactions used to produce borides are carried out on these droplets. TiO_2 particles doped with Nb, Ta or B on their surface were also made by similar a technique of this type [81]. Emulsions deserve a special section which is presented in chapter 6.

3.3 - POWDERS NOT SYNTHESIZED IN LIQUID MEDIUM

OTHER POWDER SYNTHESIS PROCESSES
Fine ceramics powders can be synthesized by a variety of methods, concurrent to the sol-gel processes, such as: Physical Vapor Deposition (P.V.D.); Chemical Vapor Deposition (C.V.D.) ; Pyrolysis of precursors; Plasma-spraying; Flame spraying; Spray drying; Freeze-drying; Liquid-drying; and Aerosols hydrolysis. The particles obtained in all these processes are characterized by a hierarchical internal structure

which depends on the precursor and the technique, with primary particles associated in secondary particles, themselves aggregated in more complex units.

Physical and Chemical vapor deposition

The P.V.D. and C.V.D techniques have reached a high level of efficiency, especially when the nucleation and growth in vapor phase are monitored with a pulsed laser, such as developed by Haggerty and al. [82]. They can also be used with alkoxides as the chemical precursors [61,83]. They produce extremely fine and pure particles. Immediately after growth the particles are non agglomerated. However, agglomeration usually occurs at the exit of the furnace [1,2]. The main advantage of these techniques is to allow the production of non-oxide and metal powders, which are not feasible in liquid medium. The final powders are smaller than by precipitation.

Pyrolysis of precursors

The various pyrolysis methods have in common with the sol-gel processes the use of the same precursors, and to require a thermal treatment [48]. They make it possible to produce ZrO_2, Y_2O_3, rare earth oxides powders [84] of Gd_2O_3, Dy_2O_3, Er_2O_3 and Yb_2O_3, some compounds such as mullite $3Al_2O_3.2SiO_2$ [67], $ZrO_2.Y_2O_3$ and HfO_2-Y_2O_3 [60], and some titanates such the PLZT (lead zirconate titanate) [85]. The mechanism of decomposition is different for each precursor. Mazdiyasni et al. [86] have studied the thermal decomposition of the Mg, Al, Ti, isopropoxides in the presence of water vapor. High purity oxides (up to 99.95%) are formed according to the reaction:

$$2MO(OR)_4 \Rightarrow MO_2 + M(OR)_4 \qquad (3.3-1)$$

In the case of Zr, the decomposition of an alkoxide is somewhat different and proceeds according to the two following steps:

$$Zr(OR)_4ROH \Rightarrow Zr(OR)_4 + ROH \qquad \text{at } 125°C \qquad (3.3-2)$$
$$\text{then: } Zr(OR)_4 \Rightarrow ZrO_2 + 2ROH + \text{olefin} \qquad \text{at } 300°C. \qquad (3.3-3)$$

The particles purity is better than 99.99%, the size of 80% of them is less than 10nm. They are cubic and transform to the monoclinic stable phase at 400°C.

Spraying techniques

A variation of the co-decomposition technique consists in maintaining the various components in a solution and to decompose them by spraying onto a hot surface. It is also possible to melt a precursor in a flame or a plasma and to spray it onto a cold surface [87]. A plasma can be realized in an arc with induction coupling [86]. Compounds which were synthesized by such techniques include $(Ni,Zn)Fe_2O_4$, $PbCrO_4$ and $Cu_2Cr_2O_4$ [88,89], Magnesium-Manganese ferrites [90], $BaTiO_3$ made from acetates and lactates [91], as well as TiO_2 and SiO_2 made from alkoxides in the

Cab-O-Sil $^{®}$ process of the Cabot Corporation and the Aerosil$^{®}$ process of the Degussa Company. The size of the particles are in a range from 10 nm to 200 nm. According to other processes, a homogeneous mixture is atomized in fine droplets which are dried or frozen at low temperature [87]. In the spray-drying technique, the liquid droplets are sprayed with an ultrasonic vibrator [92], then they are dried in a cross flow of heated air. Spraying can also be achieved by spinning, which usually produces agglomerates with a size of the order of 20 μm. Pneumatic vibration produces spheres with a diameter of the order of 1 μm [87]. All kinds of precursors can be used industrially. For instance, metal salts solution are often used and they produce hollow particles when they are made from metal salts with an endothermic decomposition reaction. Complex compositions can be tailored, some produce powders which densify well , such as mullite. In other cases, the compound does not densify well, such as with spinel [93] and ferrites [94].

Freeze Drying

In the technique of Freeze-drying the constituents of small droplets are immobilized by freezing at low temperature. Freezing can be done at - 80°C in hexane, where water transforms to ice without segregation and is sublimated without passing through the liquid state. This technique was first used by Schnettler et al. [95] to prepare reactive alumina, Al_2O_3-Cr_2O_3, spinel $MgAl_2O_4$, and Sm_2O_3. With some salts such as $Fe_2(SO_4)_3$ or with some materials such CaO stabilized ZrO_2, the freezing point is too low [96]. In this case, the addition of NH_4OH according to a method by Jaeger [97], makes it possible to obtain colloidal solutions which can be freeze-dried. This way has made possible to synthesize carbides from colloidal suspensions of graphite in ammoniac metallic solutions [98].

Liquid Drying

The method of drying in a liquid medium has been initially developed in England to synthesize radioactive pellets, such as UO_2 and ThO_2 [99]. Then it has been extended to other materials such as ZrO_2 [99],conducting powders of SnO_2-In_2O_3 [100], Al_2O_3 and $(MgMn)Fe_2O_4$ [81,97]. In this technique, a solution is atomized in the vortex of a rapidly stirred hygroscopic liquid such as acetone, methanol or isofluoropropanol, with the help of emulsifying agents [79,80]. Long chain amines make it possible to adsorb the nitrate anions. Then the emulsion is heated under vacuum, or in a microwave oven, to evaporate the water . The particles size can be monitored in a wide range; for instance Matthews and Swanson [101] have developed a process to obtain a mixture of microspheres with different size: 15 μm, 100 μm and 300 μm.

Aerosols hydrolysis

Monodispersed powders can be made by the technique of aerosols where the chemical reactions are carried out on droplets in a vapor phase [10]. For instance, powders of TiO_2 [37], Al_2O_3 [102], and various other compounds [103], have been obtained. The aerosols make it possible to aim at a given size and purity and they provide an excellent atomic homogeneity in compounds [8].

ADVANTAGES OF THE SOL-GEL PROCESS

The powders obtained by the previous methods often have a small size (100-1000 nm), a narrow size distribution and the variety of their shapes. They can also be obtained very pure when made from organic precursors such as the alkoxides. At last they could be extended to sulfides, carbides [104] and borides, once the proper investigations have been carried out.

The main advantages of sol-gel powders are [105]: (1) an easier control of aggregation; (2) a good stability in time, often for years; (3) in the case of multicomponent oxides, a good dispersion of the cations on a fine scale; (4) a good attenuation of the contamination problems due to confinement of the powders in a liquid which is often water; (5) reversibility of the sol-gel transition for some systems, which makes it possible to make corrections. For all these reasons it is possible to say that the sol-gel processes are often more convenient than other techniques [106].

3.4 - SOLS

PEPTIZATION

Colloidal particles, that is to say submicron size particles, are submitted to Brownian motion [107] when they are dispersed in a liquid time. These colloidal particles are therefore bound to collide with each other and two types of phenomena can occur during a collision: the two particles can either remain associated with each other or rebound and remain independent. In the first case, aggregation occurs and a precipitate forms. In the second case, the particles remain dispersed in the liquid, they are said to be peptized. This is an important state, as it makes it possible to control the powder transformation in further processing steps.

Hence peptization can be defined as an action of dispersing colloidal particles in a liquid medium, so that this dispersion remains stable. The stable powder dispersion can be termed a colloidal suspension, or a sol. The stability which is involved is kinetic, not thermodynamic. As a matter of fact, the surface to volume ratio of spherical particles of radius r is :

$$\frac{S}{V} = \frac{3}{r} \tag{3.4-1}$$

This ratio goes to infinity as r goes towards zero. That is to say, colloidal particles have a high surface to volume ratio and a high total surface energy. Therefore, the only stable thermodynamic state consists in sintering all of them in a single monolith. Aggregation of these colloidal particles is a first step in this direction. The kinetic stability of a sol is due to the fact that this first step can be very slow. In practice, the dispersion must remain stable for a sufficient time. This depends on interaction forces between particles. Most often, the dilution in the solvent is sufficient to consider that only interactions between near neighbor particle pairs are significant. These interactions can be divided into: (1) Van der Waals interactions, which mostly introduce attractive forces between the particles; (2) electrostatic interactions which introduce repulsive forces between the particles; (3) steric interactions which occur between the solvent and organic macromolecules adsorbed on the surface of the particles; (4) in some systems magnetic interactions.

When only the first two types of interactions are important, the sols are termed "electrostatic" according to Overbeek. A theory developed by Derjaguin, Landau, Verwey and Overbeek, known as D.L.V.O theory [108,109,110], makes it possible to derive the total interaction energy between two particles. This theory can be used to a variable degree of refinement and possibly takes into account chemical complexation reactions occurring in the liquid medium, such as in a study by Matijevic [111].

When the third type of interaction dominates, the stabilization is termed "steric" [112]. As for the magnetic interactions they concern only a few materials.

ELECTROSTATIC SOLS

Van der Waals interactions at the molecular level
Van der Waals interactions have their origin in the dipolar nature of all atoms and molecules. As shown in Figure 3.4-1, the interaction force between two electrical dipoles is usually attractive. However, several types of contributions participate in the Van der Waals interactions, they are summarized in Table 3.4-1.

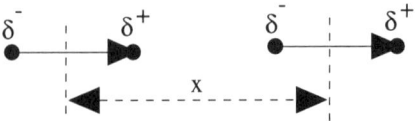

Figure 3.4-1 - Attractive force between two electrical dipoles

These various contributions sum up to the following interaction energy [113]:

$$\Sigma_A = \xi x^{-12} - \beta x^{-6} \quad \text{with } \alpha, \beta > 0 \qquad (3.4-2)$$

The term ξx^{-12} corresponds to a repulsion force which dominates at hard contact between two molecules, when their external electronic orbitals start to interpenetrate. If $x > \dfrac{c}{\nu}$ (c velocity of light, ν vibration frequency of the dipoles) the term βx^{-6} must be replaced by a term δx^{-7} which corresponds to a modified attraction term known as the retarded effect. It is due to a time lag in the transmission of dipole vibrations between two molecules; it becomes dominant at distances large enough, practically when $x > 10$ nm [114]. As a good approximation, except at hard contact, the overall Van der Waals interaction energy between two molecules can be represented by the relationship:

$$\Sigma_A = -\beta x^{-6} \quad \text{with } \beta > 0 \qquad (3.4-3)$$

Table 3.4-1 - Contributions to the Van der Waals interaction.
Adapted from Hiemenz [113]

Type	Interaction energy $\Phi = k\, x^{-n}$	Effect	Name
ion - ion	$\Phi = k\, x^{-1}$ $k > 0$ or < 0	attraction or repulsion depending on the ion charges	Coulomb
ion - permanent dipole	$\Phi = k\, x^{-2}$ $k > 0$ or < 0	attraction or repulsion depending on the ion sign and dipole orientation	Coulomb
permanent dipole - permanent dipole	$\Phi = k\, x^{-3}$ $k > 0$ or < 0	attraction or repulsion depending on the dipoles orientation	Coulomb
permanent dipole - permanent dipole	$\Phi = -\,\beta_K\, x^{-6}$ $\beta_K > 0$	attraction , by rotation of dipoles	Keesom
permanent dipole - induced dipole	$\Phi = -\,\beta_D\, x^{-6}$ $\beta_D > 0$	attraction	Debye
induced dipole - induced dipole	$\Phi = -\,\beta_L\, x^{-6}$ $\beta_L > 0$	attraction	London
or induced dipole - induced dipole, retarded	$\Phi = -\,\delta\, x^{-7}$ $\delta > 0$	attraction	Casimir and Polder
valence shell interpenetration	$\Phi = \xi\, x^{-12}$ $\xi > 0$	repulsion	

Van der Waals interaction between colloidal particles.
To derive the Van der Waals interaction between two macroscopic particles placed in
a liquid medium, the molecular interaction as expressed in equation (3.4-3) must be
summed up for all pairs of molecules composed of one molecule in each particle, as
well as to all pairs of molecules with one molecule in a particle and one molecule in
the solvent. Important contributions to this field were made by Hamaker [115] for
discrete molecules, and by Lifshits who replaced the discrete atoms by a continuous
medium. Integration of Σ_A over two spherical particles of radius r separated by a
distance S_o (Figure 3.4-2) gives the interaction energy [113]:

$$\Phi_A = -\frac{A}{6}\left(\frac{2r^2}{S_o^{\,2}+4rS_o} + \frac{2r^2}{S_o^{\,2}+4rS_o+4r^2} + \ln\left(\frac{S_o^{\,2}+4rS_o}{S_o^{\,2}+4rS_o+4r^2} \right) \right) \qquad (3.4\text{-}4)$$

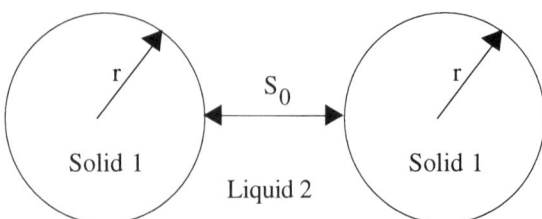

Figure 3.4-2 - Pair of particles used to derive the Van der Waals interaction

A is a positive constant termed the Hamaker constant, which depends on the polarization properties of the molecules in the two particles and in the medium which separates them. Hamaker constants A_i have the dimension of an energy and typical values from 3.5 to 8 x 10^{-20} J. A few values are provided in Table 3.4-2.

If A_1 designates the Hamaker constant for the matter inside a particle 1, A_2 for the matter inside the liquid which separates two particles, and A_3 for the matter inside a particle 3, the Hamaker constant A_{123} for particles 1 and 3 separated by liquid medium 2 which is not polar (not water) can be estimated by [116]:

$$A_{123} \approx (\sqrt{A_1} - \sqrt{A_2})(\sqrt{A_1} - \sqrt{A_3}) \qquad (3.4\text{-}5)$$

This can be a positive or a negative constant. In the latter case, a repulsion occur between the two particles. This formula also gives the Hamaker constant A_{121} for two particles 1 dispersed in the liquid medium 2 (except water):

$$A_{121} \approx (\sqrt{A_1} - \sqrt{A_2})^2 \qquad (3.4\text{-}6)$$

In this case the Hamaker constant is always positive, an attraction between the two particles always occurs.

For a separation S_o between the particles such that $\dfrac{S_o}{r} < 1$, a good approximation of equation (3.4-4) is [107]:

$$\Phi_A = -\frac{AL}{12S_o}\left(1 + 2\frac{S_o}{L}\ln\frac{S_o}{L} - \frac{15}{16}\left(\frac{S_o}{L}\right)^2 - \frac{3}{32}\left(\frac{S_o}{L}\right)^3\right) \qquad (3.4\text{-}7)$$

where $L = R + \dfrac{3S_o}{4}$

The most simple expression of this Van der Waals interaction when $\dfrac{S_o}{r} \ll 1$ is

$$\Phi_A \approx -\frac{Ar}{12S_o} \qquad (3.4\text{-}8)$$

Other simple expressions of the Van der Waals attraction are provided in Table 3.4-3.

Table 3.4-2 - Hamaker unretarded constants A_i for a few common materials. Adapted from Masliyah [116]

Materials	A_i $(10^{-20}J)$
Metals	16.2-45.5
Gold	45.3
Oxides	10.5-15.5
Al_2O_3	15.4
MgO	10.5
SiO_2 (fused)	6.5
SiO_2 (quartz)	8.8
ionic crystals	6.3-15.3
CaF_2	7.2
Calcite	10.1
Polymers	6.15-6.6
Polyvinyl chloride	10.82
Polyethylene oxide	7.51
Water	4.35
Acetone	4.20
Carbon tetrachloride	4.78
Chlorobenzene	5.89
Ethyl acetate	4.17
Hexane	4.32
Toluene	5.40

Table - 3.4-3 - Simple formulas for the Van der Waals attraction between two particles. Adapted from Hiemenz [113]

Particles	Φ_A	Conditions
Two spheres of same radius r	$-\dfrac{Ar}{12S_o}$	$r \gg S_0$
Two spheres with unequal radii r_1 and r_2	$-\dfrac{Ar_1r_2}{6S_o(r_1+r_2)}$	r_1 and $r_2 \gg S_0$
Two parallel plates with thickness δ Interaction per unit area	$-\dfrac{A}{12\pi}\left(\dfrac{1}{S_o^2}+\dfrac{1}{(2\delta+S_o)^2}+\dfrac{1}{(\delta+S_o)^2}\right)$	
Two blocks Interaction per unit area	$-\dfrac{A}{12\pi S_o^2}$	

Electrical double layers around colloidal particles

When solid particles are dispersed in a liquid medium which contains an electrolyte, some specific ions are often preferentially adsorbed on the surface of the particles. That is to say, the surface of these particles carry a fixed electrical charge density. Or, the surface of these particles is brought to an electrical potential Ψ_o. The ions which are adsorbed on the particles are termed electric potential-determining ions. For oxide particles, in most cases, these potential determining ions are H^+ and OH^-, so that the pH of the liquid medium where the particles are dispersed is very important. Each oxide is actually characterized by a particular pH for which the particles are not charged, known as the "zero-point charge" or z.p.c. At pH > z.p.c the particles adsorb more OH^- anions than H^+ ions, so that they are negatively charged. On the other at pH < z.p.c, the particles are positively charged. For instance in the case of alumina and its hydroxides the adsorption reactions of electric potential determining ions can be written:

$$
-\underset{H}{\overset{H}{O}}{}^+ + H_2O \underset{H_3\overset{+}{O}}{\xleftrightarrow{\hspace{1cm}}} -OH \underset{OH^-}{\xleftrightarrow{\hspace{1cm}}} -O^- + H_2O \qquad (3.4\text{-}9)
$$

$$
\text{pH} < 9.4 \qquad\qquad \text{pH} = 9.4 \qquad\qquad \text{pH} > 9.4
$$

The condition pH = 9.4 corresponds to neutral particles and defines the z.p.c of alumina. For pH> 9.4 alumina particles are negatively charged. For pH< 9.4 they are positively charged.

As pointed out by Verwey [117] the charges adsorbed on oxide particles are also O^2, OH^- and H^+ in non aqueous solvents such as aprotic acetone or protic alcohols. Hence it is possible to stabilize electrostatically a sol in these solvents. Similarly, in the case of a nitride powder such as Si_3N_4 dispersed in an aqueous medium, Shaw and Pethica [118] have shown that a z.p.c. existed at a pH between 5 and 6. Such a value indicates that the surface of Si_3N_4 does not look like that of silica which has a z.p.c ≈ 2.5. A possible explanation is that the surface of nitride particles can carry terminal amine NH_2 groups, instead of OH groups for the oxides. Another possibility is that a thin oxynitride layer can cover the particles, in agreement with experimental observations made by TEM, XPS and SIMS [106] which showed the occurrence of such an oxynitride layer, 3 to 5nm thick. In practice, it is possible to obtain stable Si_3N_4 sols at high pH, beyond pH = 11.

The ions in the liquid medium which are not absorbed on a particle are termed "indifferent" or electric potential "non determining" ions. Since a sol is electrically neutral, the ions with a charge opposite to that of the particles are in excess around the particles (Figure 3.4-3), they are termed counterions. These ions are not adsorbed on the particles. They move constantly by Brownian motion but they are statistically attracted by the particles, so that they constitute a diffuse layer around the particles. Their concentration is the highest immediately close to a particle and it decreases progressively at an increasing distance from a particle, so as to reach the average concentration in the liquid medium far from each particle.

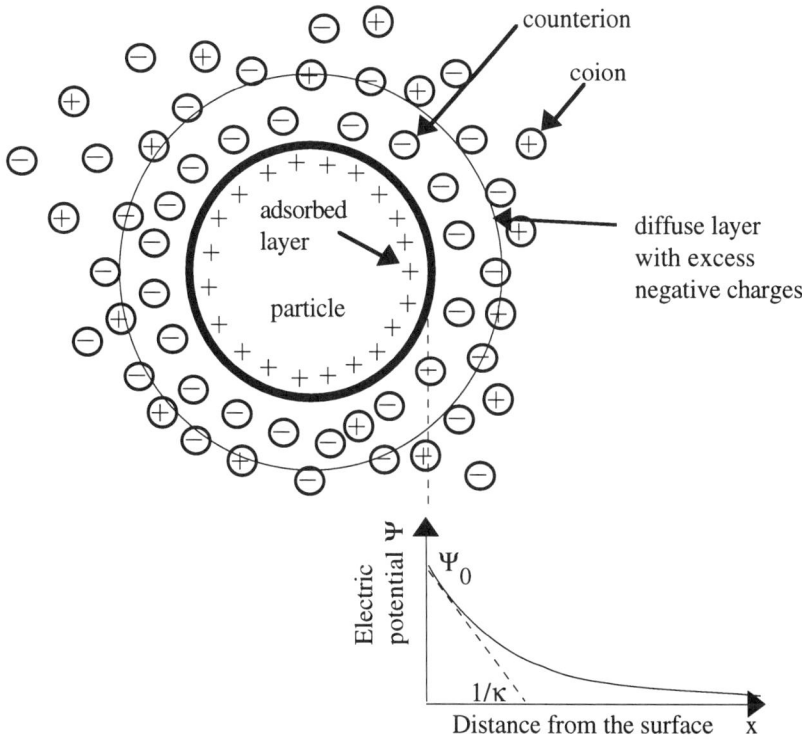

Figure 3.4-3 - Electrical double layer around a colloidal particle.

On the other hand, the ions in the liquid medium which have the same sign as the ions adsorbed on the particles are termed the "coions". They are statistically repulsed by the surface charges on the particles, so that their concentration is the lowest close to the surface of the particles, and it increases when the distance from a particle increases, so as to reach the average value in the liquid medium far from a particle.

When an electric field is applied to a sol, colloidal particles carrying a positive electric charge move in the same direction as the electric field . Experiments show that the counterions the closest to the particle move with it. This has led to define a zeta potential ζ which is the electric potential at the shearing surface which separates the liquid electrostatically moving with a particle from the immobile liquid.

The zeta potential of a particle in a liquid medium can be derived from an observation of the motion of particles in an optical microscope. For this purpose, a small quantity of very diluted sol is placed in electrophoretic cell. This device comprises two electrodes with an electric potential difference between the two electrodes such that the sol is submitted to an electric field E. If the colloidal particle has a positive ζ value, it moves in the same direction as the electric field with a velocity V (Figure 3.4-4) and the ratio

$$\mu = \frac{V}{E} \qquad\qquad (3.4\text{-}10)$$

defines the electrophoretic mobility of the particle.

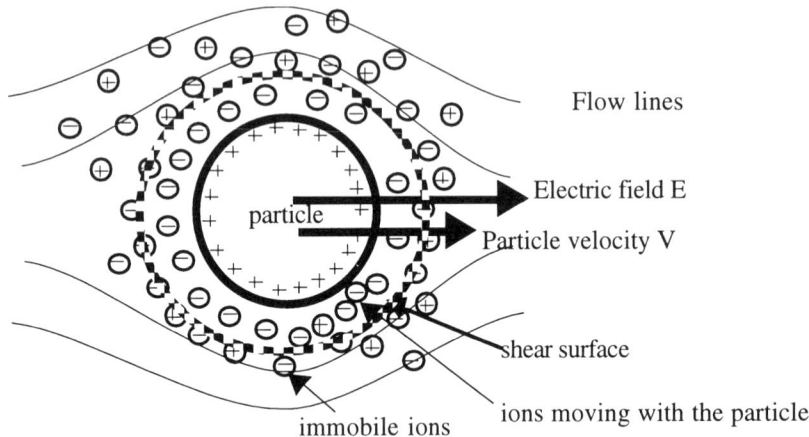

Figure 3.4-4 - Electrophoretic mobility of a colloidal particle.

The isoelectric point (i.e.p.) is defined as the pH for which the mobility of the particles is zero, that is to say when $\zeta=0$. In practice the i.e.p and the z.p.c are often considered as synonymous. Values for oxides have been investigated by Parks [119], Yoon et al. [120] and Hunter [121], a list is gathered in Table 3.4-4. Parks showed that the adsorption of aqua groups H_2O or hydroxo groups OH on the particles, lowers the z.p.c. by approximately 2 units. A better crystallized oxide particle also has a lower z.p.c. than an amorphous or poorly crystallized particle. The relationship between the zpc, the hydration energy and the order of adsorption of cations is discussed in a further paragraph.

Surface electric potential Ψ_0 of a particle
Several models have been developed to describe the electric potential $\Psi(x)$ in the liquid which surrounds a colloidal particle [113]. In the most frequent applications of the electrostatic theory, the electric-potential Ψ_0 at the surface of the charged particle is considered to be constant, for a given pH.
If the activity coefficients of the protons adsorbed on the particles surface can be considered to not depend on their surface concentration, this surface potential can be simply related to the pH according to thermodynamics considerations, by the Nernst equation:

$$\Psi_o = \frac{2.303}{F} RT \left((z.p.c) - pH \right) \qquad\qquad (3.4\text{-}11)$$

where F is the Faraday (96487 coulombs), R is the universal gas constant $(8.3143 \text{ J K}^{-1} \text{ mole}^{-1})$ and Ψ_o is in volt.

Table 3.4-4 - Heat of immersion in water, z.p.c. and simplified order of the ease to adsorb cations adsorption, of a few oxides. Completed from Dumont and al. [122].

Solid	Heat of immersion in water [123] $J/m^2 x 10^3$	ease of adsorption of cations [119]	z.p.c. [reference]
WO_3		$Cs^+ > Li^+$	0.5 [124]
V_2O_5		$Cs^+ > Li^+$	1 - 2 [124]
$\delta\text{-}MnO_2$		$Cs^+ > Li^+$	1.5 [125]
SiO_2	88 - 182	$Cs^+ > Li^+$	2.5 [121]
SiO_2(quartz)		$Cs^+ > Li^+$	3.7 [126]
TiO_2(calcined)		$Cs^+ > Li^+$	3.2 [127]
SnO_2			4.5 [121]
Al-O-Si		$Li^+ > Cs^+$	6 [128]
TiO_2	550	$Li^+ > Cs^+$	6 [129]
ZrO_2			6.7 [39]
FeOOH			6.7 [121]
$\beta\text{-}MnO_2$		$Li^+ > Cs^+$	7.3 [125]
ZnO		$Li^+ > Cs^+$	8 [130]
Cr_2O_3	650	$Li^+ > Cs^+$	8.4 [121]
Fe_2O_3	532	$Li^+ > Cs^+$	8.6 [121]
Al_2O_3	773	$Li^+ > Cs^+$	9 [121]
MgO			12 [121]

At 25°C $\qquad\qquad \Psi_o \approx 0.06 \left(\text{(z.p.c)} - pH \right)$ $\qquad\qquad$ (3.4-12)

More generally for an oxide, the two following charge formation reactions can be considered, with their equilibrium constants [131]:

$$-MOH_2^+ \Leftrightarrow -MOH + H_0^+ \qquad K_0^+ = \frac{[-MOH][H_0^+]}{[-MOH_2^+]} \qquad (3.4-13)$$

$$-MOH \Leftrightarrow -MO^- + H_0^+ \qquad K_0^- = \frac{[-MO^-][H_0^+]}{[-MOH]} \qquad (3.4-14)$$

where $[\overset{+}{H_0}]$ designates the proton concentration near the surface of the particle. According to Boltzmann's statistics

$$[\overset{+}{H_0}] \quad = \quad [\overset{+}{H_\infty}] \ \exp\left(\frac{-e\Psi_0}{k_bT}\right) \qquad (3.4\text{-}15)$$

where $[\overset{+}{H_\infty}]$ is the proton concentration far from the surface of the particle

$$\text{Hence} \quad K_0^+ = \frac{[\,\text{-MOH}\,]\ [\overset{+}{H_0}]}{[\text{-MOH}_2^+]} \quad = \quad K^+ \ \exp\left(\frac{-e\Psi_0}{k_bT}\right) \qquad (3.4\text{-}16)$$

$$\text{and} \quad K_0^- = \frac{[\,\text{-MO}^-\,]\ [\overset{+}{H_0}]}{[\text{-MOH}]} \quad = \quad K^- \ \exp\left(\frac{-e\Psi_0}{k_bT}\right) \qquad (3.4\text{-}17)$$

$$\text{where} \quad K^+ \ = \ \frac{[\,\text{-MOH}\,]\ [\overset{+}{H_\infty}]}{[\text{-MOH}_2^+]} \qquad (3.4\text{-}18)$$

$$\text{and} \quad K^- \ = \ \frac{[\,\text{-MO}^-\,]\ [\overset{+}{H_\infty}]}{[\text{-MOH}\,]} \qquad (3.4\text{-}19)$$

K^+ and K^- are intrinsic constants which characterize the acidity of the surface groups when the surface potential $\Psi_0 = 0$. The difference in magnitude between these intrinsic constants is :

$$\Delta pK = pK^- - pK^+ \quad = \log_{10} \ \frac{[\,\text{-MOH}]^2}{[\text{-MO}^-]\ [\text{-MOH}_2^+]} \qquad (3.4\text{-}20)$$

ΔpK can also be expressed as

$$\Delta pK = \quad \log_{10} \ \frac{1 - 2\theta}{\theta} \qquad (3.4\text{-}21)$$

$$\text{where} \ \theta = \frac{[\text{-MOH}_2^+]}{[\text{-MOH}_2^+] + [\,\text{-MOH}] + [\text{-MO}^-]} = \frac{[\text{-MO}^-]}{[\text{-MOH}_2^+] + [\,\text{-MOH}] + [\text{-MO}^-]} \qquad (3.4\text{-}22)$$

is the fraction of charged sites on the oxide surface.

When $\Delta pK > 2$, the fraction of charged adsorption sites is low and the concept of a zero point charge is valid. This is the case with most oxides and the z.p.c. is given by:

$$\text{z.p.c} = \frac{pK^- + pK^+}{2} \qquad (3.4\text{-}23)$$

From this expression of the z.p.c. and from the expressions of K_0^+, K_0^-, K^+ and K^-, in (3.4-16) to (3.4-19), it is possible to derive the following expression of the surface electric potential Ψ_0 :

$$\Psi_0 = \frac{2.303}{F} RT \left((z.p.c) - pH + \frac{1}{2} \log_{10} \frac{[-MO^-]}{[-MOH_2^+]} \right) \tag{3.4-24}$$

On the other hand, the fraction of ionized sites is high when $\Delta pK < 2$ and in this case the concept of i.e.p. is more useful.

Electric potential $\Psi(x)$ at a distance x from a planar surface - Gouy Chapman model [113,116]

According to the Poisson's equation, the electric potential profile Ψ is the solution to the differential equation:

$$\nabla^2\Psi = - \frac{\rho}{\varepsilon_r \varepsilon_0} \tag{3.4-25}$$

where $\nabla^2\Psi \equiv \frac{\partial^2\Psi}{\partial x^2} + \frac{\partial^2\Psi}{\partial y^2} + \frac{\partial^2\Psi}{\partial z^2}$, ρ is the electric charge density at local

position (x,y,z), ε_0 is the dielectric permittivity of the vacuum ($\frac{1}{4\pi}$ in u.e.s.c.g.s

units, 8.854×10^{-12} Farad.m^{-1} in the international system); ε_r is the relative dielectric constant of the liquid medium (78.4 for water at 25°C).

For a charged plane with its face perpendicular to the x axis, this equation simply becomes

$$\frac{d^2\Psi}{dx^2} = - \frac{\rho}{\varepsilon_r \varepsilon_0} \tag{3.4-26}$$

In the Gouy-Chapman model, the counterions layer is considered to be entirely diffuse (Figure 3.4-5) and the local charge density at x is

$$\rho = \sum_i (n_i z_i e) \tag{3.4-27}$$

Where n_i designates the local concentration (number of ions per m^3) of ionic species i with charge number z_i (positive for a cation and negative for an anion) and e is the electronic charge (1.60209×10^{-19} coulomb). The summation is carried over all ion types i. According to Boltzmann's statistics

$$n_i = n_i(\infty) \exp\frac{- z_i e\Psi}{k_b T} \tag{3.4-28}$$

where $n_i(\infty)$ is the concentration of ion species i (number of ions per m^3) far from the surface, k_b is Boltzmann's constant (1.38054×10^{-23} J K^{-1}); e is the electronic charge (1.60209×10^{-19} coulomb); T is the temperature in K.

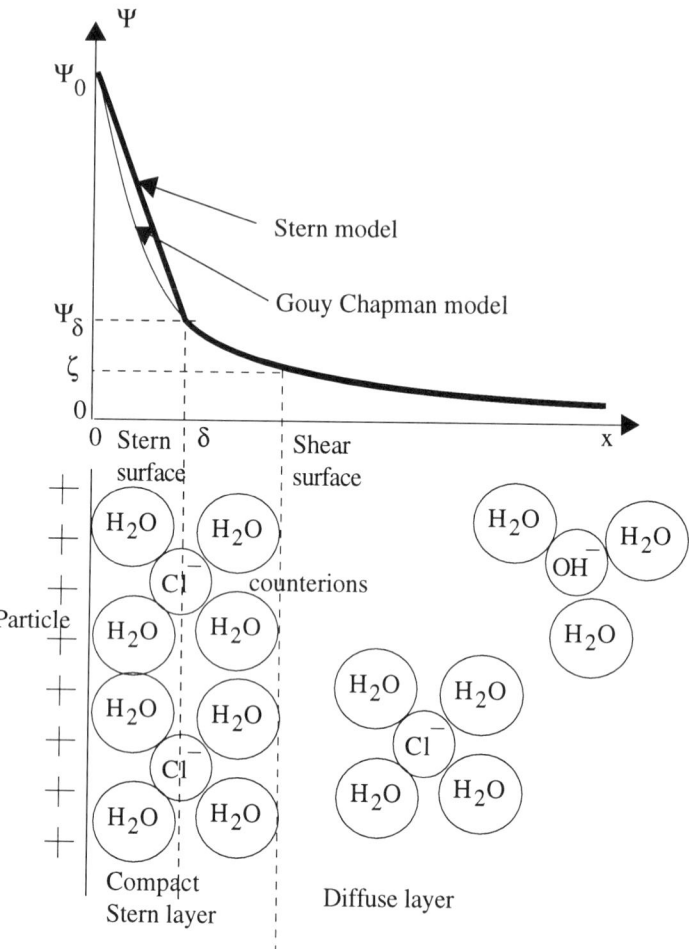

Figure 3.4-5 - Gouy Chapamn and Stern models for the electrical double layer.

If the electrolyte in the liquid medium where the particles are dispersed contains a "z:z" electrolyte (cations and anions with the same charge number z), Ψ is the solution to the differential equation

$$\frac{d^2\Psi}{dx^2} = -\frac{\sum_i (n(\infty)\,z\,e)}{\varepsilon_r \varepsilon_0}\left(\exp\left(\frac{-ze\Psi}{k_bT}\right) + \exp\left(\frac{ze\Psi}{k_bT}\right)\right) \qquad (3.4\text{-}29)$$

The boundary conditions are:

$$\Psi = \Psi_0 \text{ at } x = 0 \tag{3.4-30}$$

$$\Psi = 0 \text{ and } \left(\frac{d\Psi}{dx}\right)_\infty = 0 \text{ at } x = \infty \tag{3.4-31}$$

This equation can be exactly integrated. The solution is the following function of the distance x from a planar surface:

$$\Psi(x) = \frac{2k_b}{e} \frac{T}{z} \ln \frac{1 + \Gamma_0 \exp(-kx)}{1 - \Gamma_0 \exp(-\kappa x)} \tag{3.4-32}$$

with
$$\Gamma_0 = \tanh \frac{e\Psi_0 z}{k_b T} \tag{3.4-33}$$

$$\kappa = \left(\frac{2e^2 n(\infty)z^2}{\varepsilon_r \varepsilon_0 k_b T}\right)^{1/2} \tag{3.4-34}$$

In the above formula, $\dfrac{k_b}{e}$ can be replaced by $\dfrac{R}{F}$ in which R is the universal gas constant and F a Faraday. In spite of the fact that the electric-potential gradually attenuates as the distance x increases, κ^{-1} gives a measure of the rapidity of this attenuation. It is a parameter which has the dimension of a distance, termed the electrical double layer thickness or the Debye-Hückel length.

At large distances x when $\Gamma_0 \exp(-\kappa x)$ is small, (3.4-32) can be approximated by

$$\Psi(x) = \frac{4k_b}{e} \frac{T}{z} \Gamma_0 \exp(-kx) \tag{3.4-35}$$

A more simple expression for the electrical double layer thickness κ^{-1} is:

$$\kappa^{-1}(nm) = \frac{0.304}{z\sqrt{C}} \tag{3.4-36}$$

where C is the electrolyte concentration in mol. L^{-1} (or molarity).
A few values of κ^{-1} for z = 1 are given in Table 3.4-5 below:

Table 3.4-5 - Electrical double layer thickness for
a few concentrations of a "1:1" electrolyte

C (mol. L^{-1})	κ^{-1} (nm)
10^{-6}	304
10^{-4}	30.4
10^{-2}	3.04

The surface charge σ_0 (in coulomb m^{-2}) on the particle can be derived by considering that it must be exactly balanced by the excess counterions in the diffuse layer. Hence

$$\sigma_0 = - \int_0^\infty \rho \, dx = - \int_0^\infty \varepsilon_r \varepsilon_O \frac{d^2\Psi}{dx^2} dx = \varepsilon_r \varepsilon_O \left(\left(\frac{d\Psi}{dx}\right)_\infty - \left(\frac{d\Psi}{dx}\right)_0 \right) \quad (3.4\text{-}37)$$

Considering the boundary conditions (3.4-31)

$$\sigma_0 = - \varepsilon_r \varepsilon_O \left(\frac{d\Psi}{dx}\right)_0 = 2[2\varepsilon_r \varepsilon_O \, k_b T \, n(\infty)]^{1/2} \, \sinh\left(\frac{ze\Psi_0}{2k_b T}\right) \quad (3.4\text{-}38)$$

Debye-Hûckel approximation [113,116]
In the Debye-Hückel approximation

$$\frac{ze\Psi}{k_b T} \ll 1 \quad (3.4\text{-}39)$$

$$\text{so that } \frac{d^2\Psi}{dx^2} = \kappa^2 \, \Psi \quad (3.4\text{-}40)$$

$$\text{and } \Psi = \Psi_0 \, \exp(-\kappa x) \quad (3.4\text{-}41)$$

In the case of a positively charged surface ($\Psi_0 > 0$), the coions and counterions concentration profiles are illustrated in Fig. 3.4-6 and given by the functions:

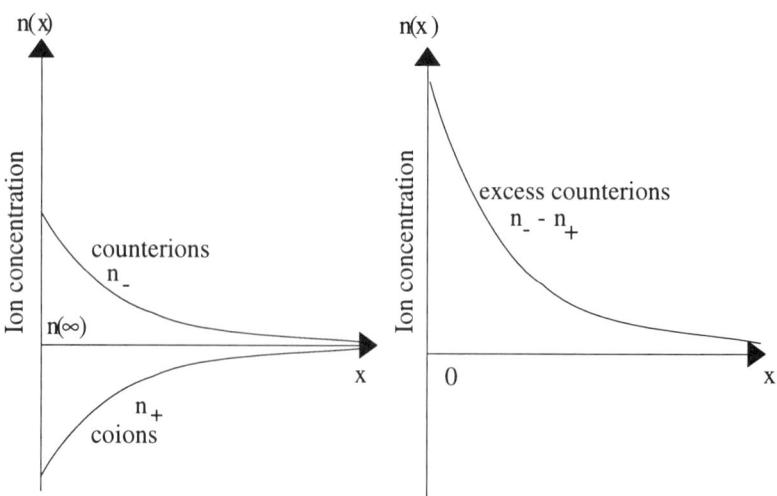

Figure 3.4-6 - Distribution of the counterions and coions as a function of the distance x from a positively charged surface.

$$\text{coions } n_+ = n(\infty) \exp\left(-\frac{ze\Psi}{k_bT}\right) \qquad (3.4\text{-}42)$$

$$\text{counterions } n_- = n(\infty) \exp\left(\frac{ze\Psi}{k_bT}\right) \qquad (3.4\text{-}43)$$

The surface charge density is

$$\sigma_0 = \frac{ze\Psi_0}{k_bT} [2\varepsilon_r\varepsilon_O k_bT n(\infty)]^{1/2} = \varepsilon_r\varepsilon_O\kappa\Psi_0 \qquad (3.4\text{-}44)$$

Stern model [113,116]
In the Stern model (Figure 3.4-5), the counterion layer itself is divided into an external diffuse part plus an inner compact layer called the Stern layer. In this case the preceding formulas for the diffuse layer can be applied by replacing Ψ_o by the Stern potential Ψ_δ while the distance x is measured from the Stern's surface instead of from the particle surface. The Stern potential Ψ_δ is actually different from the zeta-potential ζ. However for most practical cases, these two potentials are very close to each other.

Electric potential $\Psi(x)$ created by a charged spherical particle [116]
For a charged spherical particle, the differential equation which relates the electric potential Ψ to the distance x from the center of the sphere is

$$\frac{1}{x^2}\frac{d}{dx}\left(x^2\frac{d\Psi}{dx}\right) = \frac{2\,e\,z\,n(\infty)}{\varepsilon_r\varepsilon_O}\sinh\left(\frac{ze\Psi}{k_bT}\right) \qquad (3.4\text{-}45)$$

With $\Psi = \Psi_0$ at $x = r$, the particle radius.
Let $X = x\Psi$, differential. In the Debye-Hückel approximation where $\dfrac{ze\Psi}{k_bT} \ll 1$, differential equation (3.4-43) becomes

$$\frac{d^2X}{dx^2} = \kappa^2X \qquad (3.4\text{-}46)$$

The solution of the above equation is

$$X = A\,e^{-\kappa x} + Be^{\kappa x} = x\Psi \qquad (3.4\text{-}47)$$

The conditions $\Psi_\infty = 0$ gives $B = 0$
If the double layer is much thicker than the particle radius, i.e. $\kappa r \ll 1$, that is to say as $\kappa \to 0$, the electric potential around the particle is given by the potential around an isolate electric charge. Hence,

$$\text{Lim}\Psi_{\kappa \to 0} = \frac{Q_0}{4\pi\varepsilon_r\varepsilon_Ox} \quad \text{which gives } A = \frac{Q_0}{4\pi\varepsilon_r\varepsilon_Ox} \qquad (3.4\text{-}48)$$

$$\text{Hence } \Psi = \frac{Q_0}{4\pi\varepsilon_r\varepsilon_0 x} \exp(-\kappa\, x) \qquad (3.4\text{-}49)$$

Let $x = r_\zeta$ be the radius corresponding to the shear surface where the ζ potential is measured. A large value of κ^{-1} corresponds to a dilute solution, the shear surface is very close to the surface of particle, that is to say

$$r \approx r_\zeta \qquad (3.4\text{-}50)$$

$$\zeta = \frac{Q_0}{4\pi\varepsilon_r\varepsilon_0\, r_\zeta} \exp(-\kappa\, r_\zeta) \;\approx\; \frac{Q_0}{4\pi\varepsilon_r\varepsilon_0\, r} \; \exp(-\kappa\, r) \qquad (3.4\text{-}51)$$

$$\text{and } \Psi \approx \frac{r\zeta}{x} \exp(-\kappa\,(x - r)) \qquad (3.4\text{-}52)$$

As $\kappa r \ll 1$

$$\zeta \;\approx\; \frac{Q_0}{4\pi\varepsilon_r\varepsilon_0\, r} \qquad (3.4\text{-}53)$$

From (3.4-50)

$$\left(\frac{d\Psi}{dx}\right)_{x=r} = -\left(\frac{1 + \kappa r}{r}\right)\zeta \qquad (3.4\text{-}54)$$

The surface charge density on the shear surface is therefore

$$\sigma_\zeta = -\varepsilon_r\varepsilon_0 \left(\frac{d\Psi}{dx}\right)_{x=R} = \varepsilon_r\varepsilon_0 \left(\frac{1 + \kappa r}{r}\right)\zeta \qquad (3.4\text{-}55)$$

$$\text{As } \kappa r \ll 1, \quad \sigma_\zeta \approx \varepsilon_r\varepsilon_0 \left(\frac{1}{r}\right)\zeta \qquad (3.4\text{-}56)$$

$$\text{so that } Q_\zeta = 4\pi r^2\, \sigma_\zeta \approx 4\pi\varepsilon_r\varepsilon_0\, r\, \zeta = Q_0 \qquad (3.4\text{-}57)$$

electrophoretic mobility of a sphere [113,116]
Equation (3.4-45) makes it possible to determine an expression for the electrophoretic mobility of a particle in the case of a thick double ($\kappa r \ll 1$)   (Fig. 3.4-7). If Q_ζ is the total charge of the sphere composed of the particle plus the part of the electrical double layer inside the shear radius, the electrophoretic mobility of this sphere is

$$\mu = \frac{V}{E_{app}} \qquad (3.4\text{-}58)$$

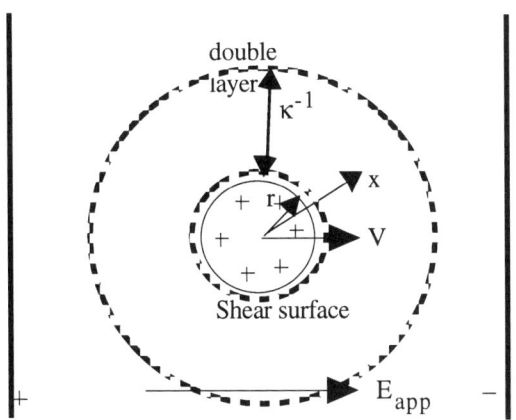

Figure 3.4-7 - Electrophoretic mobility of a charged sphere
with a thick double layer. Adapted from Masliyah [116].

where V is the velocity of the sphere and E_{app} is the electric field applied to it. The
electric force F_e acting on the sphere is :

$$F_e = Q_\zeta \, E_{app} = 4\pi \, r^2 \, \sigma_\zeta \, E_{app} \tag{3.4-59}$$

If σ_ζ is replaced by its expression in (3.4-57)

$$F_e = 4\pi \, r^2 \, \varepsilon_r \varepsilon_0 \left(\frac{1}{r}\right) \zeta \, E_{app} = 4\pi \, r \, \varepsilon_r \varepsilon_0 \, \zeta \, E_{app} \tag{3.4-60}$$

According to Stokes relation, at constant velocity,

$$F_e = 6\pi\eta V r \tag{3.4-61}$$

where η is the liquid medium viscosity. Hence, the electrophoretic mobility is

$$\mu = \frac{2 \, \varepsilon_r \varepsilon_0 \, \zeta}{3\eta} \tag{3.4-62}$$

In the case of thin double layer ($\kappa r \gg 1$) (Fig. 3.4-8), the Navier-Stokes
relationship gives

$$\eta \, \frac{d^2 V_x}{dy^2} = -\rho \, E_{app} \tag{3.4-63}$$

If this expression is combined with Poisson's equation (3.4-26)

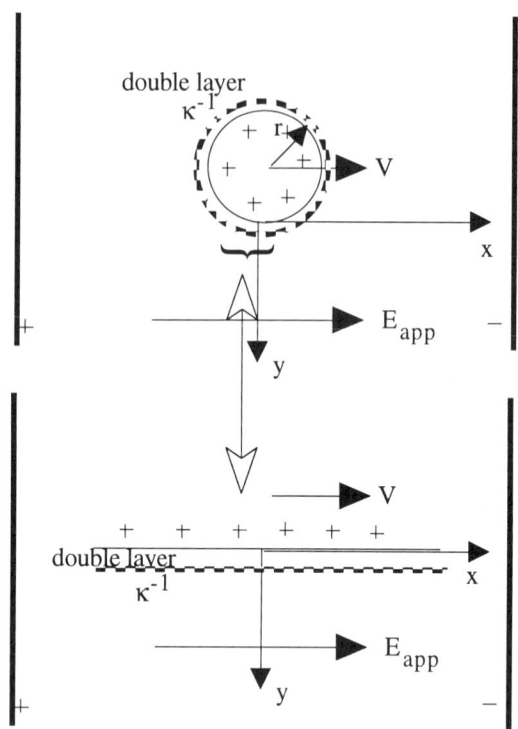

Figure 3.4-8 - Electrophoretic mobility of a charged particle with a thin electric double layer. Adapted from Masliyah [116].

$$\eta\frac{d^2V_x}{dy^2} = \varepsilon_r\varepsilon_0\ E_{app}\frac{d^2\Psi}{dy^2} \qquad\qquad (3.4\text{-}64)$$

A first integration gives $\eta\dfrac{dV_x}{dy} = \varepsilon_r\varepsilon_0\ E_{app}\dfrac{d\Psi}{dy} + C_1$ $\qquad\qquad$ (3.4-65)

Since $\dfrac{dV_x}{dy} \to 0$ and $\dfrac{d\Psi}{dy} \to 0$ as $y \to \infty$, then $C_1 = 0$

A second integration gives

$$\eta\ V_x = \varepsilon_r\varepsilon_0\ E_{app}\ \Psi + C_2 \qquad\qquad (3.4\text{-}66)$$

Since $dV_x \to 0$ and $\Psi \to 0$ as $y \to \infty$, then $C_2 = 0$

$V_x(y=0) = V$ and $\Psi(y = 0) = \zeta$ so that

$$V = \frac{\varepsilon_r \varepsilon_0 \; E_{app} \; \zeta}{\eta} \qquad (3.4\text{-}67)$$

The electrophoretic mobility is

$$\mu = \frac{\varepsilon_r \varepsilon_0 \; \zeta}{\eta} \qquad (3.4\text{-}68)$$

Electrostatic repulsion force between two charged planar surfaces
[113,116]
For the one-dimensional case, the repulsion force F_R between two plates translates in the solution which surrounds the plates, by the creation of a local hydrostatic overpressure p which depends on the abscissa x. This overpressure is itself due to an over-concentration of counterions, since from (3.4-42) and (3.4-43) and for positively charged plates

$$n_- - n_+ = n(\infty) \left(\exp\left(\frac{ze\Psi}{k_bT}\right) - \left(-\exp\frac{ze\Psi}{k_bT}\right)\right) \qquad (3.4\text{-}69)$$

$$\text{or} \quad \Delta n = n_- - n_+ = 2\, n(\infty)\, \sinh\left(\frac{ze\Psi}{k_bT}\right) \qquad (3.4\text{-}70)$$

This gradient in excess counterions induces a gradient in the osmotic pressure π of the solvent, so that this solvent is dragged in the electric double layer and creates a local hydrostatic overpressure. The gradient in local hydrostatic pressure itself induces a local hydrostatic force (Fig. 3.4-9)

$$F_h = - [p(x+dx) - p(x)] = - \frac{\partial p}{\partial x}\, dx \qquad (3.4\text{-}71)$$

Moreover, the local ions at x in a slab of volume (1. dx) are submitted to an electric force F_e given by:

$$F_e = \rho \;.1.\; dx.\; E_x \qquad (3.4\text{-}72)$$

where $E_x = - \dfrac{d\Psi}{dx}$ is the local electric field at x.

At equilibrium, the two forces F_e and F_h balance out (Fig. 3.4-9), so that

$$\frac{dp}{dx} + \rho\, \frac{d\Psi}{dx} = 0 \qquad (3.4\text{-}73)$$

If we replace ρ by its expression given in (3.4-26)

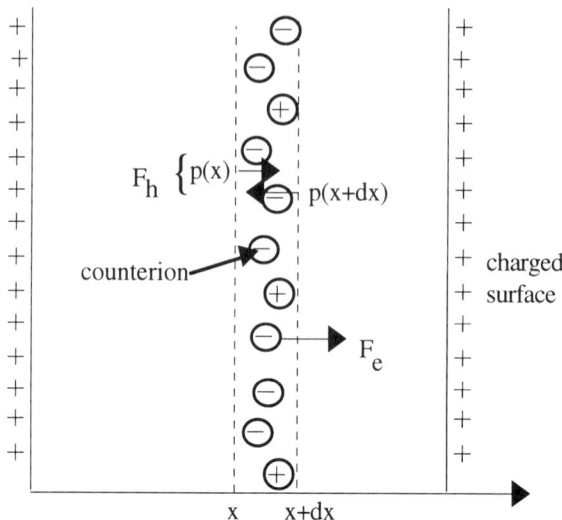

Figure 3.4-9 - Balance of forces in the liquid between two parallel plates carrying positive surface charges.

$$\frac{dp}{dx} - \varepsilon_r \varepsilon_0 \frac{d^2\Psi}{dx^2} \frac{d\Psi}{dx} = 0 \qquad (3.4\text{-}74)$$

By integration, this leads to the repulsive force per unit area between the two plates:

$$p - \frac{\varepsilon_r \varepsilon_0}{2} \left(\frac{d\Psi}{dx}\right)^2 = \text{constant} = F_R \qquad (3.4\text{-}75)$$

This equation indicates that the particles' repulsion exactly equates the local hydrostatic overpressure p_m in the solution at mid-distance between the two particles because $\left(\dfrac{d\Psi}{dx}\right)_{mid} = 0$.

In the Debye-Hückel approximation, the general solution to (3.4-39) is of the form

$$\Psi = C \cosh \kappa(x - S_0/2) + C' \sinh \kappa(x - S_0/2) \qquad (3.4\text{-}76)$$

where C and C' are two constants which must be selected to fit the boundary conditions. For two plates with different surface electric potentials Ψ_{10} and Ψ_{20}, separated by a distance S_0 :

$$C = \frac{\Psi_{10} + \Psi_{20}}{2 \cosh (\kappa S_0/2)} \qquad C' = \frac{\Psi_{10} - \Psi_{20}}{2 \sinh (\kappa S_0/2)} \qquad (3.4\text{-}77)$$

For two plates of equal surface electric potential Ψ_0

$$\Psi = \frac{\Psi_0}{\cosh (\kappa S_0/2)} \cosh \kappa(x - S_0/2) \qquad (3.4\text{-}78)$$

In this symmetrical case, if we admit that the excess counterions in the liquid between the two plates balances exactly the positive charge on the plates, the relation (3.4-38) which defines the surface charge density σ_0 is still valid, so that when the distance S_0 between the two planar surface decreases, the surface charge σ_0 on the plates decreases if the surface potential Ψ_0 is constant, as shown in Figure 3.4-10.

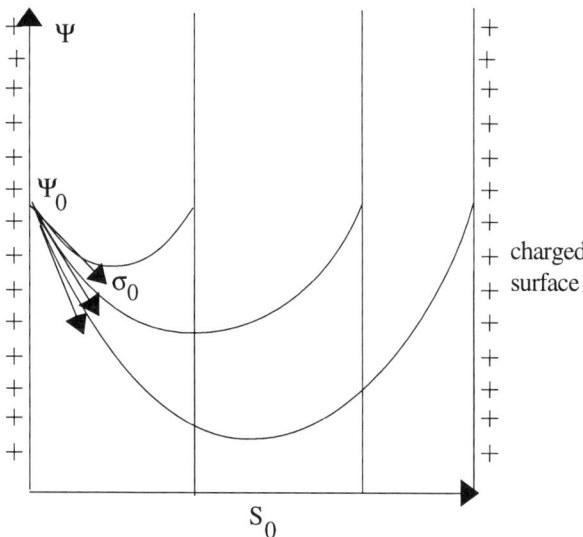

Figure 3.4-10 - Variation of the surface charge at constant surface potential, when two plates come closer to each other

Next, equation (3.4-74) can be transformed to

$$dp = -\rho \, d\Psi \qquad (3.4\text{-}79)$$

or, for a z:z electrolyte, according to (3.4-70)

$$dp = - ze \, \Delta n \, d\Psi \qquad (3.4\text{-}80)$$

$$dp = - 2n(\infty) \, ze \, \sinh \left(\frac{ze\Psi}{k_b T}\right) d\Psi \qquad (3.4\text{-}81)$$

This overpressure integrates to

$$p = 2n(\infty)\, k_b T\, [\cosh\left(\frac{ze\Psi}{k_b T}\right) - 1] \qquad (3.4\text{-}82)$$

and for $\dfrac{ze\Psi}{k_b T} \ll 1$

$$p = n(\infty)\, k_b T \left(\frac{ze\Psi}{k_b T}\right)^2 \qquad (3.4\text{-}83)$$

Making use of Debye-Hückel solution for Ψ and $\dfrac{d\Psi}{dx}$ to report in (3.4-75), the repulsion force between the particles, when the separation distance S_0 is small by comparison with the double layer thickness, that is to say when $\kappa S_0 \ll 1$, is

$$F_R = \frac{e^2\, z^2\, n(\infty)}{k_b T}\, \Psi_0^{\,2} = \frac{1}{2}\, \varepsilon_r \varepsilon_O \kappa^2\, \Psi_0^{\,2} \qquad (3.4\text{-}84)$$

When $\kappa S_0 \gg 1$

$$F_R = \frac{4\, e^2\, z^2\, n(\infty)}{k_b T}\, \Psi_0^{\,2}\, e^{-\kappa S_0} = 2\, \varepsilon_r \varepsilon_O \kappa^2\, \Psi_0^{\,2}\, e^{-\kappa S_0} \qquad (3.4\text{-}85)$$

At a large distance x when $\Gamma_o \exp(-\kappa x)$ is small and $\Psi(x)$ is given by (3.4-35)

$$F_R = 64\, n(\infty)\, k_b T\, \Gamma_0^{\,2}\, e^{-\kappa S_0}$$

In the case of two plates with different electric surface potentials Ψ_{10} and Ψ_{20}

$$F_R = \frac{e^2\, z^2\, n(\infty)}{k_b T} \left(\frac{2\Psi_{10}\, \Psi_{20}\, \cosh \kappa S_0 - 1}{\sinh^2 \kappa S_0}\right) \qquad (3.4\text{-}86)$$

The electrostatic interaction energy is

$$\Phi_R = \int_{\infty}^{S_0} -F_R\, dS \qquad (3.4\text{-}87)$$

In the case $k S_0 \gg 1$, equation (3.4-84) gives

$$\Phi_R = 2\, \varepsilon_r \varepsilon_O \kappa\, \Psi_0^{\,2}\, e^{-\kappa S_0} \qquad (3.4\text{-}88)$$

At a large distance x when $\Gamma_o \exp(-\kappa x)$ is small and $\Psi(x)$ is given by (3.4-35)

$$\Phi_R = \frac{64\, n(\infty)\, k_b T}{\kappa}\, \Gamma_0^{\,2}\, e^{-\kappa S_0} \qquad (3.4\text{-}89)$$

Electrostatic interaction between spherical particles [113,116]
The electrostatic interaction between two particles is due to the electric charges adsorbed on the particles, which can be attenuated to a variable extent by the double electric layer. This is illustrated in Figure 3.4-11, which shows that when two particles are far from each other, the counterion layers around each particle do not overlap. Hence, each particle has a virtually complete double electrical layer so that it roughly behaves as a neutral sphere (Figure 3.4-11a). In these conditions, the interaction energy between two particles is $\Phi_R \approx 0$.

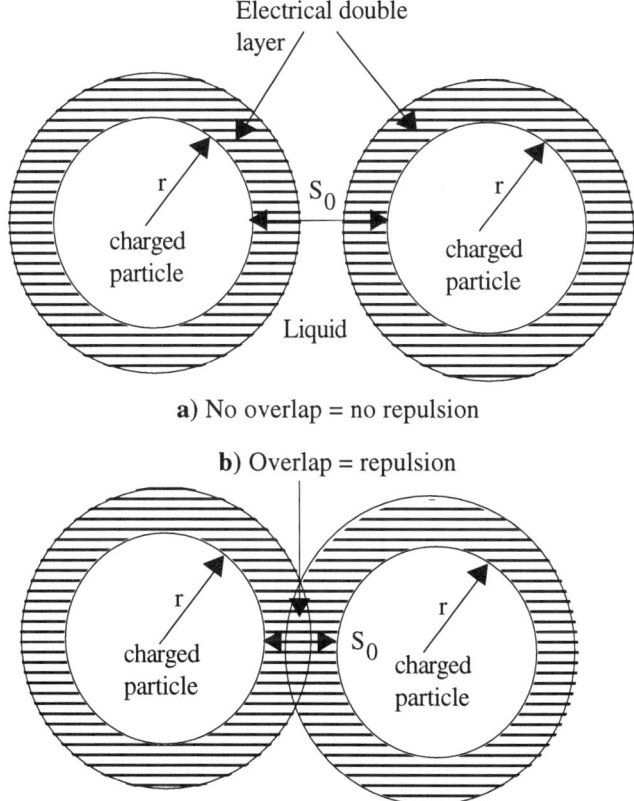

a) No overlap = no repulsion

b) Overlap = repulsion

Figure 3.4-11 - Conditions for the occurrence of electrostatic repulsion between colloidal particles.

As the particles come sufficiently close to each other, their counterion layers start to overlap. That is to say, the excess counterion concentration at mid distance is higher than what it would be from a single particle at the same distance. This increased ion concentration induces an increased osmotic flow of the solvent which itself induces a local hydrostatic overpressure in the solution between the spheres. Hence, a positive interaction energy $\Phi_R > 0$ appears. It corresponds to a repulsion force between the two particles (Figure 3.4-11b).

Several models have been successively developed to describe this electrostatic interaction energy Φ_R [113]. As mentioned previously, it is often assumed in the mathematical derivation of Φ_R that the electric potential at the surface of the particle remains a constant, Ψ_o, as the particles come close to each other. However, similar derivations have been made with the hypothesis that the surface charge density of the particles remains a constant, σ_0.

For spherical particles which have the same radius r, if the electric double layer thickness κ^{-1} is small by comparison with the nearest distance S_O between the surfaces of the particles, only the tail of the diffuse double layers overlap. Moreover, if r is large enough by comparison with κ^{-1} (Figure 3.4-12), the electric repulsion force between the two spheres can be estimated as the sum of elementary repulsion forces between parallel plates separated by a distance S such that:

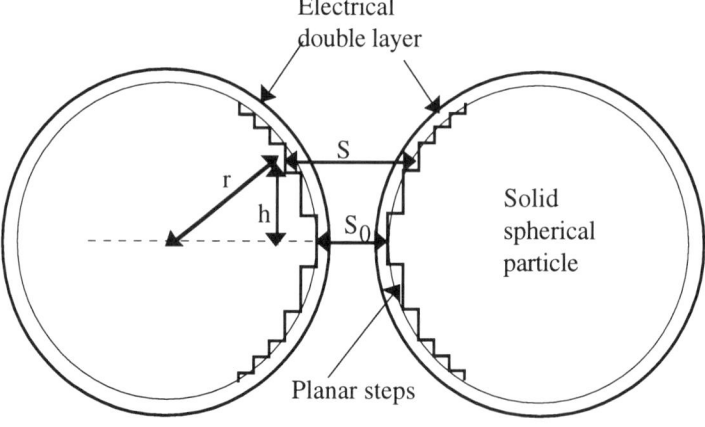

Figure 3.4-12 - Approximations made to derive equation 3.4-67. Adapted from Hiemenz [113].

$$\frac{S-S_0}{2} = r - \sqrt{r^2 - h^2} \qquad (3.4\text{-}90)$$

$$\text{Hence } dS = \frac{2h \, dh}{\sqrt{r^2 - h^2}} \qquad (3.4\text{-}91)$$

Only the central region of the spheres contribute significantly to the repulsion between the spherical particles, so that

$$dS = \frac{2h\,dh}{r} \qquad (3.4\text{-}92)$$

The corresponding repulsion force between the two elementary planar rings of the spheres is

$$dF_{RS} = F_R\,2\pi h\,dh \qquad (3.4\text{-}93)$$

where F_R can be replaced by the repulsion force between parallel planes previously calculated for $\kappa S \gg 1$ (thin double layer) , and which implicitly follows the Gouy-Chapman model and the Debye-Hückel approximation. This leads to

$$dF_{RS} = 2\,\varepsilon_r\varepsilon_0\kappa^2\,\Psi_0^2\,e^{-\kappa S}\,2\pi h\,dh = 2\,\varepsilon_r\varepsilon_0\kappa^2\,\Psi_0^2\,e^{-\kappa S}\,\pi\,rdS \qquad (3.4\text{-}94)$$

As the term $e^{-\kappa S}$ drastically decreases the contributions to the repulsion force which do not come from the central part of the spheres, an approximate integration can be done by varying S form S_0 to ∞:

$$F_{RS} \approx \pi r \int_0^\infty 2\varepsilon_r\varepsilon_0\kappa^2\,\Psi_0^2\,e^{-\kappa S}\,\pi\,rdS \qquad (3.4\text{-}95)$$

which integrates to

$$F_{RS} = 2\pi\,\varepsilon_r\varepsilon_0\kappa^2\,\Psi_0^2\,e^{-\kappa S_0} \qquad (3.4\text{-}96)$$

The electrostatic interaction energy Φ_R between the two spheres can be derived by carrying the integration of F_{RS} as this was done for plates in equation (3.4-87)

$$\Phi_R = 2\pi\,\varepsilon_r\varepsilon_0\,r\,\Psi_0^2\,\exp(-\kappa S_0) \qquad (3.4\text{-}97)$$

For unequal spheres having radii r_1 and r_2 and surface electrical potentials Ψ_{10} and Ψ_{20}, the general expression for the electrostatic interaction energy is often referred as the HHF expression [116,132]:

$$\Phi_R = \frac{\pi\,\varepsilon_r\varepsilon_0 r_1 r_2}{r_1 + r_2}\left\{ 2\,\Psi_{10}\Psi_{20}\,\mathrm{Ln}\left(\frac{1 + \exp(-\kappa S_0)}{1 - \exp(-\kappa S_0)}\right)\right.$$
$$\left. + (\Psi_{10}^2 + \Psi_{20}^2)\,\mathrm{Ln}[1 - \exp(-2\kappa S_0)] \right\} \qquad (3.4\text{-}98)$$

Equation (3.4-97) shows that the electrostatic repulsion between two particles can be experimentally modified in several ways. One method consists in modifying Ψ_0, by changing the concentration of potential-determining ions, that is to say the pH when the potential determining ions are H^+ or OH^-. This action is necessary to realize the peptization of colloidal particles made from alkoxides. A second important method consists in modifying the electric double layer thickness κ^{-1} around the particles, so as to modify the screening effect due to the counterion layer. This can be done by adjusting the concentration $n(\infty)$ of a non potential-determining electrolyte, such as Na^+Cl^- for an oxide sol. This comes from the fact that $n(\infty)$ is an important parameter in equation 3.4-34. As $n(\infty)$ increases, κ increases, the electric double layer thickness κ^{-1} decreases, so that particles can closer to each other without having a significant electrostatic interaction. It is important to note that all ions in the liquid medium where colloidal particles are dispersed, whether or not they are potential-determining, participate in the diffuse layer. For instance at pH = 7, alumina particles are charged positively. If some NaCl is added to the solution, neither Na^+ nor Cl^- modify the electric potential since they are not adsorbed. However both participate in the construction of the electric diffuse layer.

A third method consists in changing the valence z of an electrolyte which is dissolved in the liquid medium, for instance by replacing a chloride with Cl^- anions, by a sulfate with SO_4^{2-} anions or a phosphate with PO_4^{3-} anions. As z increases, equation (3.4-34) shows that κ increases drastically, hence the electric double layer κ^{-1} decreases drastically, so that particles can come much closer to each other without being submitted to a significant repulsion. This is a big effect which is at the root of the success of the DLVO theory.

Total interaction energy in electrostatic sols [113,116]
The dispersion of a fine powder in a stable sol depends on a combination of the electrostatic and Van der Waals interactions, as developed in the D.L.V.O theory. The results show that the Van der Waals attraction always dominates at small and large separation distance S_0 between the particles. The electrostatic repulsion dominates only in some instances, which depend on the double layer thickness κ^{-1}, at intermediate separation distances. The concentration $n(\infty)$ and the charge z of the ions composing the electrolyte are of great importance. The total interaction energy

$$\Phi = \Phi_A + \Phi_R \qquad\qquad (3.4-99)$$

is a function of the separation distance between two particles which is illustrated indicated in Figure 3.4-13. This figure shows an energy barrier Φ_{max} opposing to the contact between particles, which depends on the value of κ. If this energy barrier is high by comparison with the Brownian thermal agitation energy of the particles, such as $\Phi_{max} > 30$ kT for $\kappa^{-1} = 10 \times 10^{-6}$ cm in Figure 3.4-13, aggregation of the particles will not occur; the sol is kinetically stable. On the other hand, if this energy barrier is not too high by comparison with the Brownian motion energy of the particles, such as ≈ 10 kT for $\kappa^{-1} = 10^{-6}$ cm in Figure 3.4-13, it can be statistically overcome by Brownian motion. In this case, the particles will aggregate; this is the phenomenon known as flocculation or coagulation in the DLVO theory.

Coagulation [107,113,116]
To achieve coagulation, it is possible to modify Φ_A, generally a secondary effect. As indicated previously, it is more efficient to modify Φ_R by monitoring the

concentration of an electrolyte. This can be either a potential determining electrolyte which modifies the particles electric potential Ψ_0, that is to say in most case an acid or a base which modifies the pH, or a non potential-determining electrolyte such as NaCl which modifies the electrical double layer thickness κ^{-1}. In the latter case, coagulation occurs for a critical coagulation concentration, designated by C_C in this book and by C.C.C in other publications, which strongly depends on the counterions valence z.

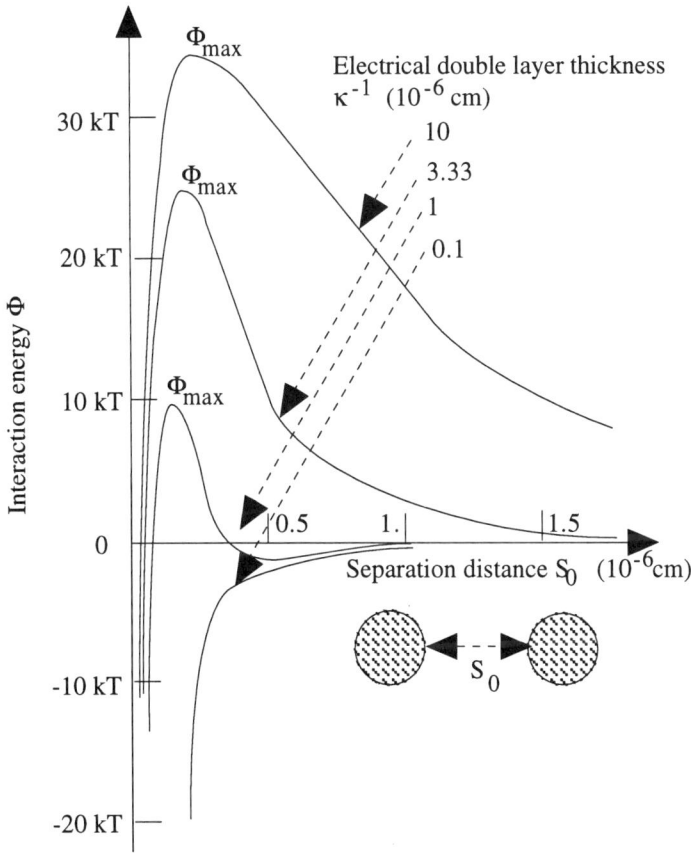

Figure 3.4-13 - Variation of the total interaction energy Φ between two spherical particles, as a function of the closest separation distance S_0 between their surfaces, for different double layer thicknesses κ^{-1} obtained with different monovalent electrolyte concentrations. The electrolyte concentration is C (mol.L^{-1})= $10^{-15}\kappa^2$ (cm^{-1}). Adapted from Overbeek [107].

For parallel plates, equation (3.4-89) for Φ_R and the appropriate formula for Φ_A in Table 3.4-3, give :

$$\Phi = \Phi_R = \frac{64\ n(\infty)\ k_bT}{\kappa}\ \Gamma_0^2\ e^{-\kappa S_0}\ -\ \frac{A}{12\pi S_0^2} \qquad (3.4\text{-}100)$$

And if Ψ_0 is high $\Gamma_0 \approx 1$
A simple way to consider when coagulation will occur is to admit that

$$\Phi_{max} \approx 0 \qquad (3.4\text{-}101)$$

while Φ_{max} is defined by $\dfrac{d\Phi}{dS_0} = 0$ $\qquad (3.4\text{-}102)$

This gives respectively

$$\frac{64\ n(\infty)\ k_bT}{\kappa}\ \Gamma_0^2\ e^{-\kappa S_0} = \frac{A}{12\pi S_0^2} \qquad (3.4\text{-}103)$$

$$\text{and}\quad 64\ n(\infty)\ k_bT\ \Gamma_0^2\ e^{-\kappa S_0} = \frac{A}{6\pi S_0^3} \qquad (3.4\text{-}104)$$

which when combined, give

$$\frac{A\kappa}{12\pi S_0^2} = \frac{A}{6\pi S_0^3} \qquad (3.4\text{-}105)$$

$$\text{or}\quad \kappa(S_0)_{max} = 2 \qquad (3.4\text{-}106)$$

Substituting this result in equation (3.4-104)

$$64\ n(\infty)\ k_bT\ \Gamma_0^2\ e^{-2} = \frac{A\kappa^3}{48\pi} \qquad (3.4\text{-}107)$$

$$\text{Hence}\quad n(\infty) \equiv C_c \approx \kappa^3 \approx z^{-6} \qquad (3.4\text{-}108)$$

If this value of $n(\infty)$ is identified with the critical coagulation concentration C_c, this result is the most outstanding success of the D.L.V.O theory as it shows that C_c is proportional to z^{-6}, in agreement with experimental data. This is an effect of large magnitude and previously known, experimentally, as the Schulze-Hardy rule. In the case of Al_2O_3 experimental data from Overbeek [110] and calculated values are compared in Table 3.4-6.

Effect of ions solvatation
A study on Fe_2O_3 precipitates made from $Fe(NO_3)_3$ precursor [122] has shown, first that the critical coagulation concentration C_C depends on the volume fraction occupied by the particles, secondly that positively charged particles (pH below the

z.p.c.) are more stable than those negatively charged (pH above the z.p.c.). The latter result cannot be explained by a simple change in the electrostatic repulsion, since the absolute magnitude of the zeta potential is the same for symmetrical pH values with respect to the z.p.c. However, it can be explained by a change in the structure of the electrical double layer due to solvatation by water molecules.

Table 3.4-6 - Critical coagulation concentrations of Al_2O_3 sols.

Valence z of counterions	Experimental values (mol. L^{-1}) [110]		Theory [113]
	$C_c(z)$	$C_c(z)/C_c(z=1)$	$C_c(z)/C_c(z=1)$
1	5.2 10^{-2}	1	1
2	6.3 10^{-4}	1.2 10^{-2}	1.56 10^{-2}
3	0.8 10^{-5}	1.5 10^{-3}	1.37 10^{-3}
4	5.3 10^{-5}	10. 10^{-4}	2.44 10^{-4}

A structuration of water occurs with many materials. In a preceding section, the properties of ions to be solvated by water molecules was presented. Actually, solid particles themselves are solvated by water molecules. Therefore, a slightly different effect of counterions on coagulation can be expected, depending the relative strength of these counterions and the solid particles, to promote self solvatation by water.

As an example, small anions such as IO_3^- and F^- are strong "structure promoter" in water where they form a solvatation shell. However, Fe_2O_3 particles themselves are strong structure promoter in water, as shown by a high hydration energy in Table 3.4-4. In practice, these anions tend to form a slightly more packed double layer close to the colloidal particles, than the larger size anions, so as to fulfill a maximum solvatation of both materials. The electrical double layer is thinner and the critical coagulation concentration C_C is lower with these small size anions. That is to say they enhance coagulation.

On the other hand, larger size anions such as NO_3^- and ClO_4^- are not much solvated by H_2O dipoles. Hence, they are rejected by a material such as Fe_2O_3. The concentration of these anions is slightly lower immediately close to the particles, the electrical double layer is thicker and the critical coagulation concentration C_C is higher. They favor the peptization of a sol.

In summary, with strongly water structure promoter particles such as Fe_2O_3 dispersed in acidic medium (positive charges on the particles), the amount of anions adsorbed decreases (and C_c increases), in the order [133]:

$$IO_3^->F^->CH_3COO^->CH_2ClCOO^->BrO_3^->SCN^->$$
$$CHCl_2COO^->Br^->NO_3^->ClO_3^->Cl^->ClO_4^- \approx I^- \qquad (3.4-109)$$

In basic conditions (negatively charged particles), a similar series occurs with cations in the order:

$$Li^+>Na^+>K^+\approx Cs^+ \tag{3.4-110}$$

In Table 3.4-4, this order is summarized by $Li^+> Cs^+$. The zpc of these colloidal particles is relatively high.

With particles which are not strong structure promoter of water, the order of adsorption of anions and cations is reversed, as indicated in Table 3.4-4.

A consequence of solvatation is that experimental z.p.c data reported for a given material are often scattered in a significant range of values, depending on the type of anions or cations in the liquid medium where the colloidal particles are dispersed, hence on their fabrication method. Very cautious work is necessary to determine the z.p.c. of pure colloids without any adsorbed foreign anions or cations. An example of such a study was made by Blesa and al. [39] on ZrO_2 particles made by hydrolysis of $ZrOCl_2.8H_2O$ and they found a z.p.c of 6.7 for pure ZrO_2 .

Electrostatic charge reversal [116]

Let us consider two plates 1 and 2, each carrying positive charges and having a positive electrical surface potential $\Psi_{10} > 0$ and $\Psi_{20} > 0$, such that $\Psi_{20} > \Psi_{10}$. The electrostatic force between these two plates is a repulsion at large separation distance S_0. However it becomes an attraction at close separation distance according to equation (3.4-86). This corresponds to an electric potential as shown in Fig.3.4-14 [6] so that the electric surface charge of plate 1 is reversed and becomes negative.

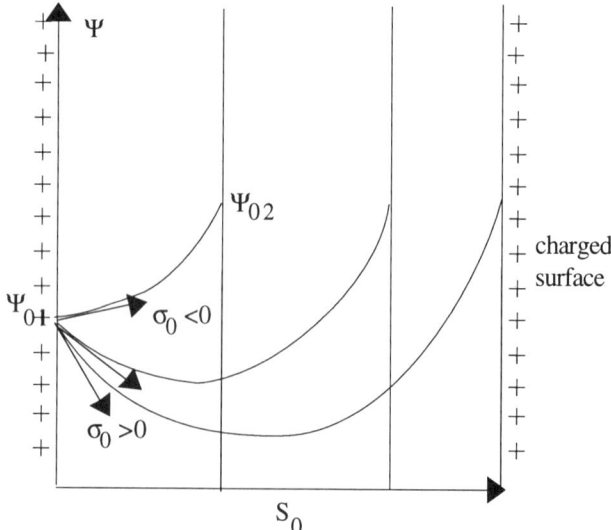

Figure 3.4-14 - Charge reversal for two plates with a different surface potential at close separation distances

However this view is very questionable. It supposes the electrical charges on the surface of the particle 1 can actually be reversed. With actual planar particles such as kaolinite, experimental behavior usually leads to retain the hypothesis that the surface electric charge σ_0, rather than the electric potential Ψ_0, is a constant [134]. Hence, repulsion always occurs between the two particles.

For oxides where the surface charge is due to the adsorption of specific ions such as H^+ and OH^-, charge reversal would mean that the $[OH^-]$ concentration in the liquid close to particle 1 is such that the pH passed on the other side of the zpc. In this case particle 2 must be of a different nature from particle 1, otherwise it would also undertake charge reversal since it has the same zpc. This in turns leads to examine interactions between particles of a different nature. Matijevic has established three-dimensional diagrams, as a function of the concentration of two components and of the pH, which show behaviors not observable with a single component. Different fields of these diagrams correspond to the possible occurrence of
- selective coagulation of one component
- mixed heterocoagulation
- mixed sols formation in which the two types of particles are randomly dispersed
-   and   sols demixion with separation in liquid domains containing a different component.

Heterocoagulation implies the formation of flocs which comprise the two types of particles. It usually occurs when particles of opposite charge are mixed [7]. Moreover, mixing these two components can be realized in various ways such as:
- (1) two sols in a nascent state
- (2) one well grown sol and one sol in a nascent state
- (3) two well grown sols.
which can lead to quite different coagulation or dispersion behaviors.

When the particles of one component are much smaller than particles of the second component, they can adsorb on this second component and change its apparent charge as well as its coagulation conditions. For instance, figure 3.4-15 illustrates spindle shaped Fe_2O_3 particles partly covered by smaller spherical Co_3O_4 particles.

When hydrolyzable metal cations are added to a sol, three types of z.p.c. of this colloid can be observed, depending on the system being considered and of the metal ions concentration [118,135]. They are:
- (1) the z.p.c. of the colloidal particles in the absence of foreign metal cations
- (2) the z.p.c. of a hydroxide from the added foreign cations, which has precipitated onto the colloidal particles.
- (3) an z.p.c. intermediate between the two previous ones due to a partial adsorption of the foreign cations on the colloidal particles.

For instance, if Mg^{2+} cation are dissolved in a Si_3N_4 sol, $Mg(OH)_2$ has an opposite charge to the particles at most pH. The hydroxide $Mg(OH)_2$ can precipitate on the Si_3N_4 particles and eventually induce heteroflocculation.

The adherence of particles to a substrate follows the same rules. Depending on the pH where precipitation or gelation is performed, the substrate and the particles may be of opposite charge (e.g., boehmite sol in a silica Becher), in which case the adherence is favored, while the inverse occurs in the case of charges of same sign.

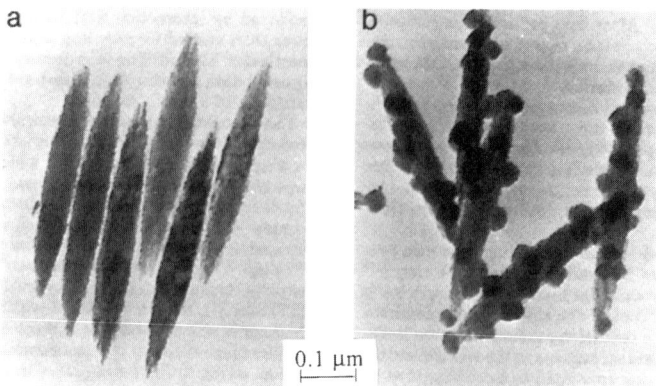

Figure 3.4-15 - Transmission electron micrographs of; (a) hematite particles; (b) hematite particles partly covered with Co_3O_4 particles. From Ishikawa and Matijevic [136] with permission.

Coagulation rate

Thermodynamically , the specific surface area of a powder has a tendency to decrease, and coagulation is one of the possible paths to achieve this evolution. The theory of Smoluchowski makes it possible to calculate a rate of fast coagulation, resulting from the Brownian motion when no energy barrier opposes to the contact between particles, such as for $\kappa^{-1} = 0.1$ in Figure 3.4-13. In this case the number n of particles per unit volume of sol decreases with time according to [107]:

$$\frac{dn}{dt} = - k_f\, n^2 \tag{3.4-111}$$

The half life time, which is the time to decrease the number of particles by half is :

$$t_{1/2,f} = \frac{3\ \eta}{4k_f\, T n_o} \tag{3.4-112}$$

in which η is the viscosity of the sol and n_o the initial number of particles.

When an energy barrier Φ exists, coagulation is slower. The half life time can be written:

$$t_{1/2,s} = \frac{t_{1/2;f}}{\alpha} \tag{3.4-113}$$

$$\text{where } \frac{1}{\alpha} = W = 2 \int_{2r}^{\infty} \exp\frac{\Phi}{k_b T} \frac{dr}{r^2} \qquad (3.4\text{-}114)$$

The efficiency of such an energy barrier to slow down coagulation increases as $\frac{\Phi}{k_b T}$

increases. Practically for $\frac{\Phi}{k_b T} > 15$, the sol remains indefinitely stable [109].

Reversibility of coagulation
When a sol has coagulated, a practical question is to know if this coagulation is irreversible, or if peptization can again be realized. Actually, re-peptization appears to be feasible in some cases only [107].
Theoretically, re-peptization appears to be possible in the case of steric coagulation, not in the case of electrostatic coagulation. As illustrated in Figure 3.4-16, the distance between two particles is large before coagulation, it can for instance correspond to the shallow minimum in (a). On the other hand, the distance between particles is short after from coagulation, it corresponds to the deep minimum in (b). If after coagulation the electrolyte concentration is modified to obtain in (c) the same energy profile as in (a), a deep energy barrier has to be overcome to again reach the shallow minimum, hence to redisperse the sol.

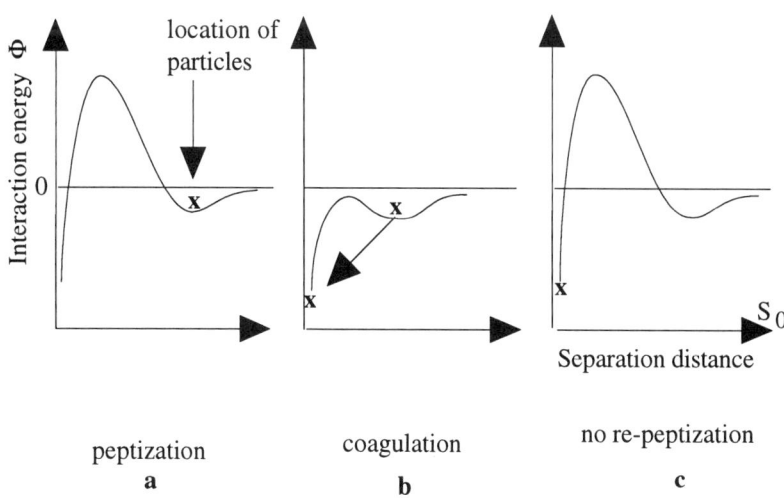

Figure 3.4-16 - Potential interaction energy diagrams for a pair of particles corresponding to: (a) a stable sol; (b) coagulation; (c) irreversible coagulation after restoration of the original peptization conditions. Adapted from Overbeek [107].

However, experiments show that repeptization is more frequent than exceptional. To explain such a behavior, Overbeek has proposed that a solvent layer of thickness 2δ remains between the particles, after coagulation. This solvent layer can eventually correspond to the Stern layer. For instance in the case of a negative AgI sol coagulated with the electrolyte KNO_3, the Stern layer is composed mainly of monovalent solvated K^+ cations.

If an electrolyte such as $Ba(NO_3)_2$ is dissolved after coagulation, the Ba^{2+} ions replace more slowly the K^+ ions in the Stern layer than in the diffuse layer. Hence the charge on the particles are not immediately modified. That is to say, the interactions between particles which were derived at constant electric surface potential before and during coagulation, must be replaced by interactions derived at constant surface charge for repeptization, at least in a transitory stage. As the new counterions in the diffuse layer have a valence 2 the hypothesis of constant surface charge results in a high ζ potential of the stern layer . In these conditions, as shown in Figure 3.4-17, one passes directly from (a) to (c) , where the lower energy barrier is favorable to redispersion of the particles. After a longer time, the Ba^{2+} end up replacing the K^+ in the Stern layer also, so that the surface is modified and the sol ends up in the situation illustrated in figure 3.4-17b, which corresponds to a stable sol.

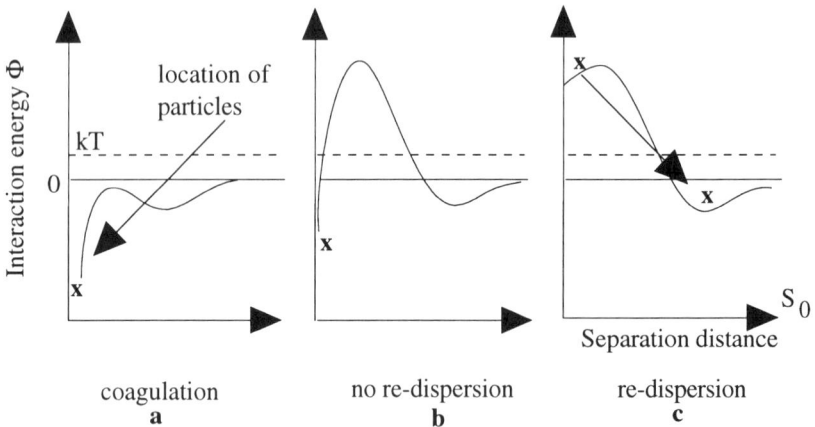

Figure 3.4-17 - Energy versus distance diagrams similar to those in figure 3.13 showing the conditions where re-dispersion from a coagulated state: (a) is impossible; (b) or possible (c). Adapted from Overbeek [107].

If the order of electrolyte substitution was reversed; that is to say first flocculation with $Ba(NO_3)_2$ followed by redissolution with KNO_3; the sol would pass directly from the situation in figure 3.4-17a to the situation in figure 3.4-17b, because the decrease in counterion valence could not produce a momentary increase of ζ. Hence, no transitory stage permitting redispersion would occur.

STERIC SOLS
Steric interactions occur when polymeric macromolecules are adsorbed on colloidal particles. These interaction are themselves a consequence of the interaction between the polymer macromolecules and the solvent. Hence, mathematical derivation of the steric interaction energy makes use of the oldest theory for polymer solutions derived by Flory and Huggins [137,138].

Polymer solutions
The behavior of a polymer as a solute in a solvent, is driven by two types of phenomena. On one hand, the Brownian motion makes each polymer macromolecule to extend in the solvent but there is a limit to this extension due to the finite length of each polymer chain and its elasticity. On the other hand, interactions such as the Van der Waals interaction occurs between the atoms in the polymer chain itself, and between the atoms in the chain and the atoms in the solvent. These interactions are summarized by the affinity of the polymer, as a solute, for the solvent.

In the Flory-Huggins theory, the total Gibbs free energy of dilution of a polymer (2) in a solvent (1) is [137,138]:

$$\Delta G_M = k_b T \; [n_1 \ln(1- \phi_2) + n_2 \phi_2 + \chi_1 \; n_1 \; \phi_2^2 \;] \tag{3.4-115}$$

In this equation n_1 and n_2 are the number of solvent and polymer molecules, $\phi_1 \equiv 1 - \phi_2$ and ϕ_2 are their volume fraction and χ_1 is a dimensionless parameter. The molar chemical potential of the solvent is:

$$\mu_1 - \mu_1^0 = R\,T\,[\ln(1- \phi_2) + (1 - \frac{1}{x}) \; \phi_2 + \chi_1 \; \phi_2^2 \;] \tag{3.4-116}$$

where x is the ratio $\dfrac{V_2}{V_1}$ of the polymer molar volume to the solvent molar volume.

In the coefficient χ_1, Flory and Krigbaum [139] distinguish a contribution of enthalpic nature h_1 and a contribution of entropic nature s_1, related to the partial enthalpy and entropy of dilution by

$$\Delta H_1 = R\,T\,h_1\,\phi_2^2 \tag{3.4-117}$$
$$\Delta S_1 = R\,s_1\,\phi_2 \tag{3.4-118}$$

These contributions are related to χ_1 by:

$$\chi_1 - \frac{1}{2} = h_1 - s_1 = - s_1 \; (1 - \frac{\theta}{T} \;) \tag{3.4-119}$$

In the latter equation, θ is a characteristics of a couple solvent-polymer. As it has the dimension of a temperature, it is termed the Flory Huggins' theta temperature.

A large value of χ_1 corresponds to a large increase of the free energy of mixing so that the solvent behaves as a bad solvent for the polymer. On the other hand, a low value of χ_1 can give a negative free energy of mixing and the solvent behaves as a good solvent for the polymer. The transition between the two situations occurs for a critical value χ_{1c}. When $\chi_1 > \chi_{1c}$ a separation of the solution in two phases occurs. For endothermic solutions ($\Delta H_1 > 0$), χ_1 decreases as the temperature increases, which favors mixing.

Temperature is an important parameter to modify χ_1. When $T = \theta$ there is no net interaction between the polymer chains and the solvent. When $T > \theta$, mixing is favored. The polymer macromolecules prefer to extend in the solvent so as to favor a good contact polymer-solvent. That is to say, the solvent behaves as a good solvent for this polymer. On the contrary, when $T < \theta$, the polymer chains contract and agglomerate with each other, so as to minimize polymer-solvent contacts. The solvent behaves as a bad solvent for the polymer.

Instead of modifying the temperature, it is also possible to modify the solute proportion. There exists a critical value of solute volume proportion ϕ_{2c} such that phase separation occurs if $\phi_2 > \phi_{2c}$. Expressions for ϕ_{2c} and χ_{1c} can be calculated from the relations:

$$\frac{d\mu_1}{d\phi_2} = \frac{d^2\mu_1}{d\phi_2^2} = 0 \qquad (3.4\text{-}120)$$

which leads to $\phi_{2c} = \dfrac{1}{1 + \sqrt{x}}$ \qquad (3.4-121)

$$\chi_{1c} = \frac{1}{2} + \frac{1}{\sqrt{x}} + \frac{1}{2x} \qquad (3.4\text{-}122)$$

Such results indicate the existence of an "upper critical" temperature (Figure 3.4-18). For high molecular weight polymers $x \gg 1$,

$$\phi_{2c} \approx \frac{1}{\sqrt{x}}. \qquad (3.4\text{-}123)$$

A comparison between the observed phase diagrams and those calculated is only qualitative but it makes it possible to understand the behavior of polymers in solution.

A more elaborate theory , known as the " State Equation theory" has also been established by Flory [140]. It rests on a derivation of the thermodynamic equation of state of the solution from a general partition function . It requires to determine experimentally 7 independent parameters. This theory presents the advantage to incorporate the individual characteristics of the components in the solution and to explain qualitatively the possible occurrence of a lower critical solution temperature in some systems. However, the agreement between the theory and the experiments remains only qualitative.

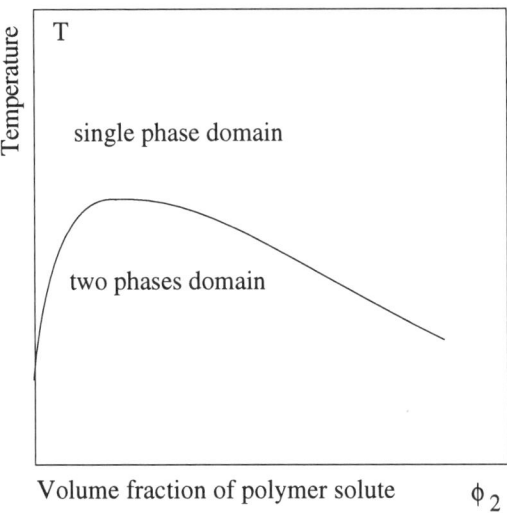

Figure 3.4-18 - Liquid liquid phase diagrams for polymer solutions showing an upper critical solution temperature . Adapted from Carpenter [140].

Steric interactions

Steric interactions operate when polymeric macromolecules are adsorbed on the colloidal particles and they are a consequence of the interaction polymer-solvent examined before. In particular, when the adsorbed polymer can only anchor by one end to a particle, it can produce either a repulsion or an attraction between two particles, depending on the temperature (Figure 3.4-19).

As indicated before, when $T > \theta$, where θ is the Flory-Huggins theta temperature of the couple polymer-solvent, the polymer macromolecules prefer to extend in the solvent so as to favor a good contact polymer-solvent. If these polymers have one end of their chains adsorbed on a colloidal particle, they make a swollen polymer shell around the particle. An osmotic diffusion flow of the solvent penetrates in between the polymer macromolecules to maintain them in the swollen state. If two colloidal particles happen to collide, the polymer in both shells tend to avoid interpenetration to maximize the contact solvent-polymer. Hence the particles remain dispersed. In terms of interaction energy, the polymer are responsible for an effective repulsion between the particles (Figure 3.4-19a).

On the other hand when $T < \theta$, the polymer chains contract and agglomerate with each other, so as to minimize polymer-solvent contacts. If two colloidal particles with adsorbed polymers happen to collide, the polymer in both shells tend to remain in contact so that the particles remain agglomerated. The polymer shells act as a cement. In terms of interaction energy, the polymer are responsible for an effective attraction between the particles (Figure 3.4-19b).

In summary, swelling of the adsorbed polymer shell and repulsion of the particles is favored by an increasing temperature T. It occurs when $T > \theta$, where θ is Flory-

Huggins' characteristic temperature of the couple solvent-polymer. Contraction of the polymer shell, hence attraction of the particles occurs when T < θ.

a - Repulsion between particles

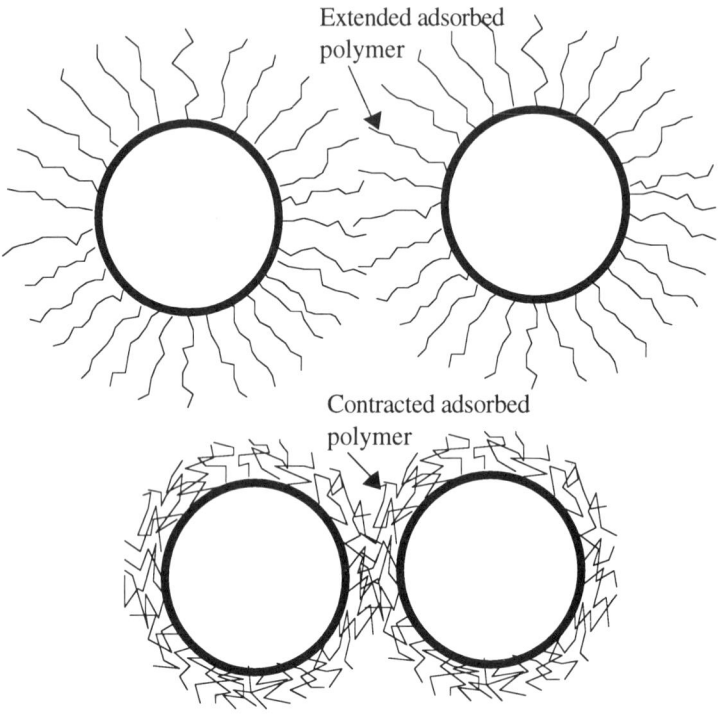

b - Attraction between particles

Figure 3.4-19 - Steric interactions between colloidal particles

Steric interaction energy

The treatment of polymer solutions by the initial Flory-Huggins theory does not apply directly to the calculation of the steric interaction energy between particles because in steric stabilization, at least one end of the polymer chains is fixed on the particles, while in the Flory-Huggins theory both ends of a polymer are free. This has led Flory and Krigbaum [139] to develop an extension of the initial polymer solution theory, appropriate for adsorbed polymers.

In this theory, the free energy of mixing $d(\Delta G_M)$ in a volume element containing dn_1 solvent molecules is estimated by,

$$d(\Delta G_M)= k_b T \, [\ln(1-\phi_2) + \chi_1\phi_2] \, dn_1 \qquad (3.4\text{-}124)$$

So that the partial free energy of dilution of the solvent becomes :

$$\Delta G_1 = - RT \left(\frac{1}{2} - \chi_1\right) \phi_2^2 \qquad (3.4\text{-}125)$$

Next, let us consider two colloidal particles, one volume element j in the polymer shell of a first colloidal particle where the polymer density is ρ_j, and one volume element k in the polymer shell of a second particle where the polymer density is ρ_k. The volume of both elements j and k can be chosen to be identical, that is to say dV. The free energy of mixing of the polymers in the solvent increases when these two volume elements are merged. An expression of this increase is :

$$d(\Delta G_M)= 2k_b T \left(\frac{1}{2} - \chi_1\right) \rho_j \, \rho_k \, V_S^2 \frac{dV}{V_1} \qquad (3.4\text{-}126)$$

where V_S represents the volume of one polymer segment .
The total interaction energy between the two particles on which the polymers are adsorbed , when they come close to each other and their polymer shells overlap, is then:

$$\Phi_S = \Delta G_M = 2k_b T \, s_1 \left(1- \frac{\theta}{T}\right) \frac{V_S^2}{V_1} \int_V \rho_j \, \rho_k \, d V \qquad (3.4\text{-}127)$$

The integration must be carried on the complete volume where interpenetration of the polymer chains occurs. The most simple solution obtained by this method is by Ottewill and Walker [141]. For two spheres of radius r, which have adsorbed a polymer layer of thickness δ, where S_o is the shortest distance between the particles surface, the steric interaction is:

$$\Phi_S = k_b T \left(\frac{4\pi C^2}{3V_1 \, \rho_2^2} s_1 \left(1- \frac{\theta}{T}\right) \left(\delta - \frac{S_o}{2} \right)^2 \left(3r + \delta + \frac{S_o}{2} \right) \right) \qquad (3.4\text{-}128)$$

In this equation, C is the polymer concentration in mol. L^{-1} in the adsorbed layer and ρ_2 the polymer density .
When $T > \theta$, these calculations show that the steric interaction energy is > 0. This corresponds to a repulsion energy between the colloidal particles, which increases very rapidly as S_0 decreases.

Mixed steric and electric interactions - case of surfactant solutions
When surfactants are dissolved in the liquid medium where a sol is dispersed, they can have an effect which is both of steric and electrostatic nature. For instance,

cationic surfactants can reduce the charge of negatively charged sols and even reverse it such as in AgI sols [142]. They can be responsible for the occurrence of a complex sequence of sol stabilization, due to the occurrence of both steric and electrostatic interactions, such as when octadecyl-ether and polyethylene-glycol are added to AgI sols [143]. Similar phenomena have also been reported when anionic surfactants are mixed in positively charged As_2S_3 sols [144].

Mixed steric and magnetic interactions - Ferrofluids[®]

The Ferrofluids[®] are special sols in which the colloidal particles have a small size ($\approx$ 10 nm), and they are made of magnetic materials. These materials include transition metals or rare earths (Fe, Co, Ni, Gd, Dy) as well as ferrites containing Mn, Co, Cu, and Mg. They were prepared for the first time in 1938 by Elmore at the Massachusetts Institute of Technology, by grinding ferrites for weeks with steel balls in oleic acid [145]. Ferrofluids[®] can also be prepared by precipitation of Fe_3O_4 particles from ferric or ferrous salts solutions, followed by extraction of the particles from the precipitate with a solution of oleic acid in a non polar liquid. In other methods, a gelatinous precipitate is first obtained with ammonia, then it is re-peptized by modifying the pH [146].

A coating layer consisting of long polar chain is adsorbed on the particle and prevents them from agglomerating by steric interaction [145,147]. Moreover, the magnetic interaction between these particles is important. These particles can move altogether when they are submitted to a magnetic field and they carry along the surrounding liquid medium with them, so that they can be used as transmission fluids.

The ferrofluids differ from the other magnetic clutch materials by having much smaller particles (10 nm instead of 1 μm). These particles are sufficiently small to be composed of one single magnetic domain, so that their sols are super-magnets with a high magnetic saturation field ($4\pi M_s = 600$ Gauss), and without any remanent field.

3.5 - OTHER PHENOMENA IN SOLS

The feasibility of producing "monodispersed" sols of various materials has led to the discovery of other phenomena . Moreover, sols undertake aging evolutions other than eventual coagulation. These phenomena are summarized in the present section

SOL DEMIXION

A phenomenon which looks like a "liquid-gas" coexistence between two dispersion states of a sol with a different content in colloidal particles, has frequently be observed. A typical case concerns small SiO_2 particles on which short alkane chains are adsorbed, in a solvent such as benzene [148,149]. A spontaneous separation in two states occurs below a temperature which depends on the solvent; from +18°C in benzene to -5°C in toluene. For each solvent, this state separation is only observed on a scale of a few °C.

The existence of such a phenomenon correlates well with the interactions between the polymer chains grafted on the particles and the solvent, as summarized in the previous section and characterized by a Flory θ temperature. With "bad solvents", small variations in temperature change the interaction. Hence they make it possible to observe a transition, which does not occur with a " good solvent" where the interaction is too strong. The length of the chains is important. The effect is more marked when short chains are adsorbed on particles which have a size below a critical size. Actually, this is an intermediate behavior between complete dispersion and classical coagulation, which can only be observed when the attraction compensates very gradually the free energy contribution of entropic origin.

If one considers a dispersion of spheres with identical size occupying a volume fraction ϕ, separation in two sol states occurs and the sphere volume fractions ϕ' and ϕ'' in each sol can be calculated by considering that the osmotic pressure π of the solvent, as well as its chemical potential μ, must be the same in the two sols [150]. Numerical simulations have shown that the separation can occur by spinodal decomposition in some cases [149]. For bi-dispersed sols (i.e; sols with two sizes of particles), the most dilute phase is enriched with the smallest particles.

The observation of a such phenomenon is also possible with the electrostatic theory. Aksay and Kikuchi [151] have shown that, when the electric ζ potential of colloidal particles decreases, a sol can undertake an order transition, from a gas-like to a liquid-like, then to a solid-like state.

In a solid-like state, the colloidal particles are densely packed although they are not in contact and the sol remains fluid. This solid-like state often occurs inside small domains themselves dispersed inside the remaining sol which has a lower colloidal particle concentration. When the colloidal particles inside a domain show no ordered packing, these small domains are termed coacervates and illustrated in Figure 3.5-1 [152].

SOLS LIQUID CRYSTALS

The solid-like state of a sol can itself be crystalline-like or amorphous-like. Crystalline-like transitions are usually observed with monodispersed sols [38]. For instance, opal is a natural example in the dry state of a solid composed of monodispersed silica microspheres and obtained from a crystalline-like sol [153]. The corresponding sols show an effect known as the higher order Tyndall effect. This effect consists in a diffraction of the sol by monochromatic light, showing the existence of a perfect "crystalline" type ordering where colloidal spheres have replaced the atoms in usual crystals . Such particular crystal types are often called colloidal crystals, or crystalloids. This type of behavior seems to be quite general when the particles size dispersion of the colloidal particles is sufficiently narrow and it is not limited to spherical shapes. However, the reasons behind the formation of such ordered structures remain poorly understood.

The phenomenon of coexistence between two sol states mentioned before and the phenomenon of liquid sol crystals formation, can occur altogether. In such a case, the state which has the higher concentration in colloidal particles consists of liquid sols crystals, which are formed inside small domains themselves dispersed inside the less concentrated state. Instead of being termed coacervates, these small domains are known as tactoids and they are illustrated in Figure 3.5-1 [152]. The hematite particles in figure 3.4-15 are an example of tactoids. During drying, these coacervates or tactoids give powders where each powder granule has a characteristic shape, for instance hexagonal in the case of CeO_2 particles [154].

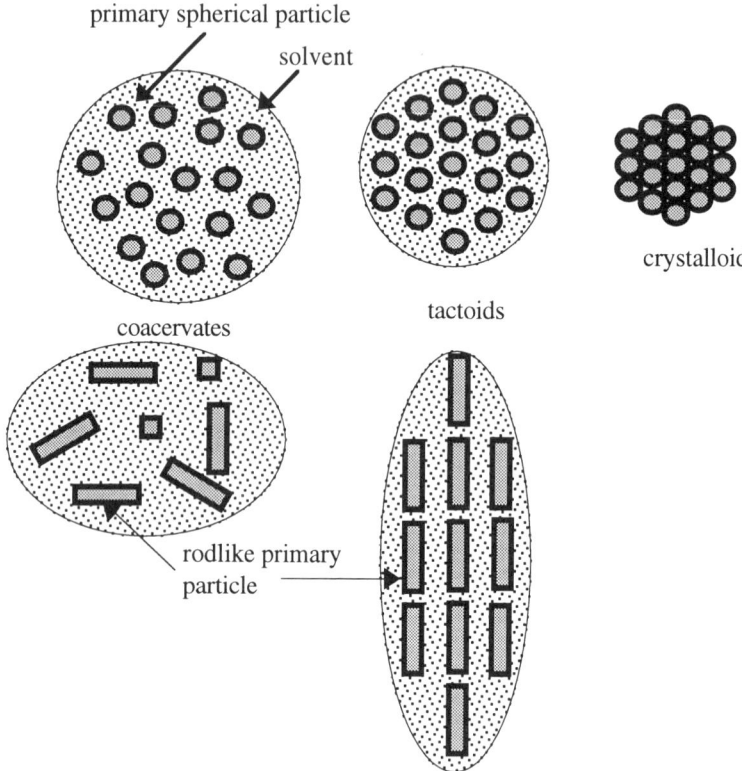

Figure 3.5-1 - Coacervates and tactoids. After Heller [152].

AGING EVOLUTION OF SOLS
Colloidal particles which are dispersed for a long time in a liquid medium often undertake a slow aging evolution, other than precipitation. This can be the formation of new particles consisting of another solid phase, or it can be an evolution in the size distribution of the colloidal particles.

Solid phase re-crystallization
Phase transformation can occur by dissolution of the particles, followed by re-precipitation, so as to produce a more stable crystalline phase.

In particular, equation (3.2-7) shows that the critical free energy for nucleation of solid particles ΔG_c decreases as the interfacial tension γ solid-solution decreases. Hence, the more soluble solid phases have a tendency to form first. As the thermodynamically more stable solid phases of a given compound are usually less soluble [155] than the less stable ones, a metastable solid phase often forms first. However, this not a general rule. A metastable phase could never recrystallize form a stable phase according to thermodynamics. However, a stable phase can sometimes form directly from a solution, without an intermediate metastable phase. It actually depends on the kinetics of the reactions of hydrolysis and condensation, and on the nature of soluble species which reach first their critical concentration for nucleation. Also, re-crystallization to a more stable crystalline phase is usually enhanced by foreign species such as the anions. Hence , these ions either modify the reaction kinetics of the various solute complexes, or act as heterogeneous nucleation centers. An example of recrystallization concerns akaganeite β-FeOOH. Particles of this phase can be precipitated from $FeCl_3$ solutions in a water/ethanol mixture and these precipitates can recrystallize to α-Fe_2O_3 after heating [23]. Schneider [156] has

shown that the particles obtained in the presence of Cl^- ions in a first stage have a very anisometric shape, such as needles 70 nm and 3 nm wide. During aging the dissolution-reprecipitation mechanism replaces the oxo bridges by double ol-bridges which themselves dehydrate further on to form Fe_2O_3. Amorphous FeOOH gels known as ferrihydrites re-precipitate also either to goethite α-FeOOH or to hematite. A similar phenomenon occurs with nickel ferrites [30] or cobalt ferrites [31]. The nature of the anions is important, as well as the pH. For instance Fe_2O_3 sols made from $Fe(NO_3)_3$ can be maintained without any noteworthy evolution and without coagulation at 100°C when the pH is < 6 as the z.p.c of hematite is 8.2 [133]. Actually, the structure of these particles depends on the fabrication pH. If the latter pH is comprised between 3 and 4, the particles are yellow, small and the crystallographic phase is goethite ($Fe_2O_3.H_2O$ or α-FeO(OH)) At a pH between 1 and 2 the particles are bigger, of a red-brick color and the crystallographic structure is hematite. If a chelating agent such as EDTA is added, it also modifies the aging process because it adsorbs on the hematite particles to which it strongly binds. Redissolution of these particles requires a pH > 9.5, as well as a temperature of at least 70°C [68].

Hematite particles themselves recrystallize to magnetite Fe_3O_4 when heated at 250°C in a strongly alkaline medium containing hydrazine and oxidized by reactants containing nitrate ions [27].

In the case of aluminum precursors, the gels and sols made from alkoxides transform to crystalline boehmite above 80°C, while below 80°C bayerite forms by dissolution-recrystallization. The conversion time increases with the size of the alcohol molecule [157].

Ostwald ripening

Another possible evolution when aging a sol, consists in a re-dissolution of the smaller particles in favor of the larger ones which grow according to the process of

Ostwald ripening [5]. This is for instance what occurs for the spherical hematite particles considered earlier.

The mechanism of Ostwald ripening is due to the fact mentioned in section 3.2 that the equilibrium solute concentration $C_s(r)$ with a particle of radius r, increases as r decreases in agreement with the Thompson Freundlich equation (3.2-19)

$$C_s(r) = C_s \exp \frac{2\gamma V_m}{RTr} \qquad (3.2\text{-}19)$$

where γ is the surface tension solid-solvent and V_m is the molar volume of the solute molecules in the solid.

Hence, when a sol comprises a distribution of particles with different sizes, a small particle of radius r_s has a higher equilibrium solute concentration $C_s(r_s)$ than a larger particle with radius $r_L > r_s$; i.e. $C_s(r_L) < C_s(r_s)$. Consequently, the solute concentration in the liquid medium of a sol maintains at an average value $C_s(r_L)$ $< C_s(Liq) < C_s(r_s)$. Dynamically, as shown in Fig. 3.5-2, the smaller particle keeps dissolving to try raising the solute concentration to a value $C_s(r_s) > C_s(Liq)$ near its surface. On the other hand, the solute precipitates on the larger particle to try reaching a solute concentration $C_s(r_L) < C_s(Liq)$ near its surface. Hence, a solute gradient concentration establishes in the liquid so that this solute undertakes a diffusion flow from the smaller towards the larger particles. This process keeps going on as long as the dissolution kinetics of the smaller remaining particles is significant.

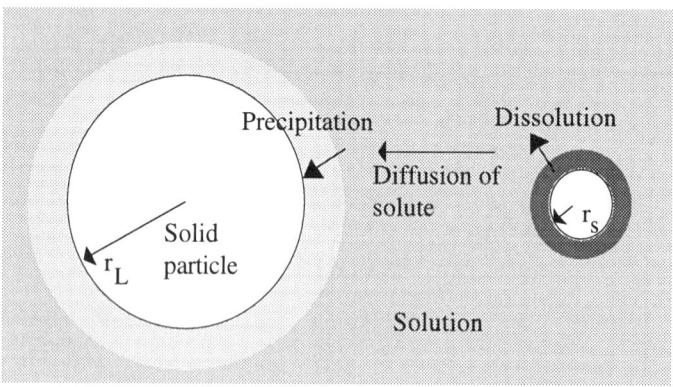

Figure 3.5-2 - Ostwald ripening mechanism.

The border between pure ripening and recrystallization with structure change is sometimes very tenuous. Very often, growth of the bigger particles does not proceed simultaneously with a change in crystallographic nature of the phases. Simply bigger crystals with less defects are formed and simultaneously, the particle change from a spherical characteristics to a more crystalline geometrical shape. For instance, whiskers, which are very elongated fiber like-monocrystals, grow on the surface of the initial particles, as this can occur with aluminum hydroxide particles [28].

3.6 - REFERENCES

1 - Haruta M., Delmon B., J. de Chimie Physique, 83 (1986) 859-868.
2 - Bowen H.K., Mater. Sci. Eng., 44 (1980) 1-56.
3 - Ewell R.H., Insley H., J. Res. Natl. Bur. Standards. 15 (1935) 173-186.
4 - Cahn J.W., Trans. Met. Soc. AIME, 242 (1968) 166.
5 - Kingery W.D., Bowen H.K., Uhlmann D.R., 'Introduction to Ceramics", John Wiley & Sons, 2nd Edition, New-York (1976), (a) p 330; (b) p 482; (c) p. 335.
6 - Matijevic E., Ann. Rev. Mater. Sci., 15 (1985) 483-516.
7 - Matijevic E., Pure Appl. Chem. 50 (1978) 1193-1210.
8 - Matijevic E., "Monodisperse colloids (Preparation, Properties and Applications), and interactions in mixed colloidal systems (heterocoagulation, adhesion and microflotation)", Seminar presented at the Université de Bordeaux I, 9-10 June 1987, France.
9 - La Mer V.K., Ind. Eng. Chem., 44 (1952) 1270-1277.
10 - Sugimoto T., Matijevic E., J. Inorg. Nucl. Chem., 41 (1979) 165-172.
11 - Bell A., Matijevic E., J. Inorg. Nucl. Chem. 37 (1975) 907-912.
12 - Sapieszko R.S., Matijevic E., J. Colloid Interface Sci., 74 (1980) 405-422.
13 - Sapieszko R.S., Matijevic E., Corrosion 36 (1980) 522-530.
14 - Tentorio A., Matijevic E., Kratohvil J.P., J colloid Interface Sci., 77 (1980) 418-426.
15 - Uchiyama K., Ogihara T., Ikemoto T., Mizutani N., Kato M., J. Mater. Sci., 22 (1987) 4343-4347.
16 - Seiyama T., Yamazoe N., Arai H., Sensors and Actuators, 4 (1983) 85-96.
17 - Haruta M., Lemaitre J., Delannay F., Delmon B., J. Colloid Interface Sci., 101 (1984) 59-71.
18 - Matijevic E., Wilhelmy M.D., J. Colloid Interface Sci. 86 (1982) 476-484.
19 - Nielsen A.E., "Kinetics of Precipitation", MacMillan, New-York (1964) pp 16-31.
20 - Williams R., Yocom P.M., Stofko F.S., J. Colloid and Interface Sci., 106 (1985) 388-398.
21 - Ozaki M. , Kratohvil S. , Matijevic E. , J. of Colloid and Interf. Science, 102 (1984) 146-151.
22 - Matijevic E., Scheiner P., J. Colloid Interface Sci. 63 (1978) 509-524.
23 - Hamada S., Matijevic E., J. Colloid Interface Sci. 84 (1981) 274-277.
24 - Matijevic E., Cimas S., Colloid and Polymer Sci., 265 (1987) 155-163.
25 - Ozaki M. , Matijevic E., J. of Coll. Interf. science, 107 (1985) 199-203.
26 - Watson J.H.L., Cardell Jr.R.R., Heller W., J. Phys. Chem., 66 (1962) 1757-1767.
27 - Sugimoto T., Matijevic E., J. Colloid Interface Sci. 74 (1980) 227-243.
28 - Matijevic E., Sapieszko R.S., Melville J.B., J. Colloid Interace Sci. 50 (1975) 567-581.
29 - Ishikawa T., Matijevic E., J. of Colloid and Interf. Science, 123 (1988) 122-128.
30 - Regazzoni A.E., Matijevic E., Corrosion 38 (1982) 212-218.
31 - Regazzoni A.E. , Matijevic E., Colloids and Surfaces 6 (1983) 189-201.
32 - Brace R., Matijevic E., J. Inorg. Nucl. Chem., 35 (1973) 3691-3705.
33 - Scott W.B., Matijevic E., J. Colloid Interface Sci. 66 (1978) 447-454.
34 - Demchak R., Matijevic E., J. Colloid Interface Sci. 31 (1969) 257-262.

35 - Matijevic E., Langmuir, 2 (1986) 12-20.
36 - Matijevic E. , Budnik M. , Meites L., J. Colloid Interface Sci., 61 (1977) 302-11.
37 - Visca M., Matijevic E., J. Colloid Interface Sci. 68 (1979) 308-319.
38 - Barringer E.A., Bowen H.K., Comm. Amer. Soc. (1982) c199- c201.
39 - Blesa M.A., Maroto A.J.G., Passagio S.I., Figliolia N.E., Rigotti G., J. of Mater.
       Sciences 20 (1985) 4601-4609.
40 - Mcfayden P. , Matijevic E., J of Coll. and Interf. Science, 44 (1973) 95-106.
41 - Milic N.B., Matijevic E., J. Colloid Interface Sci., 85 (1982) 306-315.
42 - Shimohira T., Tomuro N., Funtai oyobi Funmatsuyakin, 23 (1976) 137-142.
43 - Wilhelmy D.B., Matijevic E., J. Chem. Soc. Faraday Trans. 80 (1984) 563-570.
44 - Wilhelmy D.B., Matijevic E., Colloids and Surface 16 (1985) 1-8.
45 - Gobet J., Matijevic E., J. of Colloid and interf. Sci., 100 (1984) 555-560.
46 - Janekovic A., Matijevic E., J. of Colloid and Interf. Sci., 103 (1985) 436-447.
47 - Katsanis E.P., Matijevic E., Colloids Surf. 5 (1982) 43-53.
48 - Mazdiyasni K.S., Ceramics International 8 (1982) 42-56.
49 - Fegley B., Jr., Barringer E.A, Mater. Res. Soc. Symp. Proc., 32 (1984) 187-197.
50 - Stöber W., Fink A., Bohn E.,   J. Colloid Interface Sci. 26 (1968) 62-69.
51 - Jubb N.J., Bowen H.K., J. Mater. Sci. 22 (1987) 1963-1970.
52 - Heistand II R.H. , Chia Y.-H., Mater. Res. Soc. Symp. 73 (1986) 93-109.
53 - Ogihara T. , Ikemoto T., Mizutani N. , Kato M., Mitarai Y., J. of Mat. Sci. 21 (1986)
       2771-2774.
54 - Brown L.M., Mazdiyasni K.S., J. Amer. Ceram. Soc. 53 (1970) 590-594.
55 - Rhodes W.H., Haag R.M., "High purity fine particulate stabilized zirconia
       (Zyttrite®)" Report No. AFML-TR-70-209 prepared by Avco Systems Division
       for the U.S. Air Force Materials Lab., Wright-Patterson AFB, Ohio, (1970).
56 - Economos G., J. Amer. Ceram. Soc. 42 (1959) 628-632..
57 - Gallagher P.K., Schrey F., J. Amer. Ceram. Soc. 47 (1964) 434-437.
58 - Morgan P.E.D., J. Amer. Ceram. Soc. 57 (1974) 499-500.
59 - Brown L.M., Mazdiyasni K.S., J. of Am. Ceram. Soc. 55 (1972) 541-544.
60 - Mazdiyasni K.S., Brown L.M., J. Amer. ceram. soc. 53 (1970) 585-589.
61 - Mazdiyasni K.S., Dolloff R.T., Smith II J.S., J. Amer. Ceram. Soc. 52 (1969) 523-
       526.
62 - Smith II J.S., Dolloff R.T., Mazdiyasni K.S., J. Amer. Ceram. Soc. 53 (1970) 91-95.
63 - Mazdiyasni K.S., Brown L.M., J. Am. Ceram. Soc. 55 (1972) 548-552.
64 - Higuchi K., Naka S., Hirano S.S., Advanced Ceramic Materials 1 (1986) 104-107.
65 - Rowell R.L., Kratohvil J.P., Kerker M., J. Colloid Interface Sci., 27 (1968) 501-506.
66 - Santacesaria E., Tonello M., Storti G., Pace R.C., Carra S., J. Colloid and Interf.
       Sci., 111 (1986) 44-53
67 - Johnson C.E., Hickey D.K., Harris D.C., Mat. Res. Soc. Symp., 73 (1986) 785-789.
68 - Matijevic E., Eds L.L.Hench et D.R.Ulrich, Wiley, New-York, (1984) 334-352.
69 - Fojtik A., Weller H., Koch U., Henglein A., Ber. Bunsen-Ges. Phys. Chem. 88
       (1984) 969-977.
70 - Lume-Pereira C., Baral S., Henglein A., Janata E., J. Phys. Chem. 89 (1985) 5772-
       5778.
71 - Sanchez C., Ribot F., New J. Chem., 18 (1994) 1007-1047.
72 - Chatry M., In M., Henry M., Sanchez C., French Patent n°91-11-633 (1991)
73 - Beer H.B., Planer G.V., Brit. Communications and Electronics, 5 (1958) 939-941.
74 - Glaister R.M., Allen N.A., Hellicar N.J., Proc. Brit. Ceram. Soc., 3 (1965) 67-80.
75 - Henisch H.K., "Crystal growth in gels", The Pennsylvania State University Press,
       University Park (1970).
76 - Raman G., Gnanam F.D., Ramasamy P., J. of Crystal Growth 75 (1986) 466-470.
77 - Raman G. , Gnaman F.D. , Ramasamy P., J. of Crystal Growth 78 (1986) 155-158.

78 - Abdullah J., Baird T., Brateman P.S., J. Chem. Soc., 3 (1986) 256-257 .

79 - Lufimpadio N., Nagy J.B., Derouane E.G., "Preparation of colloidal iron boride particles in the CTAB-n hexanol-water reversed micellar system", in "Surfactants in Solution", Vol.3, Eds. Mittal K.L. et Lindman B., Plenum Press, New-York, 1984, 1483-1497.

80 - Nagy J.B., Gourgue A., Derouane E.G., "Preparation of monodispersed nickel boride catalysts using reversed micellar systems", in "Preparation of catalysts III", Ed. G. Poncelet G., Grange P., Jacobs P.A. , Eds., Elsevier, Amsterdam,Stud. Surf. Sci. Cat. 16 (1983)193-202.

81 - Barringer E.A., Jubb N., Fegley B., Pober R.L., Bowen H.K., "Processing Monosized Powders", in "Ultrastructure Processing of Ceramics, Glasses, and Composites", Eds. Hench L.L. et Ulrich D.R., Wiley, New-York (1984) 315-333.

82 - Marra R.A., Haggerty J.S., Ceram. Eng. and Science Proceed, Aug (1981) 3-19.

83 - Graham H.C., Tallan N.M., Mazdiyasni K.S., J. of the Amer. Ceram. Soc., 54 (1971) 548-553.

84 - Mazdiyasni K.S., Brown L.M., J. Amer. Ceram. Soc. 54 (1971) 479-483.

85 - Brown L.M., Mazdiyasni K.S., Anal. Chem. 41 (1969) 1243-1250.

86 - Mazdiyasni K.S., Lynch C.T., Smith II J.S., J. Am. Ceram. Soc. 48 (1965) 372-375.

87 - Wheat T.A., J. Canad. Ceram. Soc. 46 (1977) 11-18.

88 - Sadler A.G., "Ferrites, general description and fabrication of torroids", Mines Branch Technical Bull. TB-29, Canada Centre for Mineral and Energy Technology, Ottawa, Jan., 1962.

89 - Malinofski W.W., Babbitt R.W., Sands G.C., J. Appl. Phys. 33 (1962) Suppl. 1206-1207.

90 - Wenkus J.F., Levitt W.Z., "Preparation of ferrites by the atomizing burner technique", Proc. of the 1956 Conf. on Magnetism and Magnetic Materials, Pub. AIEE, (1957) 526-530.

91 - Nielsen M.L., Hamilton P.M., Walsh R.J., "Ultrafine metal oxides by decomposition of salts in a flame", in "Ultrafine Particles", Kahn W.E. Ed., J.Wiley & Son, New-York, (1963) 181-195.

92 - McColm J., Clark N.J., "Forming, Shaping and Working of High-Performance Ceramics", Blackie, London, 1988.

93 - Stuijts A.L., "New Fabrication methods for advanced materials", Sciences of ceramics Vol. 5, Eds. C.Brosset et E.Knopp, Pub. Swedish Institute for Silicate Research, Gothenburg, 1970, pp 335-362.

94 - Lau J.G.M., Amer. Ceram. Soc. Bull. 49, 572-574, (1970).

95 - Schnettler F.J., Monforte F.R., Rhodes W.W., "A cryochemical method for preparing ceramic materials", in "Sciences of ceramics" Vol. 4, Pub. Brit. Ceram. Res. Assoc., (1968) 79-90.

96 - Wheat T.A., "Synthesis of mullite by a freeze drying process", Mineral Sciences Laboratories Report MRP/MSL77-55(TR), Canada Centre for Mineral and Energy Technology, Ottawa 1977.

97 - Jaeger R.E., Miller T.J., Williams J.C., Amer. Ceram. Soc. Bull., 53 (1974) 850-852.

98 - Roehrig F.K., Wright T.R., J. Amer. Ceram. Soc., 55 (1972) 58.

99 - Wymer R.G. , Coobs J.H. , Proc. Br. Ceram. Soc. 7 (1967) 61-69.

100 - Woodhead J.L., Segal D.L., Br. Ceram. Soc. Proc. 36 (1985) 123-128.

101 - Matthews R.B., Swanson M.L., Am. Ceram. Soc. bull. 58 (1979) 223-227.

102 - Ingebrethsen B.J., Matijevic E., J. Aerosol Sci. 11 (1980) 271-280.

103 - Ingebrethsen B.J., Matijevic E., J. Colloid Interface Sci. 100 (1984) 1-16.

104 - Lemaitre J., Vidick B., Delmon B., J. Catal. 99 (1986) 415-427.

105 - Segal D.L., J. Non- Crystalline Solids 63 (1984) 183-191.

106 - Rahaman M.N., Boiteux Y., DeJonghe C., Am. Ceram. Soc. Bull., 68 (1986) 1171-1176.

107 - Overbeek J.T.G., J. Colloid and Interface Sci. 58 (1977) 408-422.

108 - Derjaguin B.V., Landau L.D., Acta Physicochim. URSS, 14 (1941) 633-662.

109 - Verwey E.J.W., Overbeek J.T.G., "Theory of the stability of Lyophobic Colloids", Elsevier, Amsterdam (1948).

110 - Overbeek J.T.G., "Colloid Science", Vol. 1, Ed. Kruyt H., Elsevier Amsterdam (1952).

111 - Matijevic E., J. Colloid Interface Sci. 43 (1973) 217-245.

112 - Evans R., Napper D.H., Kolloid Z. Z. Polym. 251 (1973) I: 409-414; II: 329-336.

113 - Hiemenz P.C., "Principles of Colloid and Surface Chemsitry". Marcel Dekker, New-York (1977).

114- Casimir H.B.G., Polder D., Phys. Rev. 73 (1948) 360-372.

115 - Hamaker H.C., Rec. Trav. Chim. 55 (1936) 1015-1026.

116 - Masliyah J.H., "Electrokinetic Transport Phenomena", Aostra Technical Publication series # 12, Aostra Edmonton, Canada (1994)

117 - Verwey E.J.W., Rec. Trav. Chim., 60 (1941) 625-633.

118 - Shaw T.M., Pethica B.A., J. Am. Ceram. Soc. 69 (1986) 88-93.

119 - Parks G.A., Chem. Rev., 65 (1965) 177-198.

120 - Yoon R.H., Salman T., Donnay G., J.of Colloid and Interface Science, 70 (1979) 483-493.

121 - Hunter R.J., "Zeta Potential in Colloid Science", Academic Press, New-York, 1981.

122 - Dumont F., Dang Van Tan, Watillon A., J. Colloid and interf. Science 55 (1976) 678-687.

123 - Gregg S.J., Sing K.S.W., "Adsorption, Surface Area and Porosity", Academic Press, N.Y., 1967.

124 - Weiser H.B., "Inorganic Colloid Chemistry ", vol.2 "Hydrous Oxides and Hydroxides", Wiley, New-York.,(1935)

125 - Stumm W., Huang C.P., Jenkins S.R., Croat. Chem. Acta 42 (1970) 223-245.

126 - Malati M.A., Estefan S.F., in "General Discussion", Discuss. Faraday Soc. 52 (1971) 377-378.

127 - Depasse J., Warlus J., J. Colloid and Int. Sci. 56 (1976) 618-621.

128 - Tadros T.F., Lyklema J., J. Electroanal. Chem. Interfacial Electrochem. 22 (1969) 9-17.

129 - Berube Y.G., De Bruyn P.L., J. Colloid Interface Sci. 28 (1968) 92-105.

130 - Blok L., DeBruyn P.L., J. Colloid Interface Sci., 32 (1970) 533-538.

131 - Jolivet J.P., "De la solution à l'oxyde", InterEditions /CNRS Editions, Paris, 1994

132 - Hogg R., Healy T.W., Fuersteneau D.W., Trans. Faraday Soc. , 62 (1966) 1638-51

133 - Dumont F., Watillon A., Discuss. Faraday Soc. 52 (1971) 352-360.

134 - Newman A.C.D., Editor, "Chemistry of Clays and clay Minerals", Longman Scientific & Technical, Mineralogical society, Harlow, England (1987)

135 - James R.O., Healy T.W., J. Colloid Interface Sci., 40 (1972) 42-59.

136 - Ishikawa T., Matijevic E., Langmuir 4 (1988) 26-31.

137 - Flory P.J. , J. Chem. Phys., 10 (1942) 51-61.

138 - Huggins M.L. , J. Am. Chem. Soc. 64 (1942) 1712-1719.

139 - Flory P.J. , Krigbaum W.R. , J. Chem. Phys. 18 (1950) 1086

140 - Carpenter D.K., "Solution properties", in "Encyclopedia of Polymer Science and Technology", Eds. H.F.Mark, N.G.Gaylord, N.M.Sikales, Interscience, New-York. vol. 12 (1970) 627-659.

141 - Ottewill R.H. , Walker T.W., Kolloid Z. Z. Polymer, 227 (1968) 108-116.

142 - Glazman Y.M., Blashchuk Z., J. Colloid Interface Sci., 62 (1977) 158-164.

143 - Nakao Y., Kaeriyama K., J. Colloid Interf. Sci. 110 (1986) 82-87.

144 - Ottewill R.H., Rastogi M.C., Trans. Faraday Soc. 56 (1960) 866-892.
145 - Elmore W.C., Phys. Rev. 54 (1938) 309-310.
146 - Massart R., IEEE Trans. Magn. 17 (1981) 1247-48.
147 - Papell S.S., US Patent 3,215,572 , Nov 2 (1965)
148 - Edwards J., Everett D.H., O'Sullivan T., Pangalou I., Vincent B., J. Chem. Soc. Faraday Trans. 1, 80 (1984) 2599-2607.
149 - Jansen J.W., De Kruif C.G., Vrij A., J. of Colloid and Interface Science 114 (1986) 481-491.
150 - Jansen J.W., De Kruif C.G., Vrij A., J. of Colloid and Interface Sci. 114 (1986) 471-480.
151 - Aksay I.A. , Kikuchi R., "Structure of colloidal Solids", in "Science of Ceramic Chemical Processing", L. L. Hench and D.R. Ulrich Eds., (Wiley, New-York, 1986) 513-521
152 - Heller W. , "Ordered and disordered aggregation of colloidal particles and macromolecules", in "Polymer Colloids 2", Ed. R.M. Fitch (Plenum, New-York, 1980) 153-207
153 - Iler R.K, "The Chemistry of Silica", Wiley, New-York, 1979.
154 - Hsu W., Ronquist L., Matijevic E., Langmuir, 4 (1988) 31-37.
155 - Nielsen A.E., Söhnel O., J. CCrystal Growth 11, (1971) 233
156 - Schneider W., Comments Inorg. Chem. 3 (1984) 205-223.
157 - Yoldas B.E., J. Appl. Chem. Biotechnol. 23 (1973) 803-809.

GELATION

4.1 - INTRODUCTION

Gelation is a process according to which a sol, or a solution, transforms to a gel. It consists of establishing links between the sol particles, or the solution molecules, so as to form a 3-dimensional solid network. However, this is very different from the solidification of a melt, since the solid structure remains completely impregnated with the liquid of the sol, or solution. This is also a very general type of transformation and it is often considered that any solid material can be transformed to a gel.

From the theoretical point of view, two types of approaches permit to understand the fundamental aspects of such a phenomenon; they are addressed in the two first parts of this chapter, the first one calls for a global description which can be presented within the framework of the thermodynamics of critical phenomena. The second one consists in a kinetic approach resting on growth models.

A third part of this chapter summarizes the methods which permit to study gelation, while the last section consists of a review of the various gelation mechanisms of inorganic materials.

4.2 - GELATION AND PERCOLATION MODELS

FLORY-STOCKMAYER MODEL

The thermodynamic understanding view of gelation is the oldest and results from developments by Flory [1,2] and Stockmayer [3], which concerns organic materials. In these first studies, gelation designates the transformation of a liquid to a solid by chemical reactions. Considering monomers with a functionality $Z>2$, calculations by Flory have shown that the condensation of monomers yields an infinitely large polymer structure, in a mathematical sense. Practically, the solid is limited in extent only by the size of the container where the reaction is performed

Gel point

Gelation occurs when the extent of polymerization reactions ξ reaches a critical value ξ_c. This precise critical stage, when for the first time a polymer of infinite size is formed, by comparison with the molecular scale, defines the Gel Point (GP).

In a practical manner, at this point, the product resulting from polymeric condensations transforms suddenly from a viscous liquid to a material with elastic properties. In the liquid state a viscosity can be measured and its value increases towards infinity when coming near the gel point. In the solid state, an elastic modulus can be measured and its value starts from zero at the gel point. In this model of gelation, called the Flory-Stockmayer model (or FS model), the infinite polymer is built by the successive addition of branches, all of which are constructed from the initial monomers (Figure 4.2-1a).

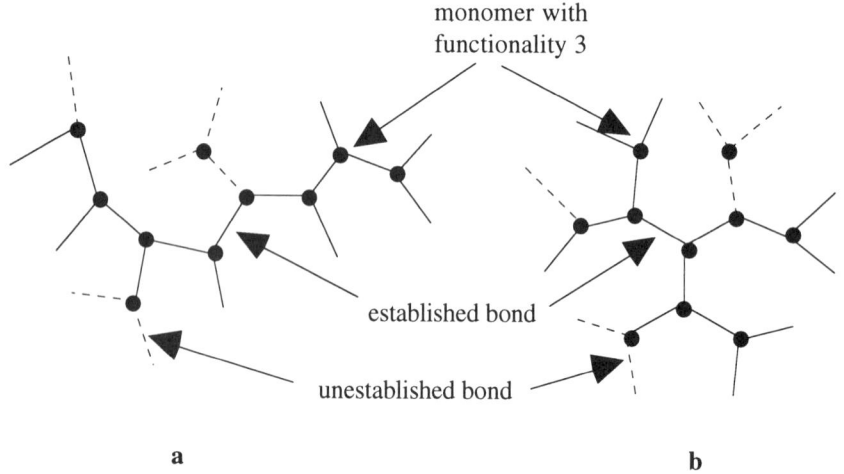

Figure 4.2-1 - (a) Flory-Stockmayer model, adapted from Flory [1]; (b) Bethe lattice, adapted from Fisher and Essam [4].

Characteristics of the Flory-Stockmayer model

The FS model explains well the main characteristics of gelation. However, instead of describing a given local aggregate, it focuses onto general parameters which characterize macroscopically a polymer solution, such as the functionality Z and the extent of reaction ξ. It is therefore not surprising that it can be described in a more abstract fashion within the framework of critical phenomena in thermodynamics, in particular by the percolation theories.

PERCOLATION MODELS

The development of the percolation theories was initiated in 1957 by Hammersley [5]. Since that time, a large number of percolation models have been studied and applied to the description of many phenomena in physics, chemistry, and the life sciences. They go from theoretical models as in the works of Stauffer et al. [6], to computer simulations like Leung and Echinger [7].

Site and bond percolations

The most simple of the percolation models are isotropic, such as the bond percolation and the site percolation. In these models, either a bond is established, or a site is occupied, with a probability p, in a completely random fashion throughout a geometrical network, as illustrated in Figure 4.2-2 on a planar square network

occupied site   established bond

empty site   unestablished bond

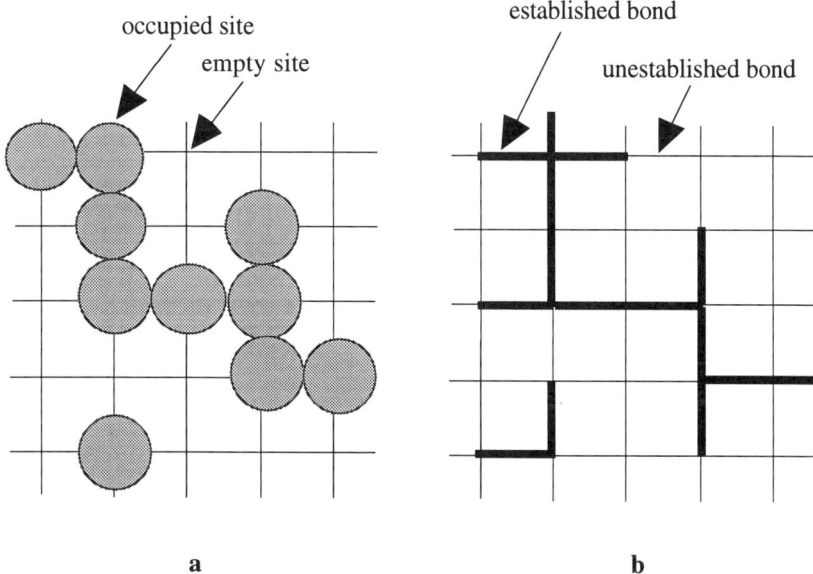

a   b

Figure 4.2-2 - Site percolation and bond percolation on a square bi-dimensional network.

Percolation threshold, critical exponents and scaling laws

These percolation theories show first the existence of a critical probability p_c, termed the percolation threshold, such that if $p>p_c$ an infinite continuous cluster of bonds or of sites exists. The threshold corresponds to the GP in the case of gelation, but its significance is more general.

Moreover, the mathematical formalism of percolation theories introduces some mathematical functions of the probability p which diverge mathematically when p comes close the GP. The divergence of these functions is equivalent to those of viscosity and elastic modulus in a gel, but with a stiffness mathematically described by well defined critical exponents.

More precisely, when coming close to the percolation threshold from the $p<p_c$ side , which is the equivalent of coming close to the GP in a sol, two characteristic properties can be defined (Figure 4.2- 3):

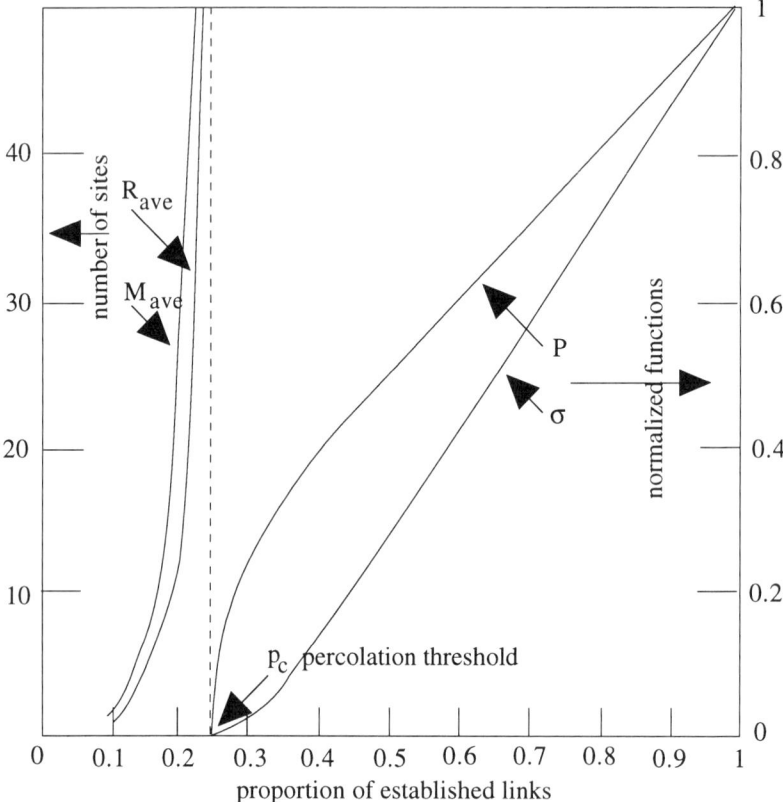

Figure 4.2-3 - Variation with the fraction p of established bonds, of characteristic properties of bond percolation on a three- dimensional cubic lattice: normalized conductivity $\sigma(p)$ after Kirkpatrick [8]; percolation probability $P(p)$ after Frisch and al. [9]; average cluster mass $M_{ave}(p)$ and average radius $R_{ave}(p)$. Adapted from Zallen [10].

- The average mass of clusters, defined as the number of connected sites in one cluster, or by the connected bonds. It diverges according towards infinity as p comes close to p_c according to the scaling law (4.2-1).

$$M_{ave}(p) \approx (p - p_c)^{-\gamma} \qquad (4.2-1)$$

-The average value of any characteristic distance, such as the cluster diameter diverges according towards infinity as p comes close to p_c according to the scaling law (4.2-2).

$$R_{ave}(p) \approx (p - p_c)^{-\nu} \qquad (4.2-2)$$

When p comes close to the percolation threshold p_c from the $p>p_c$ side, that is to say on the gel side in the case of gelation, two other characteristic properties can be defined (Figure 4.2- 3):
- The probability P(p) that a bond, or site, belongs to the infinite cluster and therefore participates in the elastic or conduction properties of a gel. It is given by the scaling law (4.2-3).

$$P(p) \approx (p - p_c)^{\beta} \qquad (4.2-3)$$

- Any transport property of this cluster, such as its elastic shear modulus G(p), or its electric conductivity $\sigma(p)$ when the bonds conduct electricity, is a function of p given by equation (4.2-4).

$$G(p) \approx \sigma(p) \approx (p - p_c)^{t} \qquad (4.2-4)$$

At last, at the percolation threshold where $p=p_c$, which is the equivalent of the GP in the case of gelation, two more important characteristics are:
- The probability P(M) for a cluster to have a mass M. It is given by equation (4.2-5). This is due to the fact that, at the percolation threshold, a whole population of clusters exists, and only one is mathematically infinite in size.

$$P(M) \approx M^{-\tau;} \qquad (4.2-5)$$

-The geometry of the unique infinite cluster is such that, starting from any site or bond, the mass inside a radius R follows the equation 4.2-6.

$$R(M) \approx M^{1/f} \qquad (4.2-6)$$

The exponent "f" is termed the fractal or mass fractal or Haunsdorf dimension. It was introduced by Mandelbrot [11] in 1977. It provides a mathematical description of the lace geometry of the solid gel network.
It is also possible to relate the radisu R of an aggregate to its surface S by a relationship of the type :

$$R(S) \approx M^{1/f_s} \qquad (4.2-7)$$

where f_s is the surface fractal dimension. If an aggregate is such that $f_s > 2$, this is a surface fractal aggregate.
Each of the exponents f, t, τ, β, ν and γ has the important property that they only depend on the Euclidean dimension d of the space in which the critical phenomenon occurs. They are termed "critical exponents". The Euclidean dimension is d=3 for actual gelation and all three-dimensional percolation models, d=2 for all planar

percolation models. It is possible in theory to consider percolation inside Euclidean spaces with dimensions other than 2 or 3. The corresponding critical exponents as a function of d are plotted in figure 4.2-4. This remarkable mathematical result has permitted to define the notion of a "universality class", which gathers all the phenomenon described by the same critical exponents, and for which the main characteristic properties obey the same mathematical laws. For instance, the theories of percolation and gelation belong to the same universality class and the equivalence between the various functions are gathered in table 4.2-1.

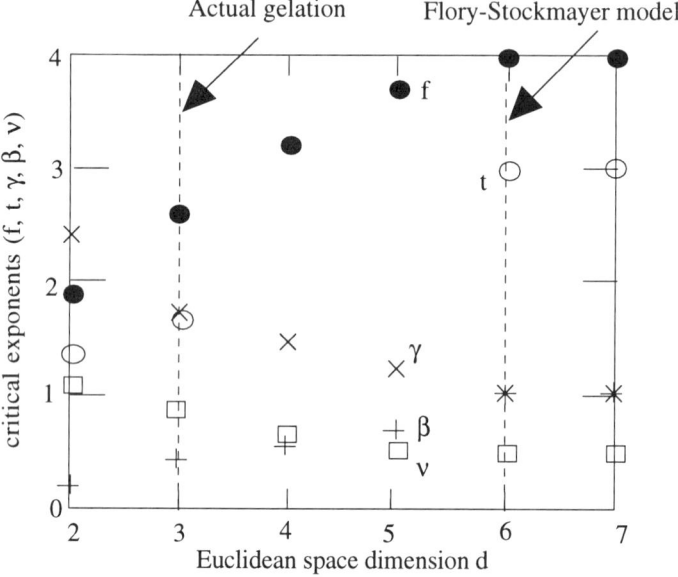

Figure 4.2-4 - Variation of the critical exponents for percolation, as a function of the Euclidean space dimension d. Adapted from Zallen [10].

Mean field or Effective medium theory
Figure 4.2-4 shows the existence of a Euclidean dimension d* beyond which the critical exponents level out to a constant value. This dimension is called the marginal dimension. For percolation and gelation d*=6. In this special case, all mathematical laws coincide with those resulting from another type of model known as the "mean field" theory, also called "effective medium" theory, which was developed independently by Bruggeman [12]. In this model, the environment around a given polymer (around a given site in percolation) is well described by replacing all other solvent or polymer molecules (all other occupied or unoccupied sites in percolation), by identical "fictitious" molecules with properties intermediate between the properties of the solvent and those of the polymers (equivalent to "partially

occupied" sites in percolation). The percolation theories demonstrate, therefore, that provided the dimension d of a Euclidean space is sufficiently high, each site is surrounded by a large number of neighboring sites, so that the details of occupied or unoccupied sites is unimportant: only the average occupancy matters.

Table 4.2-1 - Comparison of bond percolation and gelation.
Adapted from Zallen [10].

BOND PERCOLATION	GELATION
- Network coordination number Z	- Polymer functionality Z
- Bond Connectedness probability p	- Extent of reaction ξ
- Percolation threshold p_c	- Gel Point
- Percolation probability P	- Proportion of monomers belonging to the gel, or gel fraction P(x)
- Mean cluster radius R_{ave}	- Mean polymer radius R_{ave}
- Mean cluster mass M_{ave}	- Mean polymer mass M_{ave}
- Network conductivity σ	- Shear modulus G
- Bethe lattice	- Flory-Stockmayer model

Actually, the FS model examined previously corresponds to a bond percolation model on a lattice known as a Bethe Lattice [4]. This is a lattice in which each node is connected to Z other nodes, without making any closed loop, as shown in Figure 4.2-1b where Z=3. This coordination number Z in this model plays exactly the same role as the functionality Z in the FS model and the probability p to establish a bond is equivalent to the extent of reaction ξ. However, it has been demonstrated mathematically that a Bethe lattice cannot be extended indefinitely in a Euclidean space of finite dimension d, because sooner or later an overcrowding problem will occur. An infinite Bethe lattice can only exist in a Euclidean space of dimension d = ∞, a very abstract space only mathematically conceivable. Such a value for d is therefore above the marginal dimension d*=6; as a result, the FS model is both a special percolation model and an "effective medium" theory model, and it has the same critical exponents as the latter model. In particular, its fractal dimension is f=4, instead of f≈2.6.for all percolation models in the Euclidean space d=3.

Other critical parameters in percolation
In addition to the universal critical exponents introduced before, the percolation threshold probability p_c is another important critical parameter. However its value depends, in addition to the Euclidean dimension d, either on the coordination number Z in the case of a regular site network, or on the volume fraction ϕ occupied by the sites if they are replaced by hard spheres packed in a random fashion. For the Euclidean space d=3 for instance, figure 4.2-5 shows that a quasi-linear relationship

exists between p_c and either Z or ϕ. The latter relationship has not been
mathematically demonstrated but has only been found by computer simulations.
Therefore, while these percolation models have in common some universal
characteristics, they also differ from each other by many aspects. The FS is one of
the most simple model; its critical probability p_c is related to the coordination
number (or functionality) Z by:

$$p_c = \xi_c = \frac{1}{(Z-1)} \tag{4.2-8}$$

Its main default is to neglect the formation of closed loops, contrary to what occurs
in real gelation.

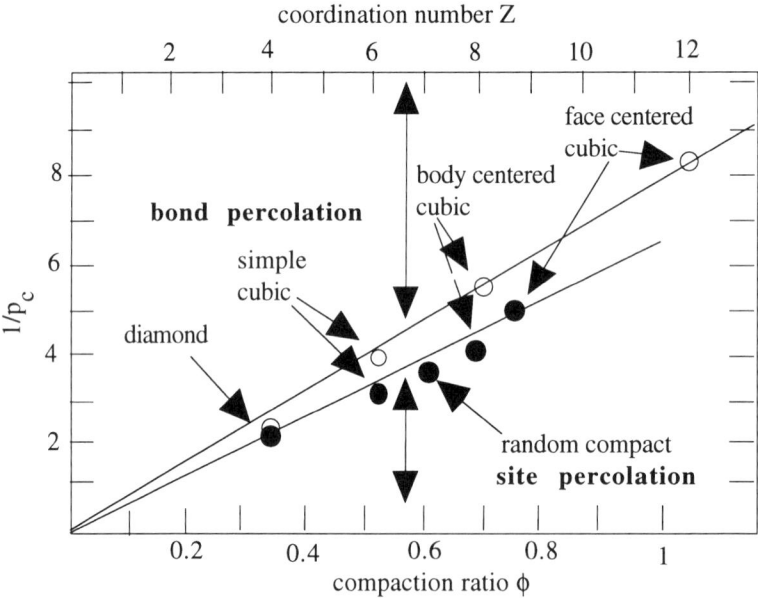

Figure 4.2-5 - Empirical correlation between the percolation threshold and the
lattice coordination for various three-dimensional structures.
Adapted from Zallen [10].

Other percolation models
A more accurate description of gelation in a real system requires percolation models
more complex than the simple site or bond percolation. For instance a gelation
model for silica should involve the different functionality Z=4 for the silicon and Z=2
for the oxygen. In many cases of heterogeneous gelation, products of a different
nature are necessary to build the gel network; such as an alkoxide and a binding agent
such as water in a third solvent. In this case, a special percolation model termed site-

bond percolation has been developed. It combines sites, occupied by the species to link, and bonds which take the place of the binding agent. For each concentration of one species, one critical concentration for the other species exists. Hence a line of GP exists which is a function of the concentrations of both species.

Moreover in many experimental cases, one has to consider colloidal particles which are no longer spherical but which can be small fibers or plates, a shape very difficult to take into account in theoretical developments. Computer simulations and experimental studies on the conductivity of such systems [13] have shown that the percolation threshold can be reached with a small volume fraction of particles, of the order of 3% for fibers, therefore much lower than for spheres (of the order of 25%). Furthermore, it is not known whether the critical exponents keep the same universal values as in isotropic percolation. The derivation of mathematical models remains therefore an open research domain.

4.3 - GROWTH-GELATION MODELS

MAIN DIFFERENCES BETWEEN PERCOLATION AND GROWTH-GELATION MODELS

Instead of forming a gel by a uniform phase change inside a fluid, it is quite possible to obtain a gel structure by the successive addition of sol particles (or of polymeric molecules in solution), to an initial particle (or an initial polymer molecule). Gelation studies which have adopted this point of view give up a macroscopic description of the sample to favor the local study of one aggregate.

Such an approach presents two advantages. First, it uses models in which the parameters can be changed progressively, so that it can show that chemical species or colloidal particles dispersed in a liquid can associate in dense aggregates in some conditions, while they build much more open structures in other conditions. The transition between both situations is quite progressive. Dense aggregates correspond to particles such examined in chapter 3, but with a more complex internal structure. Open structures are typically gel-like.

Secondly, when open structures are obtained, they are the result of kinetics processes which are out of equilibrium. It is still possible to measure their fractal dimension, but the values differ from those indicated by the theory of critical phenomenon.

In common with percolation models, most growth-gelation models describe the liquid medium by a site network. However, each site is occupied by a chemical species which keeps moving with time. Hence, this is a kinetic approach. Typical examples are described in the following sections

EXAMPLE OF GROWTH-GELATION MODELS

Polymerization-growth model by Mannerville and de Seze [14]

In this description, some sites of a network are attributed to monomers with a functionality 2, other sites to monomers of functionality 4 and the remaining sites to solvent molecules. The proportions of each type of sites is respectively C_2, C_4 and $C_S = 1 - C_2 - C_4$. (Figure 4.3-1).

Next a fraction C_I of chemical species acting as bonding initiators are initially distributed on the monomer sites, and they are let free to move by random walk from their initial position. Several things can occur. Each time an initiator jumps from a site occupied by a monomer to another site occupied by a monomer, the two monomers are bonded to each other to form a dimer. When it jumps to a solvent site, no bonding occurs. At last, when two initiators jump on the same site they annihilate each other. The process stops when all the available sites to the initiators are either monomers which have reacted, or solvent molecules. Fractal aggregates with critical exponents different from percolation are obtained.

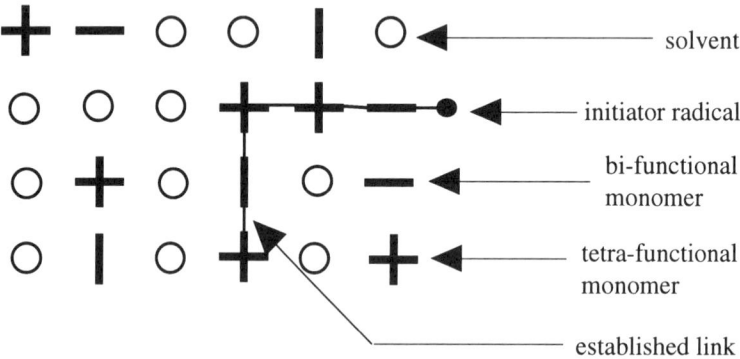

Figure 4.3-1 - Addition polymerization model. Adapted from Manneville and De Seze [14].

Invasion percolation growth model
In the invasion-percolation growth model by Chandler et al. [15], a random number in the range from 0 to 1 is attributed to each site. Next, one site is selected as the initial nucleus and growth consists in adding the perimeter site with the lowest random number to this nucleus, at each step. Hence the focus is on the growth of one particular aggregate, rather than on the average growth of all aggregates such as in percolation. The aggregates which are formed in this process have the same fractal dimension a classical bi-dimensional percolation, that is to say $f \approx 1.89$.

Eden model and other unfractal growth models
In the "Eden" or "cancer-growth" model [16,17], growth also proceeds from one initial site which is selected as the initial nucleus. The sites which are added to this nucleus are then selected at random amongst all the perimeter sites of this nucleus. The resulting structures have holes as illustrated in figure 4.3-2, but their mass fractal dimension is identical to the Euclidean dimension (f=d). They are therefore compact non-fractal structures. However they are surface fractal aggregates.
Other growth models which produce compact solids have been developed. One of them is the ballistic model [19] where the particles which are added to a nucleus

arrive in straight line. Another one is the multiple aggregation model [20,21] which consists in the simultaneous aggregation of a finite density of particles.

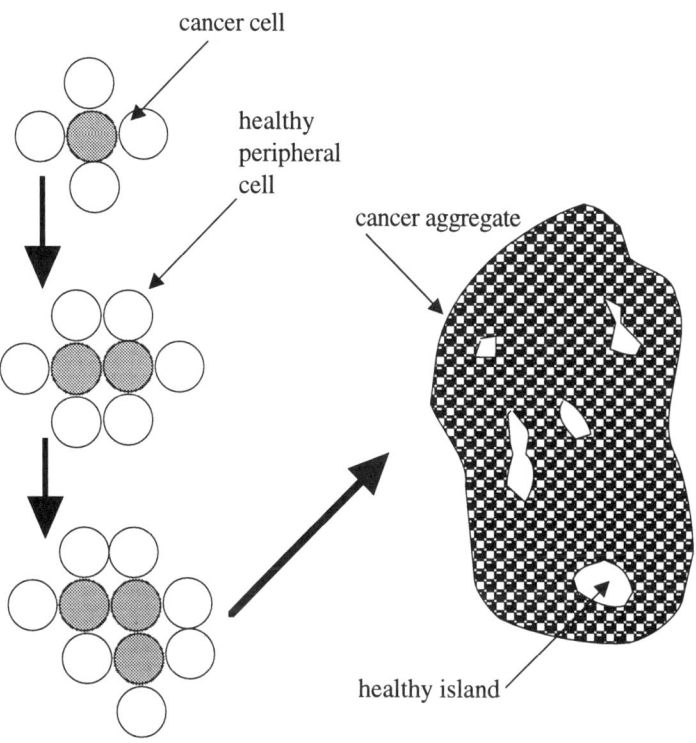

Figure 4.3-2 - Eden's cancer growth model. Adapted from Stanley et al. [18].

Rikvold crystallization model [22]
This model is a modification of the cancer-growth model, in which a new peripheral site added to a nucleus can diffuse on a length L_d along the cluster perimeter, so as to simulate surface diffusion. The results have shown that, starting from a low value L_d (negligible diffusion) which produces compact structures, the growth figure becomes more and more ramified with increasing L_d (Figure 4.3-3).

The "electric breakdown" model
This model, studied by Sanada [23], simulates the electric breakdown of capacitors. It is derived from the cancer-growth model by weighing the perimeter sites with a factor R when they are close to a tip of an aggregate. These aggregates remain compact if R<<35, but become ramified if R>>35 .

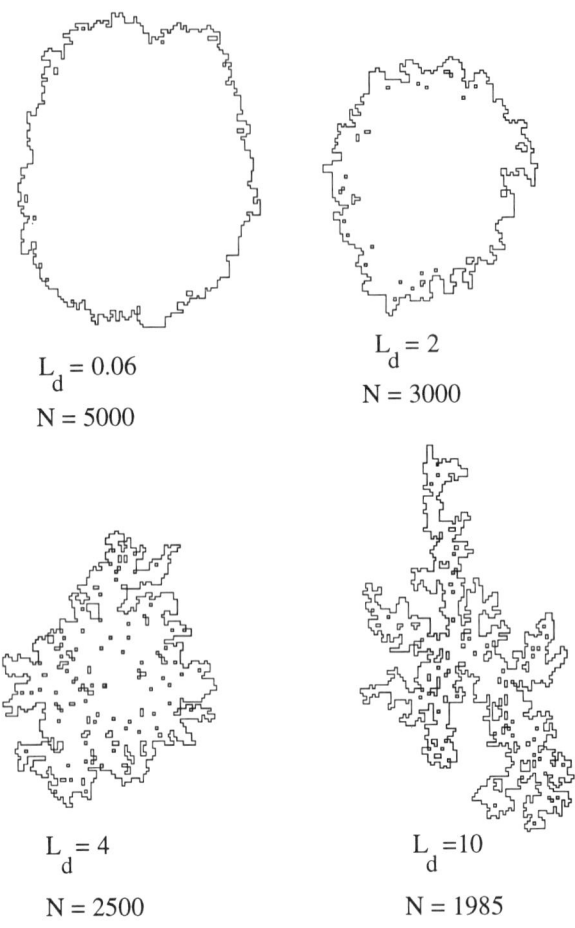

$L_d = 0.06$

$N = 5000$

$L_d = 2$

$N = 3000$

$L_d = 4$

$N = 2500$

$L_d = 10$

$N = 1985$

Figure 4.3-3 - Examples of aggregates in the Rikvold crystallization model for different values of the diffusion length parameter L_d. N designates the number of sites. Adapted from Rikvold [22].

Diffusion limited aggregation models (DLA model) [21,24]

The sites at a given distance from the original nucleus are allowed to walk randomly, so as to simulate the Brownian motion of the particles in a sol. They may therefore either avoid meeting a nucleus, or reach one of them and participate to its growth. As soon as some branches of the aggregate statistically start to grow around this nucleus, the possibility for sites to diffuse up the corridors without hitting a side branch becomes low. These branches have therefore a tendency to keep growing so that the aggregates becomes even more ramified than in the percolation models. The fractal dimensions of bidimensional models are 1.67 for DLA and 1.89 for percolation and respectively 2.5 (DLA) and 2.6 (percolation) for tridimensional models. The characteristics of the corresponding aggregates are illustrated in Figure 4.3-4.

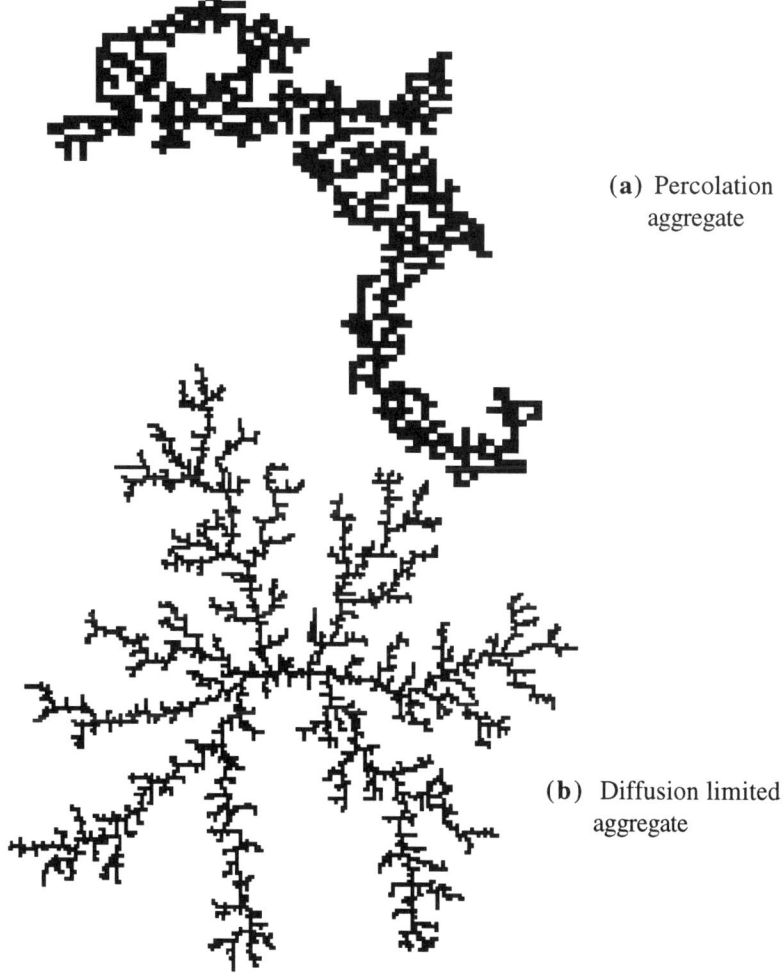

(a) Percolation aggregate

(b) Diffusion limited aggregate

Figure 4.3-4 - Bi-dimensional aggregates on a square lattice with different fractal dimensions: (a) aggregate near the percolation threshold in site percolation, adapted from Stanley et al. [25] and (b) diffusion-limited aggregation cluster comprising 3600 particles, adapted from Witten and Sander [21].

A refinement of the DLA model [26-28] considers several nucleation centers. Each nucleus grows by diffusion and simultaneously diffuses slowly, by Brownian motion. Hence, primary fractal DLA aggregates grow and finally aggregate themselves in a more open structure. The fractal dimension of such hierarchical DLA, also termed "cluster-cluster" is f=1.4 for bidimensional aggregates and f = 1.80 for tridimensional ones if the linkage of the primary aggregates is the rate limiting step. If the diffusion of the primary aggregates is the rate limiting, f = 2.09 in tridimensional models .

Onoda and Toner hierarchical model
The preceding hierarchical DLA aggregation-growth with its "Russian dolls" looks like a hierarchical model by Onoda and Toner [29]. According to this model, a population of primary spherical particles with identical size, aggregate to form a secondary spherical particle. The packing ratio p, defined as the volume fraction of the secondary sphere which is occupied by the primary spheres, and the ratio   S $= \dfrac{\text{secondary sphere radius}}{\text{primary sphere radius}}$, are fixed. The secondary particles themselves aggregate in tertiary spherical particles, characterized by the same packing ratio p and radius ratio S. This operation can be repeated several times (Figure 4.3-5), and it can be shown that it builds a fractal structure with the fractal dimension:

$$f = d + \frac{\ln p}{\ln S} \qquad (4.3\text{-}1)$$

As an example for S = 10, p= 0.5 and d=3,   f=2.7

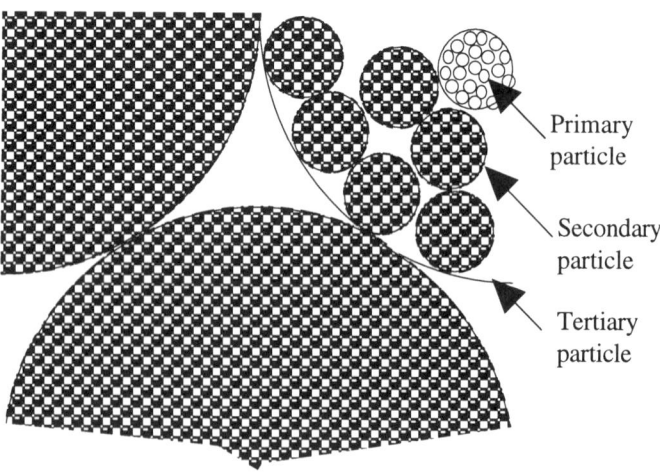

Figure 4.3-5 - Hierarchical generation of aggregates in a powder.
Adapted from Onoda and Toner [29].

Other mathematical studies on growth-gelation

All previous aggregation-growth models have in common the posibility to be studied by Monte-Carlo simulation methods on a computer. On the other hand, pure mathematical analytic studies in continuous media have been attempted, although they have not been very successful [30,31].

GELATION AND THE DLVO THEORY

Ramified structures such as described in the previous paragraphs can be observed when a large number of primary particles have aggregated. At a scale of a few particles, it is not obvious to make a difference between compact agglomerates and fractal aggregates. However, a linear bonding between colloidal particles is possible in some conditions as soon as the first contacts occur and this phenomenon can be described within the DLVO theory presented in chapter 3.

Critical electrolyte concentration for gelation

In fact it has been known that the gelation of sols obeys the experimental Schulze-Hardy rule which concerns the influence of the concentration of a non-potential determining electrolyte on the coagulation of a sol. In chapter 3, the existence of a critical electrolyte concentration C_c for the occurrence of rapid coagulation was introduced. The existence of a critical concentration for gelation C_g has also been known or a long time [32]. If the concentration C of non-potential determining electrolyte is $C < C_g$, a sol is stable. If $C_g < C < C_c$, the sol slowly gels. At last if $C > C_c$, rapid coagulation occurs instead of gelation.

An increasing concentration of non-potential determining electrolyte has the effect of decreasing the energetic barrier which must be overcome by Brownian motion to realize the aggregation of two particles. As shown in chapter 3, if $C < C_g$, the energy barrier to overcome is too high, no linkage occurs. For $C_g < C < C_c$, the energy barrier which opposes coagulation is still significant, but not as high as when $C < C_g$. Hence a low, but not negligible probability to overcome this barrier exists. With time contacts between particles occur and they form pairs of particles, in a first step. In some conditions, DLVO computations show that a third particle has a smaller energy barrier to overcome to add up linearly to a two-particles aggregate, than to add up laterally. Hence, a fibrous structure made of strings of particles starts to build up slowly from the very beginning of the coagulation process; it really constitutes a nascent gel (Figure 4.3-6). At last, if the concentration of non-potential-determining electrolyte is $C > C_c$, the energy barrier which a third particle has to overcome to link with a two-particles aggregate, is low. It can occur at random laterally as well as linearly, hence the coagulation of dense, randomly packed aggregates occurs.

Electrostatic conditions of gelation

The conditions to realize a gel, according to the DLVO theory, are reached when the thickness of the electric double layer κ^{-1} (chapter 3) is of the same order as the diameter of the colloidal spherical particles. In this case, a noticeable population of counterions are excluded from the volume occupied by two adjacent spheres in a string of particles.

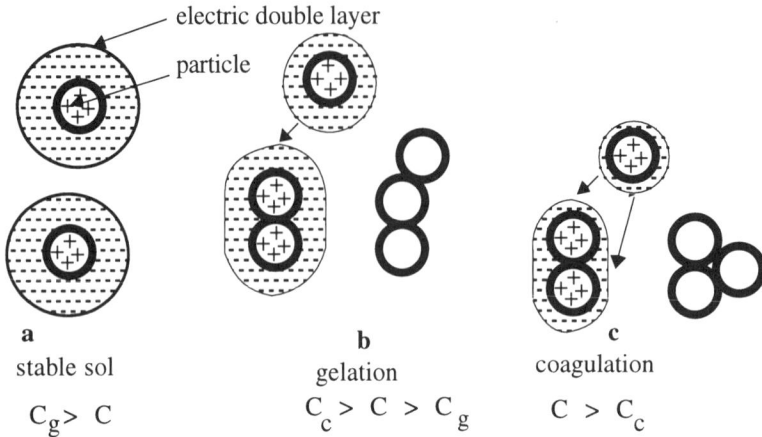

Figure 4.3-6 - Conditions for the formation of: (a) a stable sol; (b) gelation; (c) coagulation, as a function of the non-potential determining electrolyte C.

Moreover, the electric double layers of these adjacent spheres partly cover each other [33] (Figure 4.3-7). With respect to electrostatic repulsion, a chain of colloidal spherical particles is therefore perceived as a smooth cylinder by a free spherical particle which approaches the chain on its side. In these conditions, the hydrostatic overpressure which builds up by osmotic diffusion in between the coming particle and the existing chain of particles, is strong on a larger front for a lateral approach than for a linear approach. Computation by the method of Derjaguin shows that the electrostatic repulsion is stronger for a lateral approach than for a linear approach, by a coefficient of the order of $\sqrt{2}$, while the Van der Waals interaction is modified in a much lower ratio [34]. Hence, linear strings of particles keep growing. Lateral linkage can statistically occur with a lower probability than linear linkage, which insures the formation of branches and crosslinking between the chains of spheres.

Example of gel structure according to DLVO theory
A possible example of structure is the following; let us consider some primary particles of diameter 1 nm, with a surface electric potential of 27.5 mV at pH 3, in a material with z.p.c 2.5. The energy barriers to overcome are 2.85 and 4 kT respectively for linear and for lateral approach. These two values are much lower than 15 kT, therefore they can easily be overwhelmed by Brownian motion and the primary particles agglomerate in relatively compact spherical secondary particles (Figure 4.3-8). When the secondary particles reach a diameter of 5 nm the linear and lateral energy barriers become respectively 15 kT and 21 kT. In these conditions, linear linkage becomes much more probable than lateral linkage; it is a slow process which constructs chains made of these secondary spherical particles. The overall

hierarchical structure illustrated in Figure 4.3-8 is very similar to that actually admitted for silica gels [35].

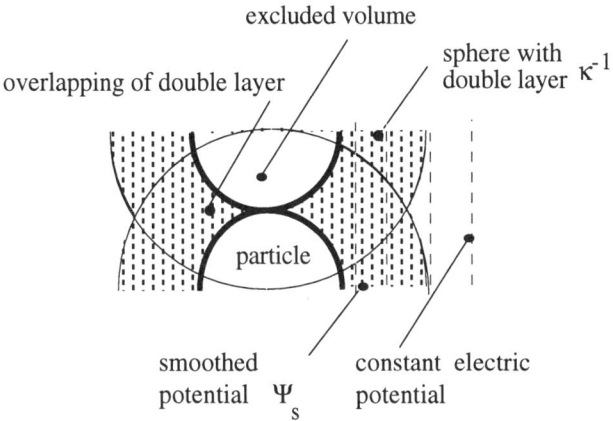

Figure 4.3-7 - Smoothed electric potential around a string of colloidal particles with a thick double electrical layer. Adapted from Pierre [34].

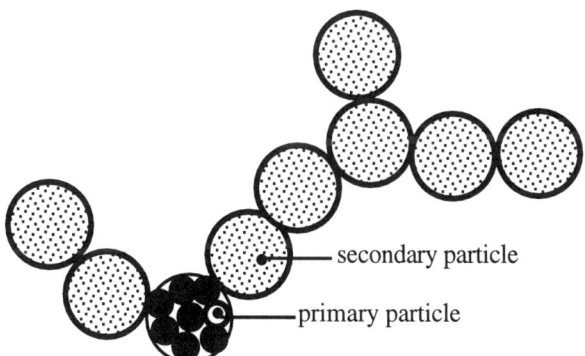

Figure 4.3-8 - Possible structure of a gel obtained by electrostatic destabilization. Adapted from Pierre [34].

SUMMARY OF THE GELATION THEORIES

Critical phenomenon theories, such as percolation, offer a powerful mathematical formalism to show that a sol-gel transition produces fractal structures which can be described within the framework of classical thermodynamic.

On the other hand, growth simulations on computer such as the DLA models, show that fractal aggregates can also be formed by kinetic, non equilibrium processes. In this case, the existence of a powerful mathematical formalism is missing. The fractal dimension of aggregates cannot be derived by mathematical analysis, but deduced by the computer from the relationship between the computed aggregate mass and size.

Computer growth models, as well as percolation theories, give a macromechanistic view of gelation. The DLVO theory brings a micromechanistic view, as it explains how linear strings of particles can also be build from the beginning of aggregation, when appropriate chemical conditions are satisfied.

Overall, gel-like solid networks can have a very different origin. It is now necessary to examine the gelation behavior of actual materials, in the light of all the theories introduced in the present section.

4.4 - EXPERIMENTAL STUDY OF GELATION

RHEOLOGICAL METHODS

Steady flow curves

The most traditional manner to study a sol-gel transition makes use of a viscometer [36]. Before the gel point, a sol or a solution is submitted to a shear flow. The viscosity is measured as a function of time, which is itself directly related to the extent of the chemical reaction, until the limit of the apparatus is reached. Above the gel point, a gel behaves as a solid; it is submitted to a torsion strain and the steady state shear modulus is measured as a function of time. The gel point itself is only accessible by extrapolation (Figure 4.4-1). Such a method has been applied to colloidal boehmite sols [37], to SiO_2 polymeric solutions, or even to multicomponent systems, for instance comprising the elements Si, Al, B, Ba and Na [38]. Such measurements are tedious. On the other hand, a more simple and very qualitative method consists in tilting a sol and its container. Before gelation, the sol flows. After the gel point, the container can be turned upside down and the sample does not flow.

Oscillatory shear flow [36]

Another rheological technique consists in applying an oscillatory shear, of small amplitude and of frequency ω, to a sample during gelation [36]. The instantaneous shear modulus is recorded; for a given frequency ω, it is a periodic function G(t) of time which has the same frequency ω as the oscillatory shear, but which is not in phase with it. Hence, the instantaneous shear modulus can be considered as a complex modulus and decomposed in a component G'(w) termed the storage modulus which is in phase with the oscillatory shear modulus, and a component G"(w) termed the loss modulus which is out of phase with the oscillatory shear. The storage and loss modulus can be obtained by the following Fourier transforms:

$$G'(\omega) = \omega \int_0^\infty G(t)\sin(\omega t)dt \qquad (4.4-1)$$

$$G''(\omega) = \omega \int_0^\infty G(t)\cos(\omega t)dt \qquad (4.4-2)$$

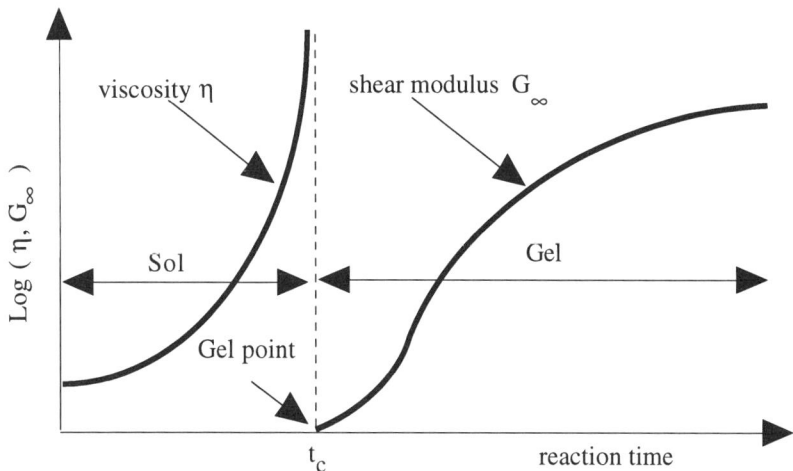

Figure 4.4-1 - Schematic representation of the steady state shear viscosity and of the equilibrium elastic modulus of a polymer during a sol-gel transition. Adapted from Winter and Chambon [36].

This method was applied to an organic compound for which gelation was well known; the polydimethylsiloxane. It has shown that during gelation the storage modulus increased faster than the loss modulus as the time increased and that a crossing point existed (Figure 4.4-2).

The time necessary to reach the intersection depended on the oscillation frequency ω [39]. However, this frequency could be changed by a factor up to 105 without affecting the following equality at gel point:

$$G'_c(\omega) = G''_c(\omega) \qquad (4.4-3)$$

Therefore, this intersection could be used to define the experimental gel point. Moreover, if the temperature was modified within a large range from -50°C to +180°C, all intersection data points for G'_c and G''_c could be gathered on a single curve on which they were plotted as a function of $k_T \omega$ and the coefficient k_T was a function of the temperature only. At last it was possible to define a characteristic constant C for each material, which was independent of the temperature T and of the frequency ω, such that:

$$G'_c = G''_c = C\, k_T\, \omega \qquad (4.4-4)$$

This definition of the gel point is not contradictory with the previous traditional definition which corresponds to an oscillatory frequency $\omega = 0$. It can be shown that the viscosity η_{GP} and shear modulus G_c at the gel point can be deduced from the loss and storage moduli according to equations (4.4-5) and (4.4-6). As shown in these two equations, they take values which are in agreement with the characteristic values of the viscosity and shear modulus at the percolation threshold:

$$\eta_c = \lim_{\omega \to 0} \left(\frac{G_c''}{\omega} \right) = C \, k_T \lim_{\omega \to 0} (\omega^{-1/2}) = \infty \qquad (4.4\text{-}5)$$

$$G_c = \lim_{\omega \to 0} (G_c') = C \, k_T \lim_{\omega \to 0} (\omega^{1/2}) = 0 \qquad (4.4\text{-}6)$$

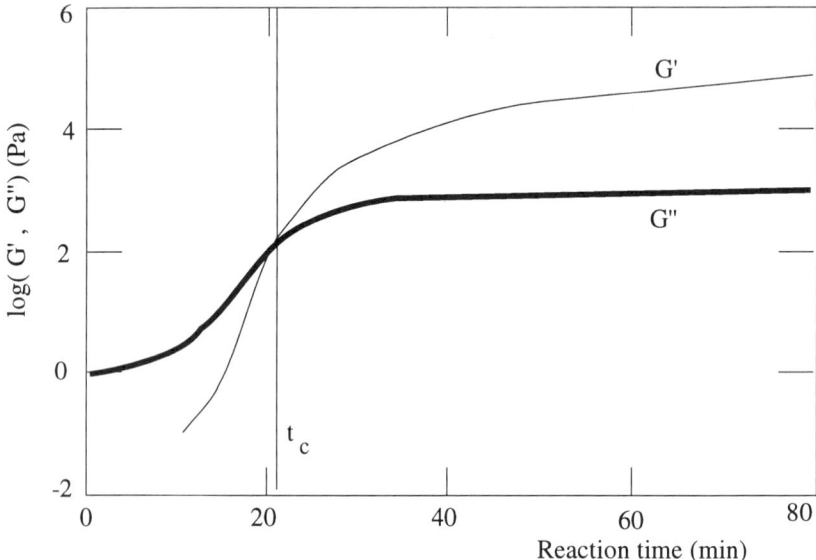

Figure 4.4-2 - Evolution of the storage modulus (G') and loss modulus (G") of a polymer during gelation, when submitted to an oscillatory shear of frequency ω . Adapted from Winter and Chambon [36].

OTHER METHODS
Techniques other than rheology make it possible to study some particular aspects of this phenomenon once the gel point has elsewhere been proven, if not to prove the existence of a gel point.

Scattering techniques
The scattering techniques can give a numerical estimate of the fractal dimension f of an aggregate, since the scattered intensity I (k) varies with the scattering vector k as [40]:

$$I (k) = k^{-f(3-\tau)} \qquad (4.4\text{-}7)$$

Where f and τ are critical exponents defined previously.

For the gelation of a colloidal sol with particles size of the same order as the visible light wavelengths, light scattering is capable of providing some information on the size and shape of the particles. During gelation, the Tyndall effect according to which a sol takes on a bluish aspect when observed in transmission, often increases. However, this can be the consequence either of the formation of bigger aggregates, or of an increasing difference between the optical density of these aggregates and that of the liquid medium which surrounds them. The latter difference actually increases when the pH is farther from the isoelectric point, because a larger number of chemical species are adsorbed on the particles [41]. As pointed by Tanaka et al. [42], Rayleigh light scattering cannot be used to demonstrate the existence of a gel point. In some instances, the diffused light cannot be observed at the gel point. On the other hand, some phenomena which do not correspond to gelation can be observed by this technique, such as the coexistence of a collapsed state and a swollen state in a gel.

Dynamic light Scattering consists of submitting a sol to a monochromatic coherent beam. The impact of the colloidal sol particles under constant Brownian motion, onto the photons, is the cause of a "red shift" of the wavelengths of these photons. The shifts at large angles (90 degrees) and at small angles, provide respectively the translational and rotational velocities of the particles and permit to determine the size and shape of aggregates. In this way, an estimate of the effective hydrodynamic radius of an colloidal aggregate r_c can be determined. Its evolution with time can be followed, as well as its divergence towards in infinite value if gelation occurs.

At a much lower size scale, another radius termed the radius of gyration r_g, can be determined by small angle X-ray scattering (SAXS) [43]. Porod [44] has shown that if $r_g k \gg 1$ and if the size of monomers r_m which constitute the primary aggregates responsible for scattering is such that $r_m k \ll 1$, the scattered intensity I(k) follows a law which depends on the packing structure of the monomers. For instance:

$$I(k) \approx k^{-4} \text{ for distinct 2-phases systems} \qquad (4.4-8)$$

$$I(k) \approx k^{-2} \text{ for random walk chain polymer} \qquad (4.4-9)$$

$$I(k) \approx k^{-2.6} \text{ for random bond percolation theory} \qquad (4.4-10)$$

The range of k can be extended to lower values, from $8. 10^{-3}$ to 8.10^{-4} nm^{-1} with the technique of small angle neutron scattering (SANS), than with the SAXS technique. Practically, the scattering techniques can be used to design a structure model for a material, which combines a measured size of primary particles given by r_g, a measured size of colloidal aggregate r_c, and a measured fractal dimension f. The correlation function $\Gamma(r)$ of the model is calculated, then Fourier transformed . The result is compared to the experimental data and the values of f, r_g and r_c which lead to the best agreement with the experiment indicate a possible correct structure model. However, this model is not unique, hence it does not constitute a definitive proof of the actual structure [45].

Microscope techniques

The technique of scanning electron microscope (S.E.M) does not usually permit to observe the texture of porous networks at the size scale of the pores in most gels.

On the other hand, when the electronic density of colloidal sol particles is high, the actual shape of the aggregates can be visually and directly studied under a transmission electron microscope (TEM). This was the case for the gold sols studied by Weitz and Huang [46], composed of spherical particles with a mean radius of 14.5 nm. These sols were made by reduction of the gold salt $Na(AuCl_4)$ by sodium citrate. Electrostatic peptization of the particles occurred by adsorption of the citrate ions on the particles surface which therefore carried a negative surface charge. Destabilization leading to gelation was done with pyridine. The corresponding molecules were neutral but they could displace the citrate ions and adsorb on the surface of the gold particles. The aggregates made in this example had a fractal characteristics with a fractal dimension f < 2.

Other physical chemistry techniques

All physical chemistry techniques can be used to study the evolution of a network during gelation or coagulation. For instance in their study of the gelation of spherical particles of an organic material (poly-4-methyl-1-pentene) in cyclohexane, Tanugami at al. [47] froze the organic gel in its medium and they separated manually the small spheres which belonged to the gel network, so that they could determine the total mass fraction X_g of the colloidal gel, as a function of time. The sol-gel transition itself can also be studied by electronic paramagnetic resonance (EPR), such as with silica [48] or vanadium pentoxide [49]. In this technique, the exchange energies between neighboring metals in a gel are measured. Their magnitude can give an idea of the type of bonding responsible for gelation. Other useful techniques comprise nuclear magnetic resonance (NMR) as in a large number of studies, Raman spectroscopy, thermogravimetric analysis (TGA), differential thermal analysis (DTA) and dilatometry .

4.5 - GELATION MECHANISMS OF CERAMIC MATERIALS

SHAPING A GEL

For technologic applications, the sol-gel transition is very important because this is the first moment when the shape of a ceramic part can be controlled. The gelation can be directed to obtain continuous fibers, thin films, or spherical powders with a given size, according to a technique which depends on the material and desired shape.

To produce spherical powders, the sol or the solution can be dispersed as liquid droplets, by stirring, in an organic solvent which is unmiscible with water. For instance Yamagishi and Takahashi [50] made ThO_2 gel microspheres with an average diameter of 1.1 mm, dispersed in CCl_4, from a sol obtained by hydrolysis of $Th(NO_3)_4$ with NH_4OH at pH of 3.05. The gelation of these droplets could be achieved by:

- adding long chain amines to the emulsion, which had the effect of dehydrating the microspheres.
- adding an ammonia precursor such as hexamethylene tetramine $(CH_2)_6N_4$, in the solution, before formation of the emulsion droplets. Ammonia was released by heating and gelation occurs
- passing a hydrophilic reactant such as 2-ethyl hexanol or gaseous NH_3 in the solution to remove the anions from the sol droplets and induce gelation by dehydration.

The spraying techniques constitute good alternate methods to make microspheres. The precursor solution can be sprayed in a chilled liquid such as xylene or liquid nitrogen. Next, the chilled microspheres are slowly heated under vacuum to extract the solvent by freeze-drying. A sol can also be sprayed as an aerosol, which consists of a mist of fine droplets, on which drying is performed [51]. Eventually, hydrolysis itself can be achieved with an aerosol of the precursor [52]: colloidal spherical particles of TiO_2 have been made in this way [53] as well as aluminum hydroxide powders [54].

All these techniques make it possible to eliminate a grinding step which requires an increasingly long time when smaller particles are desired. They also permit to synthesize powders in a more industrial fashion than by nucleation and growth in a solution. The latter technique, examined in chapter 3, must be distinguished from the synthesis method described in this section to make microspheres: these microspheres really are small spherical monoliths, with a bigger size more adapted to industrial handling than the colloidal ones. They contain a polymeric or colloidal three-dimensional network, made with the colloidal particles described in chapter 3.

For other shapes such as fibers or coatings, it is possible to spin or draw a sol near the gel point, or also to dip a surface to be coated in the sol. The gelation techniques presented above for spherical powders, can be adapted to these shapes.

INFLUENCE OF THE NATURE OF CATIONS

As mentioned before, gelation also depends on the material being considered. For instance, it is known that the elements which produce a stable hydroxide do not easily make gels from their chemical precursors [55]; in the case of Al, only the monohydroxide can easily be peptized after hydrolysis to a clear sol which gels.

Classification of the gelation behaviors

With some materials, gelation can be achieved from polymer solutions which are clear and transparent, where no colloidal particles can be observed. With other materials a more or less opaque characteristics is visible, because colloidal particles are present. Hence a first practical distinction must be made between two types of gelation, one known as polymeric gelation, the other known as colloidal gelation. Also and independently from the two previous categories of gels, the bonds which must link the polymer molecules or the colloidal particles in a gel, can be quite diverse. They comprise covalent bonds, Van der Waals or hydrogen bonds, electrostatic bonds, bonds by hydrophobic patches, ion bridging [56,57]. In practice, this diversity in the nature of bonds responsible for gelation has led to divide the oxide gels in two categories with respect to the reversibility of gelation [32]. For a first class which including some clays such as montmorillonite, boehmite AlO(OH), or the similar CrO(OH) and FeO(OH) gels, gelation is reversible. A dry gel can be diluted back to an aqueous sol. For a second class of gels which comprises SiO_2, ZrO_2 and TiO_2, reversed dilution is not possible; gelation is irreversible.

Irreversible gelation

When the bonds responsible for gelation are strong and do not hydrolyze readily, such as Si-O-Si or Ti-O-Ti, gelation usually occurs spontaneously after a sufficient time, once the appropriate extent of chemical reaction has been reached. This gelation behavior is often designated as chemical gelation. It can be realized by hydrolysis with H_2O or with H_2O_2 for instance in the oxides made from alkoxides [58], or by bubbling an H_2S gas flow in a non-aqueous solution of alkoxide [59]. Chemical gelation can be achieved from colloidal sols as well as from polymeric solutions, when strong bonds link the particles in contact. The nature of the final bonding between particles is independent from the electrostatic conditions favorable to the establishment of contact according to the DLVO theory (chapter 3).

It is often considered that the elements leading to chemical gelation are capable of producing polymeric gels and that they coincide with the glass forming elements; This is the case with a number including Si, B and Ge [60]. Al, which is considered as a glass forming element in spite of the fact that no pure alumina glass has never been obtained, does not gel spontaneously. However, there is no complete identity between the two properties of glass forming and chemical gelation. For instance Ti and Zr do undertake chemical gelation but they do not produce pure glasses.

There is no coincidence neither between the categories of polymeric gels and chemical gel forming elements. For instance, in some conditions Al form gels which can be considered to be polymeric, as they are clear, extremely amorphous and transparent. However these alumina gels are soluble and they do not belong to the category of chemical gels.

Gels termed polymeric gels always form from clear transparent solutions. A common opinion is that polymeric gelation occurs under conditions when the reactions of polymerization-condensation are faster than the hydrolysis reactions [61]. However this view is very controversial and we saw that the formation of linear polymers is favored when the hydrolysis rate is faster than polymerization in the case of silica. The gelation of polymers requires crosslinking points where bonds are established in order to build a three-dimensional continuous network. A consequence is that the gelation process is influenced by the all the factors which modify the hydrolysis and polymerization rates. As discussed in chapter 2, these factors include: the chemical nature of the reactants; the hydrolysis water ratio $r_w = \dfrac{[H_2O]}{[alkoxide]}$; the pH; the presence of a catalyst; the temperature; the concentration of the reactants and the nature of the solvents.

The influence of the previous parameters was mainly studied for silica. The corresponding precursors, when hydrolyzed in excess water, produce strongly crosslinked structures instead of linear polymers [62]. As an example, the colloidal gelation of tetramethoxysilane $Si(OCH_3)_4$ (or TMS) in solution in methanol, catalyzed by ammonia, proceeds faster with an increasing hydrolysis ratio r_w, or with an increasing precursor concentration [63], the gel transition becomes sharper and the gel more transparent. The fastest gelation rate of colloidal silica occurs at pH $\approx$ 5, while for 6 < pH < 11 a silica sol is stabilized. In the latter pH range, linear polymerization is not sufficient to favor the formation of siloxane bonds between the colloidal particles. When the pH is below 5, the overall hydrolysis rates decreases and allows linear polymerization to occur. The various polymeric or colloidal structures obtained, as a function of the pH, are reviewed in chapter 2. In summary, polymeric gelation is favored by acidic catalysis and it requires a longer time. Such a process is not modified when foreign additives which do not participate to the reaction are added, such as acetone or a phenolic resin [64].

In the chemical gelation of other colloidal sols, the electrolytic or steric effects influence the formation kinetics as well as the more or less compact structure of the gel obtained. As this was explained before in this chapter, according to the DLVO theory, adding a metal salt to a sol can reduce the electrostatic repulsion so as to favor contact between colloidal particles when they come in a linear approach. Polymeric gelation can also be favored by non aqueous solvents, as in the ZrO_2 gels made in cyclohexane by Kindu and Ganguli [65].

Reversible gelation
In some instances, the bonds responsible for gelation are weak so that physical gelation occurs. In this case, it is necessary to evaporate the solvent to favor contact and bonding between polymeric chains or colloidal particles and to transform the sol or solution to a gel [66]. Therefore, the sol-gel transition is reversible with respect to dilution in the initial solvent [67]. That is to say, a piece of dry gel will slowly swell when placed in the solvent, until it loses any solid consistency. The most simple technique to remove a solvent consists in evaporating it from a dish which contains the sol [64]. After some time, the liquid becomes more and more viscous and solidifies to a soft gel with a glassy characteristic [64]. The gelation of such soluble systems can also be achieved by addition of a liquid which is not a solvent of the system but which adsorbs water. However the gel obtained in this way will still swell and re-dissolve when immersed in its original solvent.

Physical gelation is also often thermoreversible [68], as the weak bonds responsible for gelation can be broken by thermal agitation. Ionic, Van der Waals and hydrogen bonding interactions are often involved in this type of gelation [69,70], which is frequent with organic materials. For instance non-covalent bonds can be sufficient to crystallize polymeric molecules, in some domains, and to insure the gelation. as this occurs with globular proteins. The linking can also be achieved by low-molecular weight molecules or by tensioactive compounds [41]. In all these systems, gelation depends on the steric volume and functionality of the molecules. Many organo-metallic compounds are prone to this type of mechanism in solution; an example is the paramagnetic copper complex with a disk shape which comprises eight paraffin chains, as illustrated in Figure 4.5-1. When in solution in cyclohexane, these molecules make a discotic liquid crystal mesophase [71], where each domain is constituted by a hexagonal array of columnar associated disk-like molecules. This liquid undertakes a sol-gel transition when the concentration in organo-metallic material is above 4 % by mass, at a temperature of about 20°C. Observations of the dry gel under the electron microscope have shown that the organo-metallic molecules are arranged in chains.

Classical crystallization and physical gelation sometimes look quite similar to each other [72]. During crystallization, dendritic particles may grow and build a network which looks like a gel. These structures are unstable and usually dissolve again to recrystallize to a different crystalline form. However the dissolution scheme is different from a physical gel; these dendrite packing are unable to swell to a soft gel when a small quantity of solvent is added . There is no bonding between the crystals and one cannot speak in this case of a gel. The formation temperature of these crystalline structures depends on the cooling rate. They often are observed after repeated heating and cooling [41].

The oxides particles which have a high hydration energy promote a strong water structuration immediately close to their surface. In this case, the hydration water can play an important role because the structured water dipoles can stay in between the colloidal particles after they aggregate according to the DLVO theory [73] and help to link them to each other. Moreover, they offer an easy way for the intercalation of more water when a gel is immersed in this solvent, so that it can easily swell and progressively return to a sol state. Such a behavior occurs with gels of thorium arsenate, phosphate and molybdate, and of iron, chromium and aluminum hydroxides. A number of other materials present a low hydration energy and gel irreversibly, such as the ferric arsenate, borate and phosphate; the zirconium hydroxide, molybdate and borate; the vanadium pentoxide; the manganese dioxide and the tannic hydroxide, borate and tungstate [74].

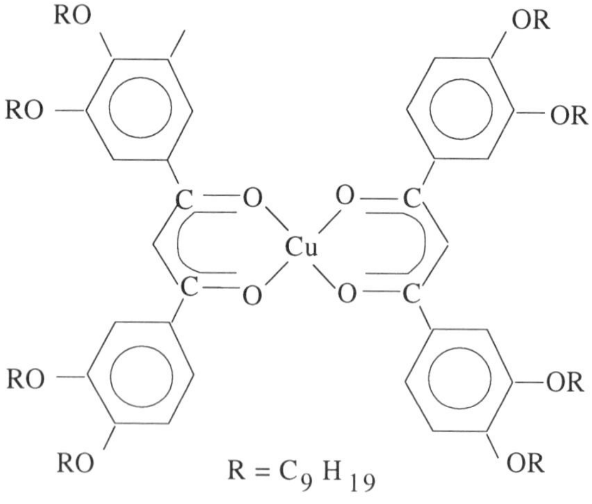

Figure 4.5-1 - Copper complex with eight paraffin chains.
Adapted from Terech et al. [57].

The early transition metals (V, Mo, W) can be made to gel by decreasing the charge of their polyanions in solution [75]. In this way, vanadium gives a decavanadic acid which polymerizes spontaneously in water, according to a self-catalytic condensation process [76,77]. The gel structure is fibrillar, however it redissolves in excess water. Aluminum is an element which typically makes physical gels. However such colloidal gels can only be obtained from the monohydroxide boehmite $AlO(OH)$ [78]. The particles in a boehmite sol have a plate-like shape, as in clay sols and, their structure is lamellar. The anisotropic growth of these particles is eased by heating above 80°C. Depending on their length and the type of electric charges on their edges and faces, an evolution in the gel network formation occurs. For instance, when an increasing concentration of an electrolyte such as HNO_3 is used to peptized a sol, the gelation mechanism depends on the acid concentration [79]. At low acid concentration, the boehmite particles have a low electric charge and they produce a gelatinous precipitate in which the particles are randomly packed. For a slightly higher acid content, a sol is formed with particles which only carry one type of charge, positive, on their flat faces. The volume at gel point is lower than previous gelatinous precipitate because these charges compel the particles to organize parallel to each other, inside domains. For still higher acid contents, negative charges are adsorbed on the edges of the particles while their faces are positively charged. It is possible to consider that gelation occurs by edge-to-face linkage and the volume at gel point becomes high (Figure 4.5-2). In summary the volume at gel point presents a minimum for a molar acid concentration ratio $\dfrac{[HNO_3]}{[Al\text{-sec-butoxide}]}$ =0.07 [80].

Shear stress versus shear curves, termed flow curves, can be experimentally determined in a viscometer for any sol. In the case of the previous boehmite sol, a flow curve for a HNO_3 molar ratio of 0.56 shows damped oscillations which are quite reproducible during successive cycles. Moreover, a yield stress must be applied to the sol to make it flow once the sol has been maintained at rest for at least one second. The same phenomenon can be observed for an acid molar ratio of 0.28, however the sol must be maintained at rest for at least one hour to again observe a yield point. This behavior is characteristic of thixotropy, which is common with particles which having a plate-like or a rod-like shape and are able to link to each other. These links are destroyed when a shear stress is applied. However, they reconstruct when the suspension is at rest and they eventually build an extended three-dimensional network, so that a gel is obtained.

Such a particular gelation mechanism is known as "thixotropic gelation". The corresponding gels are characterized by the existence of a yield stress beyond which they are plastic. Of course, their behavior depends on the previous material history, since the mechanisms of destruction-reconstruction are not spontaneous. The existence of a network reconstruction mechanism, together with the occurrence of flow beyond the yield point, constitute the main differences between a strong chemical gel, where any bond destruction is irreversible and flow does not occur, and a weak physical-gel.

The phenomenon of thixotropic gelation is observed with materials other than boehmite, in particular ferric hydroxide sols [81] and clay. The colloidal properties of boehmite gels are very similar to those in clay gels.

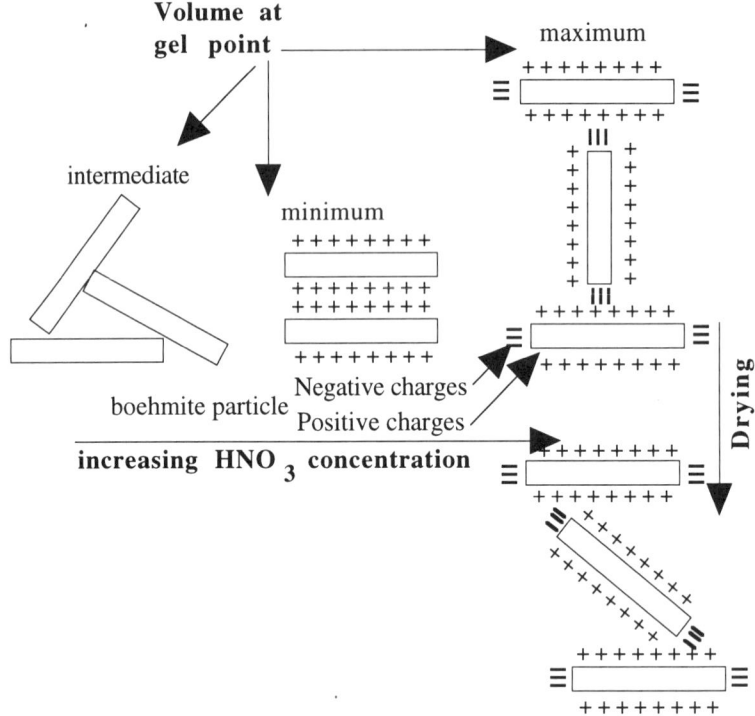

Figure 4.5-2 - Gelation of boehmite sols. After Pierre and Uhlmann [79].

The main difference between the two materials is that the sign of electric charges, on the edges and faces, are reversed [80]. The faces of clay particles are usually charged negatively while the edges can carry positive charges. Several gelation mechanisms have actually been proposed for clays, in particular gelation by formation of H_2O dipole chains with counterions as the keystones [82] and gelation by increase of the double layer thickness, which would decrease the fluidity of a sol [83]. However, the latter one could explain the development of rigidity in compression, not in tension and none of these mechanisms easily explains the existence of a minimum gelation volume. Hence, the formation of card-house structures as in boehmite, by edge to face attraction when they have opposite electric charges appears to be a credible architecture.

However such card-house packing are not really observed in an electron microscope. This fact is in part due to the shape of the particles which are not nicely drawn rectangles but very irregular platelike particles with a variable size [84-88]. On the other hand, after treatment with phosphates to provide negative charges to both the edges and faces of platelike kaolinite particles, and with electrolyte additives such as $AlCl_3$ or $FeCl_3$, random aggregation where the particles are packed with a nearly edge to edge linking modes are observed (Figure 4.5-3), in aggregates dried by the CO_2 supercritical method presented in chapter 5.

Boehmite and all clay gels are of the colloidal type, however aluminum hydroxide can also make weak chemical polymeric gels, for instance by hydrolysis of Al-sec-butoxide at room temperature in excess water, with a molar ratio of nitric acid of at least 0.28 [89]. In these conditions a solution as clear as water rather than a sol is obtained. Polymeric gelation occurs during drying and, here again, this is a weak gel which can redissolve in excess water.

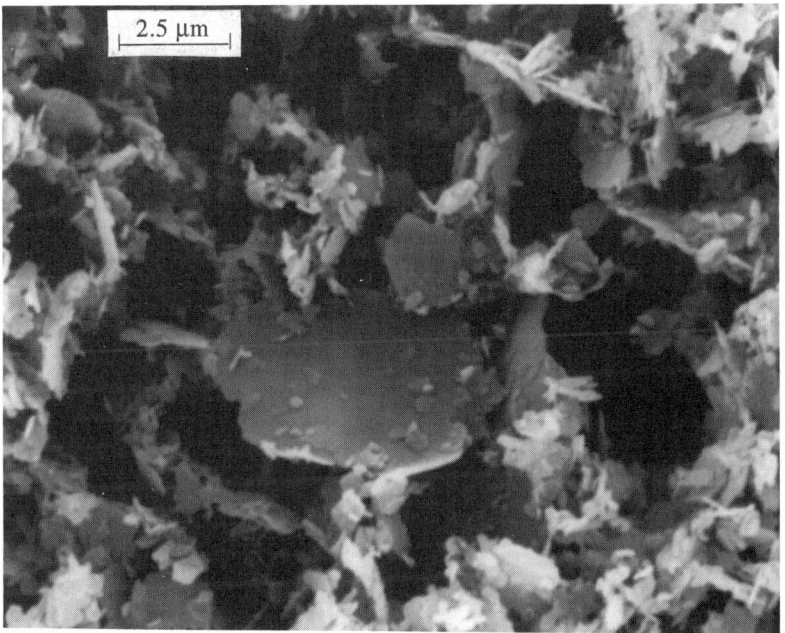

Figure 4.5-3 - Packing architecture of platelike kaolinite particles treated with $Na_4P_2O_7$, dispersed in a 2 mM $AlCl_3$ aqueous solution and dried by the CO_2 supercritical method. From Ma and Pierre [90].

Other characteristics of gelation

Physical gelation and chemical gelation both show a latent heat. Also, they are sometimes accompanied by a volume change which can be either a contraction such as with gelatin + water, or an expansion as with silicic acid + water. With silicic acid, the volume expansion begins in the same time as the reactions of hydrolysis and condensation and it keeps going during gelation. It originates in the formation of water molecules coming from the condensation reactions. In other cases, such as with iron hydroxide, no volume change has been observed.

GELATION OF ACTUAL MATERIALS AND THE PERCOLATION OR AGGREGATION MODELS

With respect to the description of gelation by percolation or by growth-aggregation a previously mentioned study by Weitz and Huang [46] on gold sols, showed that the growth-aggregation models correctly describe the gelation characteristics of colloidal sols. Overall, it appears that the fractal dimension of inorganic colloidal gels is close to f=1.8, which is in agreement with the hierarchical growth-aggregation [31]. Percolation gives a good description of truly polymeric gels, essentially organic gels. It may also possibly apply to polymeric silica gels.

Limitation by transport or by fixation of new colloidal particles on a fractal aggregate

One practical consequence of growth aggregation models is that the gelation kinetic can result from the slowest of the two following steps, on one hand the transport of solute species towards the aggregate, on the other hand the fixation of new species on the aggregate.

When an aggregate becomes big enough, the probability for a new particle to remain attached to this aggregate increases, and at some moment the kinetics become limited by the transport of new species. As the fractal aggregates themselves diffuse by Brownian motion, we therefore observe a transition in the kinetic regime from simple DLA to hierarchical DLA and this has an influence on the fractal dimension f of the final network which decreases. In the case of colloidal gold gels where the bonding between colloidal particles is due to Van der Waals forces, the growth kinetics are almost entirely dominated by the particles diffusion. The size of the aggregates is an increasing function of time, with an exponent between 0.4 and 0.6. The low value of the fractal dimension, f=1.74, is consistent with a hierarchical DLA model. Aggregates start to grow by diffusion, and they themselves aggregate with each other by diffusion.

An example of actual DLA aggregate made with colloidal kaolinite particles and observed in a scanning electron microsope is shown in Figure 4.5-4 .

Colloidal gelation and percolation models

Another previously mentioned study by Tanigami et al. [47] on a colloidal organic gels, showed that the weight fraction X_S of colloidal particles belonging to aggregates could be represented by an Avrami's law [91] :

$$X_S = 1 - \exp(-kt^n) \tag{4.5-1}$$

where k and n are two constants which depend on the material and on the growth mechanism.

In the case of the material studied by Tanigami et al., the value of n before gelation is comprised between 3 and 4, which is not too far from the value n=4 corresponding to growth limited by diffusion, in the general case. Moreover, the Avrami graphs in Figure 4.5-5 show that a kinetic transition occurred near the gel point, as approximately determined by turning the test tube containing the sol upside down. After gelation, only the particles remaining inside the holes of the gel aggregate could still add to its network. Hence the value of n fell below 2. For a given initial concentration of colloidal particles, this transition defined a gel point time t_g and a gel point aggregate mass fraction X_{sg}.

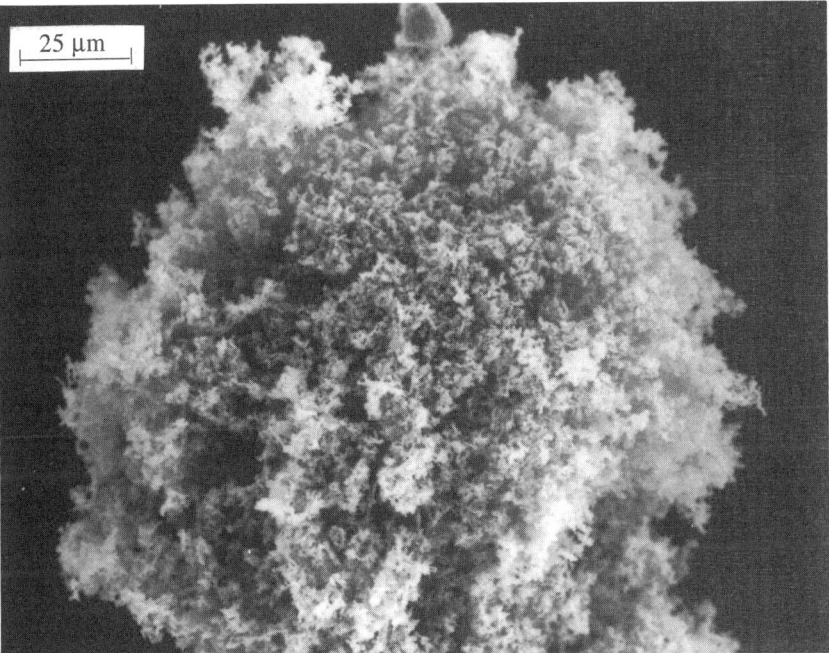

25 μm

Figure 4.5-4 - Packing architecture of platelike kaolinite particles aggregates treated with $Na_4P_2O_7$, dispersed in a 2 mM $FeCl_3$ aqueous solution and dried by the CO_2 supercritical method. From Ma and Pierre [90].

By repeating this operation for different colloidal particle concentrations, it was possible to plot t_g as a function of the initial colloidal particle concentration. The resulting graph, in Figure 4.5-6, looked like that found with a simple mixed site-bond percolation model, as this was illustrated in the section 4.2 .

GELATION OF MULTICOMPONENT SYSTEMS
The gelation of multicomponent systems shows the existence of new phenomena which can be described within the broad class of heterogeneous gelation.

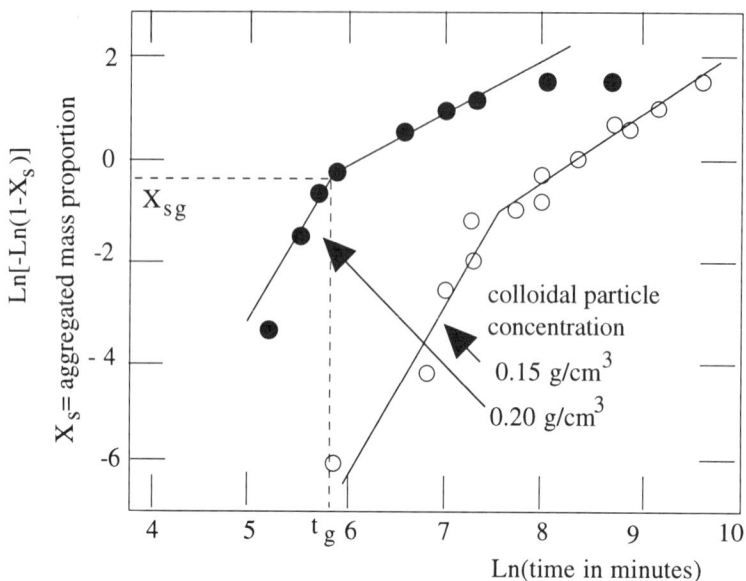

Figure 4.5-5 - Avrami plots for the growth-gelation in a colloidal organic system. Adapted from Tanigami et al. [47].

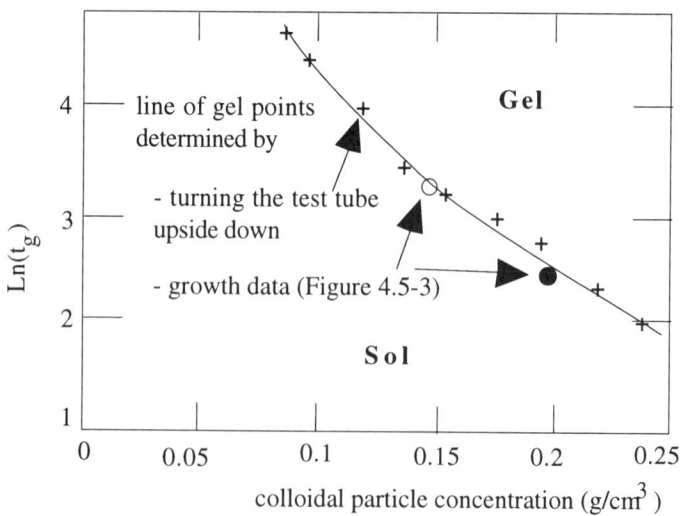

Figure 4.5-6 - Gelation time, at 50°C for the gel in Figure 4.5-3. Adapted from Tanigami et al. [47].

A combination of different bonding mechanisms is possible, depending on the individual mechanism of each component and of its state, as a sol or a solution. The co-precipitation of hydroxides mixtures can itself be considered as the formation of a heterogeneous gel-precipitate and occurs in a large number of examples [61].

When two sols are mixed, the homogeneity is at the scale of the particles which constitute the sols. Moreover the electric charges of the various components can carry electric charges with the same sign or with an opposite sign, which modifies the gelation mechanism. In the case of mullite $3Al_2O_3.2SiO_2$, the silica particles usually carry an electric charge with a sign opposite to that of boehmite particles. Depending on the relative size of the particles of each oxide, the following gelation mechanisms are possible:

- If the silica particles are much smaller than the boehmite particles, they build a shell around these boehmite particles. Consequently, the gelation of these composite particles is similar to that of silica, it is irreversible.

- If the sizes are reversed, the nature of the core and of the outer shell of the composite particles are reversed and the gelation mechanism is similar to that of boehmite, it is reversible.

In the case of mixed magnesium and aluminum systems, $Al(OH)_3$ precipitates in slightly basic solutions while $Mg(OH)_2$ requires strong basic solutions. Depending on the selected conditions, it is possible to produce a boehmite gel which encloses colloidal particles of a Mg compound, or a mixed gel roughly corresponding to a double hydroxide [92]. It is also possible to enclose the colloidal particles with a polymeric alumino-silicate gel [93].

The gelation of multicomponent systems can be considered to be polymeric when the different cations can be homogeneously dispersed at the atomic level. In this case also it could also be possible to theoretically consider two different type of gelation mechanisms; one according to which a single polymeric gel network comprising the various cations is built; the second in which two independent and interpenetrating polymeric networks are formed. In practice, the first situation is considered to prevail as in the complex silicate system SiO_2-Al_2O_3-K_2O [94]. The second situation could prevail when the condensation kinetics of the various cations is very different, in particular when it is not possible to realize a mixed alkoxide comprising these various cations.

4.6 - REFERENCES

1 - Flory P.J., J. Am. Chem. Soc. 63 (1941) I: 3083-3090; II: 3091-3096; III 3096-3100.

2 - Flory J.P, "Principles of Polymer Chemistry", Cornell Univer. Press, Ithaca, New-york. 1953

3 - Stockmayer W.H., J. Chem. Phys., 11 (1943) 45-55.

4 - Fisher M.E., Essam J.W., J. Math. Phys. 2 (1961) 609-619.

5 - Hammersley J.M., Proc. Cambridge Phil. Soc. 53 (1957) 642-645.

6 - Stauffer D., Conoglio A., Adam M., Adv. Pol. Sci. 44 (1982) 103-158.

7 - Leung Y.K., Eichinger B.E., J. Chem. Phys., 80 (1984) 3877-3891.

8 - Kirkpatrick S.,   Rev. Modern Phys., 45 (1973) 574-588.

9 - Frisch H. L., Hammersley J.M., Welsch D.J.A., Phys. Rev. 126 (1962) 949-951.

10 - Zallen R., "The physics of amorphous solids", John Wiley, New-York, 1983.

11 - Mandelbrot B., "Fractals: Form, Chance and Dimension", Freeman, San Francisco, 1977.

12 - Bruggeman D.A.G., Ann. Phys. (Leipz.) 24 (1935) 636-664.

13 - Carmona F.,   Barreau F., Delhaes P., Canet R., J. Physique-Letttres 41 (1980) L531-L533.

14 - Manneville P., De Seze L., "Percolation and Gelation by Additive Polymerization", in "Numerical Methods in the Study of Critical Phenomena", Eds., Della Dora I., Demangeot J., Lacolle B., Springer, Berlin, 1981, 116-124.

15 - Chandler R., Koplik J., Lerman K., Willemsen J.F., J. Fluid Mech., 119 (1982) 249-267.

16 - Eden M., Proceedings of the 4th Berkeley Symposium on Mathematical Statistics and Probability, Ed. Neyman J., University of California Press, Berkeley, 1961. Vol. 4 p 223. Reference quoted by Peters et al. [17].

17 - Peters H.P., Stauffer D., Holter H.P., Loewenich K., Z. Phys. B 34 (1979) 399-408.

18 - Stanley H.E., Family F., Gould H., J. of Polymer Science, Polymer Symposium 73 (1985) 19-37.

19 - Bensimon D., Domany E., Aharony A., Phys. Rev. Letters 51 (1983) 1394.

20 - Voss R.F., Bull. Am. Phys. Soc., 28 (1983) 487.

21 - Witten T.A.Jr., Sander L.M., Phys. Rev. Lett., 47 (1981) 1400-1403

22 - Rikvold P.A., Phys. Rev. A 26 (1982) 647-650.

23 - Sawada Y., Ohta S., Yamazaki M., Honjo H., Phys. Rev. A 26 (1982) 3557-3563.

24 - Deutch J.M., Meakin P., J. Chem. Phys., 78 (1983) 2093-2094.

25 - Stanley H.E., Birgeneau R.J., Reynolds P.J., Nicoll J.F., J. Phys. C, 9 (1976) L553-L560.

26 - Meakin P., Phys. Rev. Letters, 51 (1983) 1119-1122.

27 - Kolb M., Botet R., Julien R., Phys. Rev. Letters, 51 (1983) 1123-1126.

28 - Witten T. A. Jr., Meakin P., Phys. Rev. B28 (1983) 5632-5642.

29 - Onoda G.Y., Toner J., J. Am. Ceram. Soc. 69 (1986) C278-C279.

30 - Meakin P., Phys. Rev. B 28 (1983b) 6718-6732.

31 - Sander L.M.,   "Theory of Fractal growth Processes", in "kinetics of aggregation and Gelation", P.Family P., Landau D.P., Eds., Elsevier (1984) 13-17.

32 - Prakash S., Dhar N.R., J. Ind. Chem. Soc. 6 (1929) 391-409.

33 - Thomas I.L., McCorkle K.H., J. of Colloid and Interf. sciences, 36 (1971) 110-118.

34 - Pierre A.C., "The Gelation of Colloidal Gels and the Derjaguin-Landau-Verwey-Overbeek (or DLVO) Theory", In "Ultrastructure Processing of Advanced Materials", Edited by D.R. Uhlmann and D.R. Ulrich, John Wiley and Sons (1992) 103-110.

35 - Fricke J., J. of Non-Crystalline Solids 100 (1988) 169-173.

36 - Winter H.H., Chambon F., J. of Rheology 30 (1987) 367-382.

37 - Pierre A.C.,   "Ceramic Composites by Sol-Gel techniques", PhD Thesis, Massachusetts Institute of Technology, september 1985.

38 - Mukherjee S.P., "Final report on Sol-Gel Coatings for Si-Ge substrates", Sandia Laboratories (DOE Contract/DE-AC04-IG-DP00789) October 22, 1979.

39 - Tung C.Y.M., Dynes P.J., J. Appl. Pol. Sci. 27 (1982) 569-574.

40 - Schaefer D.W., Martin J.E., Wiltzius P., CannelL D.S., "Aggregation of Colloidal Silica", in "Kinetics of Aggregation and Gelation", Eds., Family F.. Landau D.P., Elsevier Science Publishers B.V., (1984) 71-74.

41 - Hermans P.H., "Reversible systems", in "Colloid Science", H.R.Kruyt, Ed., Elsevier, New-York, vol. II (1952) 483-650.

42 - Tanaka T., Swislow G.,Ohmine I., Physical Review Letters, 42 (1979) 1556-1559.
43 - Brinker C.J., Keefer D.W., Schaefer D.W., Ashley C.S., J. of Non-Crystalline Solids 48 (1982) 47-64.
44 - Porod G., Kolloid Z., I, 124 (1951) 83-114; II, 125 (1952) 51-57 et 108-122.
45 - Sinha S.K., Freltoft T., Kjems J., "Observation of power-law correlations in silica-particles aggregates by small-angles neutron scattering", in "Kinetics of Aggregation and Gelation", Eds., Family F.. Landau D.P., Elsevier Science Publishers B.V., (1984) 87-90.
46 - Weitz D.A., Huang J.S., Self-Similar Structures and the Kinetics of Aggregation of Gold Colloids", in "Kinetics of Aggregation and Gelation", Family F.. Landau D.P., Eds., Elsevier Science Publishers B.V., (1984) 19-28.
47 - Tanigami T., EN K., Yamaura K., Matsuzawa S., Polymer Journal 18 (1986) 31-34.
48 - Martini G., Burlamacchi H., J. Phys. Chem. 83 (1979) 2505-2511.
49 - Gharbi N., Sanchez C., Livage J., J. Chim. Physique 82 (1985) 755-759.
50 - Yamagishi S., Takahashi K., J. of Nuclear Materials 144 (1987) 244-251.
51 - Sane A.Y., "Refractory metal borides, carbides and Nitrides, and composites containing them", European Patent 0115745- Eltech Systems Corporation, 15. 08 (1984).
52 - Nicolaon G., Cooke D.D., Kerker M., Matijevic E., J. Colloid Interface Sci., 34 (1970) 534-544.
53 - Visca M., Matijevic E., J. Colloid Interface Sci. 68 (1979) 308-319.
54 - Ingebrethsen B.J., Matijevic E., J. Aerosol Sci. 11 (1980) 271-280.
55 - Mazdiyasni K.S., Ceramics International 8 (1982) 42-56.
56 - Ross-Murphy S.B., McEvoy H., British Polymer Journal, 18 (1986) 2-7.
57 - Terech P., Chachaty C., Gaillard J., Giroud-Godquin A.M., J. Physique 48 (1987) 663-671 (1987).
58 - Komarneni S., Roy R., Breval E., J. Am. Ceram. Soc., 68 (1985) C41-C42.
59 - Melling P.J., Am. Ceram. Soc. Bull., 63 (1984) 1427-1429.
60 - Brinker C.J. Bunker B.C., Tallant D.R., Ward K.J., J. de Chimie Physique, 83 (1986) 851-858.
61 - Zelinski B.J.J., Uhlmann D.R., J. Phys. Chem. Solids 45 (1984) 1069-1090.
62 - Sakka S, "Gel method for making glass", dans "Treatise on Materials Science and Technology", Edited by M. Tomazawa and R.H. Doremus., Vol 22 (1982) 129-167.
63 - Debsikdar J.C., Advanced Ceramic Materials 1 (1986) 93-98.
64 - Wei G.C., Kennedy C.R., Harris L.A., Amer. Ceram. Soc. Bull. 63 (1984) 1054-1061.
65 - Kindu D., Ganguli D., J. of Materials sciences Letters 5 (1986) 293-295.
66 - Partlow B.P., Yoldas B.E., J. Non-crystalline Solids 46 (1981) 153-161.
67 - Segal D.L., J. Non- Crystalline Solids 63 (1984) 183-191.
68 - Guenet J.M., Le Courrier du CNRS 63 (1985) 50-53.
69 - Taniguchi Y., Susuki K., J. Phys. Chem. 78 (1974) 759-761.
70 - Terech P., J. Colloid Interface Sci. 107 (1985) 244-255.
71 - Nguyen Huu Tinh, Dubois J.C., Malhete J., Destrade C., C.R. Heb. Sean. Acad. Sci. C 286 (1978) 463-464.
72 - Von Weimarn P.P., Rev. Gén. Colloïdes 7 (1929) 153-158.
73 - Dumont F., Dang Van Tan, Watillon A., J. Colloid and interf. Science 55 (1976) 678-687.
74 - Prakash S., Dhar N.R., J. Ind. Chem.Soc. 7 (1930) 417-432.
75 - Livage J., Lemerle J., Ann. Rev. Mater. Sci. 12 (1982) 103-122.
76 - Frey-Wyssling A., Mühlthaler K., Vierteljahrenchr Naturf. Ges., Zurich 89 (1944) 214-215.

77 - Lemerle J., Nejem L., Lefebvre J., J. Inorg. Nucl. Chem. 42 (1980) 17-20

78 - Pierre A.C., Uhlmann D.R., J. Amer. Ceram. Soc. 70 (1987) 28-32.

79 - Pierre A.C., Uhlmann D.R., Mater. Res. Soc. Symposium Proc., 73 (1986) 481-487.

80 - Yoldas B.E., J. Mater. Sci. 10 (1975) 1856-60.

81 - Cotton A., Mouton H., Ann. Chim. Phys. 11 (1907) 145-205, et 289-347.

82 - Ford R.W., "Drying", Institute of ceramics, Textbook Series; MacLaren and Sons, London, England (1964).

83 - Callaghan I.C., Ottewill R.H., Faraday Disc. of the Chemical Society, 57 (1974) 110-118.

84 - Pierre A.C., Ma K., Barker C., J. of Materials Science, 30 (1995) 2176-2181.

85 - Pierre A.C. and Ma K. , J. of Materials Sciences, 1997, accepted for publication .

86 - Ma K., PhD Thesis, Department of Mining, Metallurgical and Petroleum Engineering, The University of Alberta. July 17, 1995.

87 - Pierre A., J. Chim. Phys., 93[5] (1996) 1065-1079.

88 - Zou J. and Pierre A.C., J. of Materials Science Letters, 11 (1992) 664-665.

89 - Pierre A.C., Uhlmann D.R., J. of Non-Cryst. Solids, 82 (1986) 271-276.

90 - Ma K. and Pierre A.C., unplished data.

91 - Avrami M., J. Chem. Phys. I, 7 (1939) 1103-1112; II, 8 (1940) 212-224.

92 - Bratton R.J., Am. Ceram. Soc. Bull., 48 (1969) 759-762.

93 - Barrett W.T., Sanchez M.G., Smith J.G., Advances in Catalysis 9 (1957) 551-557.

94 - Klein L.C., Garvey G.L., J. Non-crystalline Solids 38-39 (1980) 45-50.

GELS

5.1 - INTRODUCTION

Inorganic gels constitute a group of materials which cover a large variety of structures. However any gel is composed of a solid network and a liquid matrix when in the wet state. Quite special properties result from this double liquid-solid composition. Some of them are reversible transformations which can be described within equilibrium thermodynamics, such as swelling or shrinkage inside the liquid in some cases. Other properties are irreversible transformations such as syneresis, aging and drying. Gels also are unique materials which have very interesting properties in the wet state as well as in the dry state, in particular new mechanical, optical, thermal and sound conduction properties. The present chapter addresses successively the structure of these materials, as well as their properties.

5.2 - STRUCTURE AND CLASSIFICATION OF GELS

ORIGINALITY OF GELS AS MATERIALS

At the beginning of colloid sciences, gels were considered as one state of matter, in the same manner as solids and liquids. Von Nägeli [1], in 1858, proposed a "solid-liquid" structure in which small crystallites were surrounded by highly bound water shells, constituting what was called micelles; the latter being arranged in a network inside the liquid. Finally it is in 1898 that Van Bemmelen [2] proposed a fibrillar structure, according to which solid matter constituted directly a continuous solid framework inside a liquid matrix.

The solid volume proportion can be extremely low in a gel, even lower than in some gases near their critical temperature. In an eye vitrea, the solid network occupies a volume proportion of 0.001 compared to a typical value of 0.083 in a gas near its critical point. Such a low compacting can be only be achieved if the solid adopts an extremely open fibrillar structure [3].

Later on, it appeared that the solid network could be crystalline or amorphous and that all gels were not alike.

Most often, the inorganic gels comprise two phases in the thermodynamic sense: the solid network, and the liquid matrix. However, this is not always the case: the truly polymeric gels are composed of only one thermodynamic phase as the solutions; their elastic skeleton is of molecular nature instead of being composed of solid particles with macroscopic dimensions in three directions by comparison with the size of one molecule. It is even possible for a solid network to be a polymer of the monomeric liquid component, in which case the gel is named an isogel. An example of isogel is styrene, which forms a framework of polystyrene molecules in a styrene monomer liquid matrix [4].

GEL CLASSIFICATIONS
In the field of ceramics, a large variety of gel structures has been found. However, presently no classification is universally accepted.

Flory's Classification
A general classification of both organic and inorganic gels was proposed by Flory. It comprises (Figure 5.2-1):
(a) lamellar gels such as mesophase and clays.
(b) covalent gels largely represented in organic chemistry.
(c) gels constituted by local crystallization of polymeric chains.
(d) particulate gels in which macroscopic particles of various shapes, are linked to each other to form a porous network.

Ceramists' classification
In inorganic chemistry, it is not easy to present the structure of gels within this classification because:
- the nature of the units constituting the solid network are not always accurately known .
- the nature of the bonding between the particles is uncertain also.
- some gels, such as clay or boehmite, could be considered as belonging to both classes (a) and (d) in the classification by Flory. They are composed of particles with a plate-like shape, but each particle consists of a parallel stacking of atomic layers with an interlamellar distance which can swell by a large factor.
In the present state of knowledge, ceramists have consequently adopted a more simple classification of gels comprising :
- polymeric gels.
- colloidal gels
The latter gel category is really equivalent to the particulate gels in the classification by Flory.

POLYMERIC GELS

Organic gels
The element most prone to build extended polymeric molecules which are not hydrolyzed by water, is carbon. Organic gels which comprise a polymeric carbon backbone, constitute a large class, which is not really the subject of this book. However, hybrid organic-inorganic gels are very new materials which constitute an exception and are addressed in chapter 6. Moreover, it is often useful to refer to the

large extent of knowledge gathered on organic gels, to have an insight in the structures and properties of the inorganic gels.

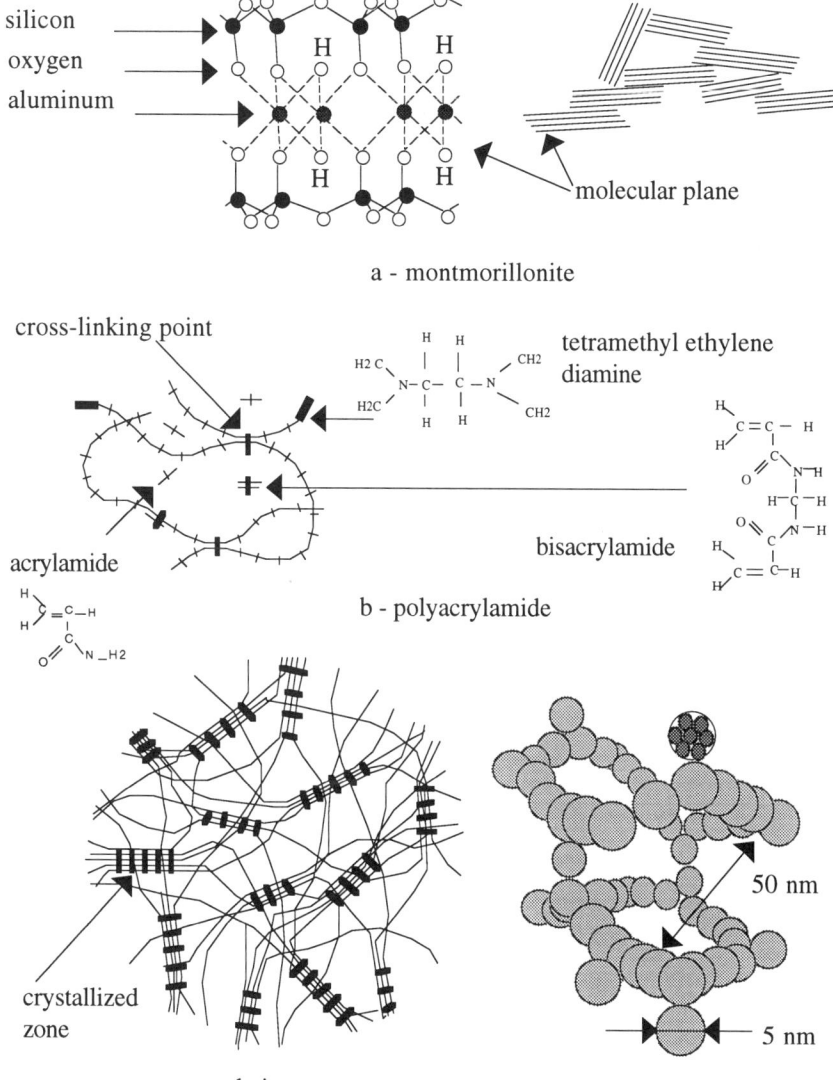

Figure 5.2-1 - The gels in the classification by Flory; (a) montmorillonite lamellar gel, adapted from Pierre and Uhlmann [5]; (b) covalent polyacrylamide gels, adapted Tanaka [6]; (c) gelatin gel comprising crystallized domains, adapted Gerngross and Hermann [7] ; (d) particulate silica gel adapted from Fricke [8].

Silica gels
The second element most prone to polymerization is silicon. The covalent nature of the Si-O-Si bonds is significant, though not as marked as that of C-C bonds. Actually, silica gels which are termed polymeric have a polymer molecular structure which is not as well defined as in the case of carbon, so that ceramists have adopted another more practical convention to define polymeric gels in a "large sense". A gel is here considered as polymeric when the solid backbone is composed of particles with a size below a given dimension: e.g., 1.5 nm for silica [9].
As indicated in chapters 2 and 4, polymerization can build unidimensional, bi-dimensional, or three-dimensional polymers depending on the conditions. Yoldas [10] established correlation between the polymerization connectivity and the oxide content in the final gel (Figure 5.2-2) and such correlation have allowed him to show that an increasing water/alkoxide hydrolysis ratio, as well as an increasing dilution, favored an increasing oxide content, that is to say an increasing connectivity of the network. Silica gels which can be classified as polymeric, are obtained in rather acidic conditions (below pH 2.5).

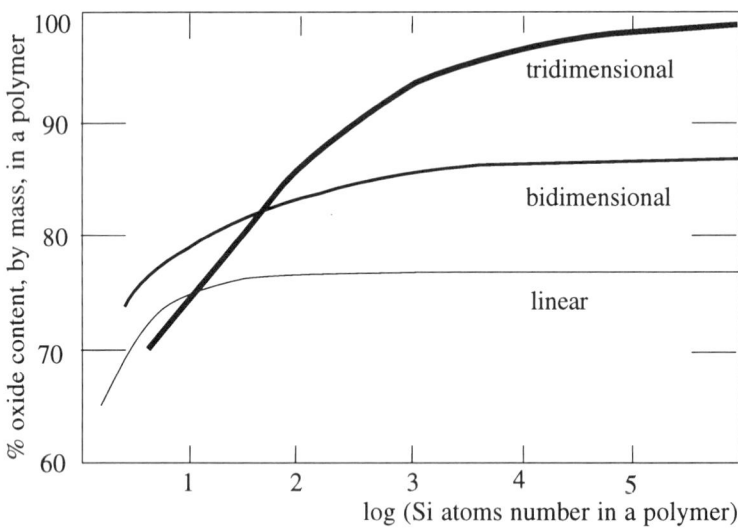

Figure 5.2-2 - Equivalent oxide content of silicon hydroxide polymers as a function of the silicon content, in linear, planar, and 3-dimensional polymerization. After Yoldas [10].

Borate gels
Another element easily forming polymeric gels is boron. A possible structural model for borate gels is based on an infinite network with B-O-B bonds in which tetra-

coordinated and tri-coordinated boron alternate [11]. All the non-bonding oxygen are terminated with an alkoxide group (figure 5.2-3). A structure of this type prevails in the alkali-borate systems $xR_2O.(1-x)B_2O_3$ where x must exceed a minimum value which is 0.2 in alcohol, or 0.15 in tetrahydrofuran (THF).

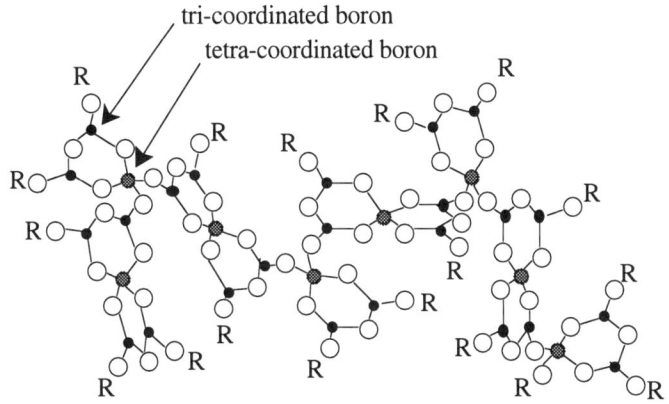

Figure 5.2-3 - Borate gel network, adapted from Brinker et al. [11].

Other polymeric gels
Other oxide gels are often described as being polymeric, although this qualification is controversial. With the oxides, this polymeric nature is related to an amorphous solid structure. The powder X-ray diffraction pattern of an amorphous polymeric gel such as silica, differs from the X-ray pattern of silica glass by the presence of a central peak which is due to the presence of pores smaller than 10 nm (figure 5.2-4). However, dense colloidal particles can also have an amorphous structure.
Amorphous oxide gels which can be considered as being polymeric in a large sense include V_2O_5 gels which are synthesized with a low hydrolysis water ratio r_w [13] and WO_3 gels made at a temperature T ≈ 20°C [14].
With aluminum hydroxide, polymeric gels can be obtained in acidic conditions and at temperature T ≈ 20°C. Marboe and Bentur [15] have mentioned such a gel termed α-gel which they made from aluminum salts and NH_4OH. Other polymeric aluminum hydroxide gels have also been synthesized by Pierre and Uhlmann [16] in cold acidic medium with HNO_3. These gels are very transparent and they have a glassy characteristics (figures 5.2-5 and 5.2-6).
Titanium dioxide TiO_2 forms gels which are categorized as being polymeric, although their polymer structure is not known in details [17]. Komarneni et al. [18] have reported that gels obtained form the titanium alkoxides contain oxygen, carbon dioxide and water, and they have a fibrillar structure which appears to be amorphous

when examined under X-rays, but micro-crystalline when examined by electronic diffraction. Hence, they actually are microfibrillar colloidal gels.

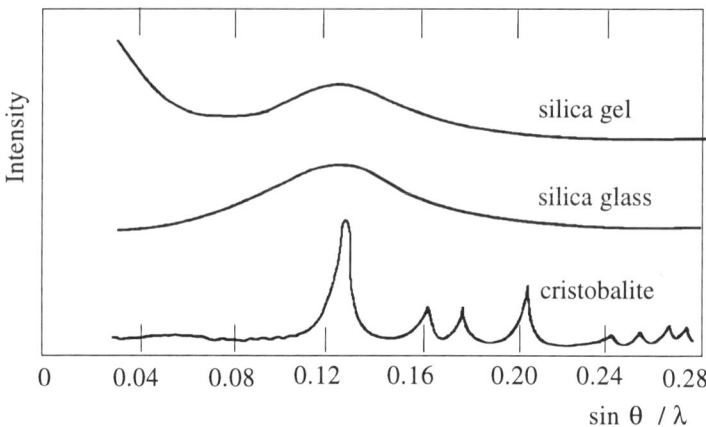

Figure 5.2-4 - X-ray diffraction pattern of; (a) cristobalite; (b) silica gel; (c) silica glass. After Warren and Biscoe [12].

Amorphous zirconium gels can be made from the zirconium alkoxides, and they have a glassy transparent characteristic. Their synthesis is favored when alkoxides with a small alkyl group are used as precursors and the water ratio r_w used for hydrolysis is low [19], which corresponds to conditions usually favorable to the formation of polymeric gels. Zirconium oxide gels which can be considered as polymeric have also been obtained by chelation with acetylacetonate [20]. Their structure appears to be amorphous in a transmission electron microscope. Infra-red investigations have shown that some absorption bands could be attributed to monoclinic zirconia (sharp bands at 425 and 415 cm^{-1}), while other bands corresponding to this phase were missing (absence of the 820 cm^{-1} band). The pore size of these gels was rather narrowly centered around 1.5 nm, which is quite compatible with a polymeric nature, in the large sense.

It is probable that polymeric gels, in the large sense, are formed in the mixed systems when SiO_2 is the major component, especially when the other components include B_2O_3 and P_2O_5 [21], or Na_2O as in "NaSiCon" gels [22]. Such polymeric gels can form provided the chemical conditions (pH, solvent dilution, hydrolysis ratio) are such that silica itself would be polymeric. If these conditions are satisfied, it is quite possible that double metaloxane bonds, such as Si-O-B in the mixed SiO_2-B_2O_3 system, can form. However, the real structure of these mixed components gels is far from being unraveled. Elements such as Na or P can also remain trapped inside the pores of the silica gel network and linked by hydrogen bonds or ionic bonds as side groups on the gel backbone.

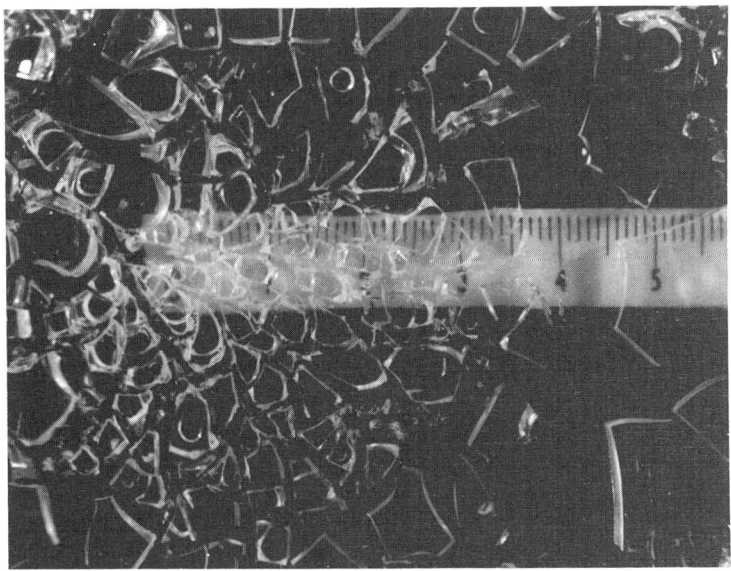

Figure 5.2-5 - Aluminum hydroxide polymeric gels made from aluminum s-butoxide, in excess water, with a molar proportion of nitric acid of 0.28 .

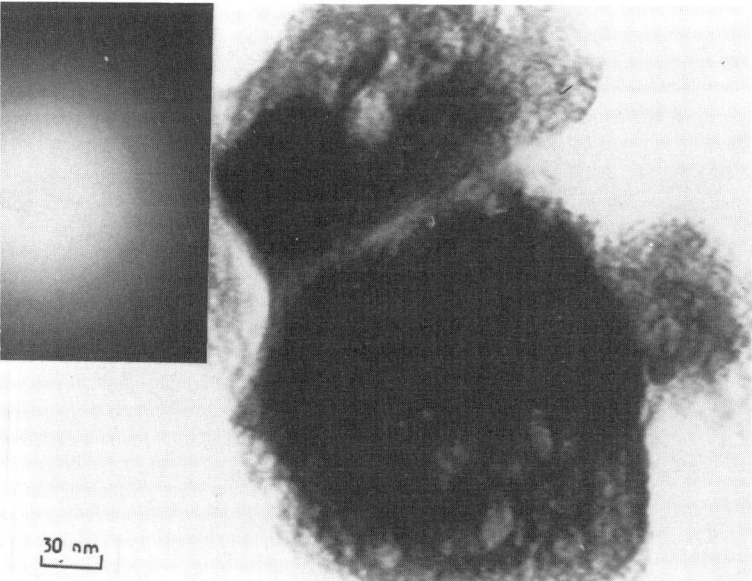

Figure 5.2-6 - (a) Characteristics under a transmission electron microscope and (b) electron diffraction of the gel in figure 5.2-5.

Overall, this makes it possible to enlarge the span of structural variations for a given cation composition, by introducing variations in the chemical process [23].

In a preceding chapter, it was indicated that alkoxides make it possible to homogenize the distribution of several metals at the atomic scale. This is a factor which helps the formation of polymeric structures, provided the double alkoxide molecule is not destroyed during hydrolysis, which would introduce again a segregation of the different metals. If double metaloxide bonds are destroyed and no mixed complex is formed, it is quite possible to obtain a polymeric gel from a single cation type, which entraps particles of the second oxide. This can occur for instance in the binary system $CaO-SiO_2$ [24] where the alkoxide $Ca(OEt)_2$ hydrolyzes faster than the Si precursors. When the second component is incorporated as a salt, it may remain as a very mobile cation in the wet gel, but it often crystallizes again as a salt, possibly with another anion than in the initial salt, during drying. As an example, Na can be added as NaCl in the ternary system $Na_2O-B_2O_3-SiO_2$ and crystallized as $NaNO_3$ in the dry gel, if nitrate ions were also introduced with another precursor [25].

In systems other than those containing SiO_2 , in particular with the titanates, the crystallographic structure of the solid phase which constitutes the gel is often not known. As an example, mixed (Ba,Ti,O) gels made from mixed alkoxides [26] are amorphous after drying, for the whole range of $\dfrac{[Ba]}{[Ti]}$ composition ratio except between $\dfrac{2}{1}$ and $\dfrac{1}{2}$ where a mixture of crystalline $BaTiO_3$ and an amorphous product is obtained. When a dry gel is amorphous, it is more convenient to describe it as polymeric. Similar results are often true, in particular when a complex alkoxide precursor could be synthesized, as with a double alkoxide of Pb and Ti which forms monolithic gel at once [27].

Moreover, polymeric structures are not limited to oxides. Germanium, which is just under silicon in the periodic classification table, makes GeS_2 gels which can be termed polymeric when they are amorphous at room temperature [28]. The crystallinity increases with the temperature and a higher solvent proportion.

COLLOIDAL GELS

Gels which can be classified as colloidal, according to the size of the solid particles which makes the tridimensional network, can differ from each other by :
- the size, shape and crystallographic structure of the colloidal solid particles which constitute them.
- the connectivity between the particles : a gel which macroscopically appears as quite fibrous may be composed of spherical particles linked in a very linear fashion.
- the type of linkage between the particles.

Pore texture

One of the best classical way to characterize the solid network in a colloidal gel consists in describing its porosity. The macroscopic data usually available for this task are:

- the apparent density ρ_a : it can be measured for instance by immersion in mercury, which does not penetrate into the micropores.
- the true density ρ_t : it can be measured in a fluid such as helium which penetrates into the smallest open micropores.
- the specific area S_a determined by adsorption of nitrogen by the Brunauer Emmett and Telling (BET) method.

These data make it possible to determine the specific pore volume V_{por} and a mean particle radius r_{par} of the gel, according to the formulas:

$$V_{por} = \rho_a^{-1} - \rho_t^{-1} \qquad (5.2\text{-}1)$$

$$r_{par} = 3(\rho_t \ S_a)^{-1} \qquad (5.2\text{-}2)$$

As for the pore radius r_{por}, it can be determined if a shape is assumed. For cylindrical pores this is:

$$r_{por} = 2 \ (\rho_a^{-1} - \rho_t^{-1}) \ S_a^{-1} \qquad (5.2\text{-}3)$$

The pore volume percent can be extremely high in a gel, up to 98% in silicate gels hydrolyzed in an excess of solvent and dried in supercritical conditions [21]. The pore structure depends in a large extent on the fabrication method. Additives named "drying chemical control additives" (DCCA) can be used to modify it. In the case of alumino-silica gels made from sodium silicate and sodium aluminate solutions for instance, Snell [29] has shown that a sodium ions content > 0.5% by mass had a marked effect on the pore structure (figure 5.2-7).

The pore size does not only depend on the size of the particles but on the aspect ratio, that is to say the ratio $\dfrac{\text{length}}{\text{thickness}}$ for a section of the solid network located between two branching point. For instance with TiO_2 , an aspect ratio ≈ 35 has been reported when the precursor was Ti ethoxide and ≈ 150 with Ti isopropoxide [18]. The pore size statistical distribution depends on the aggregation hierarchy of primary solid units in progressively bigger units. The smallest pores are associated with the smallest primary particles, and so on [30]. A finer description of a network texture usually requires techniques complementary to the BET method, such as scanning or transmission electron microscopy [31,32], small angles X-rays scattering (SAXS) [33] or light scattering [34].

The results of all these techniques can make it possible to propose a geometrical model for the pore network and to compare the resulting specific area and pore volume with the experimental data measured on a real gel.

Fractal characteristics of colloidal gel networks
It is often difficult to directly observe a gel network in an electron microscope, because an oxide gel usually undertakes a solid phase transformation when it is heated by the electron beam. This can be feasible when the cations in an oxide have a high electron density as this was done for colloidal gold gels in which the gold particles were linked by the intermediate of organic molecules. These gels did not phase transformed in the electron beam and they offered a high electronic contrast in a

microscope [35]. In this example, electron micrographs made it possible to actually observe the fractal characteristics of a solid gel network, which is illustrated in figure 5.2-8. On size scales which ranged from 50 to 500 nm, the network architecture keeps the same relative geometry. Therefore, even a colloidal gel can be described as a fractal organization, at least in a range of magnification.

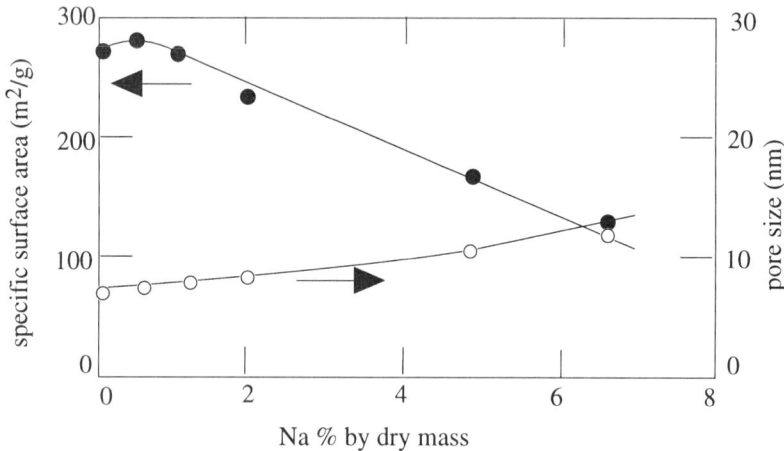

Figure 5.2-7 - Effect of the sodium ion concentration on the pore structure of alumina-silica gels. Adapted from Snel [29].

A solid network comprises a proportion of dangling ends which are linked by only one end to the main network (figure 5.2-9). These dangling branches contribute to the density, the pore size and the specific area of a dry gel. They also act as barriers to the diffusion of species such as molecules or colloidal particles inside the liquid matrix of the gel. However, they do not modify properties such as the elasticity or the electric conduction of a gel network. If all dangling branches in a gel network are suppressed, the remaining part of the network is composed of branches which are linked at their two ends to other neighboring branches. They are named active branches and constitute the active network of a gel. In the Flory-Stockmayer model of gelation for instance, the number of active branches N_a can be calculated as a function of the extent of reaction ξ ; it is given by [36] :

$$N_a = \frac{3}{2} \frac{(2\xi-1)^3}{\xi} \qquad (5.2-4)$$

Such a formula makes it possible to relate the texture of a gel to its theoretical properties, such as the osmotic swelling behavior for an organic polymeric gel and to compare it with experimental data.

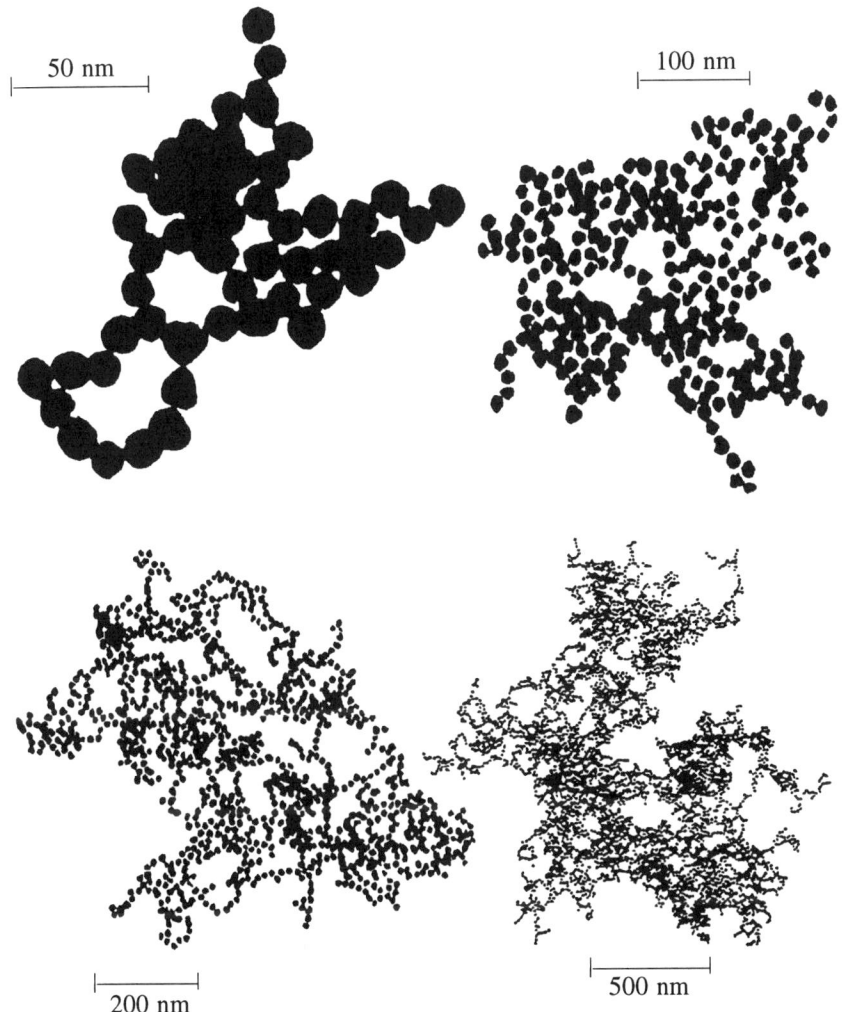

50 nm

100 nm

200 nm

500 nm

Figure 5.2-8 - Representative electron micrographs of clusters of colloidal gold gel at different scales. After Weitz and Huan [35].

Silica colloidal gels
Colloidal gels are formed with silica in a pH range from ≈ 2.5 to ≈ 6. They are composed of spherical particles linked by siloxane bonds. Preferential growth of the necks between two particles produces a cylindrical fibrillar structure during aging. Examples of silica gel made in excess water, termed xerogel because they were dried by evaporation, are show in figure 5.2-10.

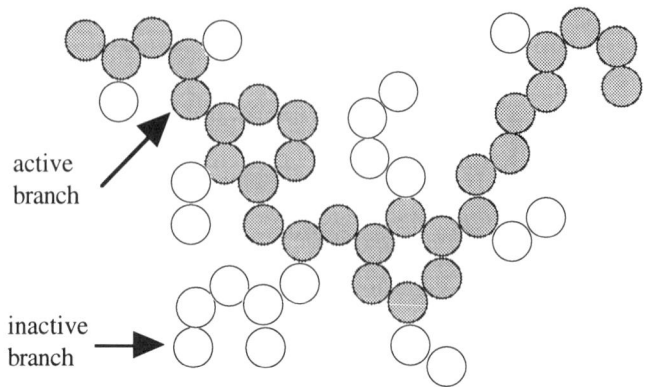

active
branch

inactive
branch

Figure 5.2-9 - Active and inactive branches of a gel network.

As this is explained in a further section, gels can also be dried in supercritical conditions, in which cases a dry gel is termed an aerogel. The solid network structure of a silica aerogel was illustrated in figure 5.2-1d. It can be described as follows [8]. On a scale of 1 nm, the structure consists of dense primary silica particles ($\rho_a = 2200$ kg.m^{-3}). These primary particles are aggregated into roughly spherical secondary particles with a radius of the order of a few nm. The internal surface of these secondary particles makes the biggest contribution to the specific surface area of silica gels. Their apparent density is of course lower than that of dense silica . The secondary aggregates are themselves arranged in chains, on length from 50 to 100 nm, which give its macroscopic apparent density to the aerogel. A silica aerogel can actually be described as a fractal construction, but only in a limited scale range.

Other colloidal gels
Colloidal gels with a fibrous structure can be made with Vanadium [13]. These gels are composed of high molecular weight polymeric species, arranged in an entanglement of fibers about 0.5 µm long and 5 to 10 nm thick [37], as illustrated in figure 5.2-11.
Similar tungsten oxide gels can be made similarly [38,39]. As for many oxides, the gel X-ray structure depends on the hydrolysis temperature. For instance with tungsten ethoxide, the gel is amorphous when hydrolysis is carried out at 20°C . On the other hand, the crystalline tungstite phase $WO_3.H_2O$ forms when hydrolysis is done at 80°C [14].
Colloidal gels obtained from aluminum precursors have a quite different structure. They comprise two types of oxo-bridges [15]:

$$-Al-O-Al- \qquad \text{and} \qquad \begin{matrix} -Al-O-Al- \\ | \quad\quad | \\ OH \quad OH \end{matrix} \qquad (5.2\text{-}5)$$

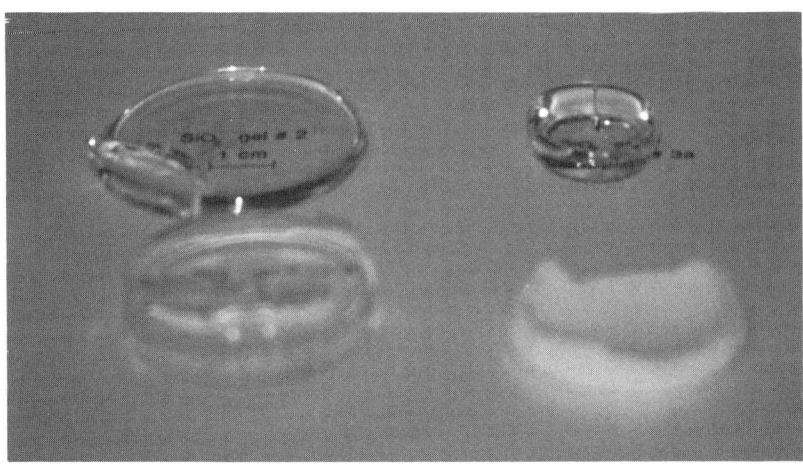

Figure 5.2-10 - Silica xerogels made from TEOS in different conditions.

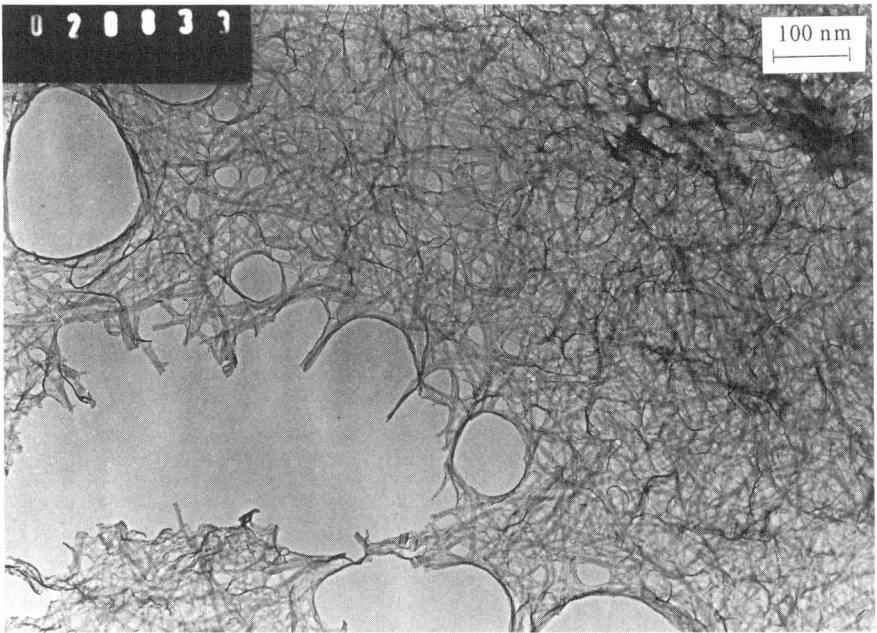

Figure 5.2-11 - Vanadium oxide gel seen under an electron microscope. From J. Livage with permission.

In acidic conditions, these types of links are responsible for the formation of colloidal boehmite particles with a platelike shape and a variable shape ratio length/thickness which depends on the growth conditions. These gels can have a very marked planar stacking of octahedral atomic layers when in the dry state [5]. For instance the characteristics of two boehmite gels are shown in figure 5.2-12; one sample has molecular layers with an orientation parallel to the flat glass surface on which it was dried ; the other one was dried over a layer of mercury and showed no preferred orientation of the its atomic layers.

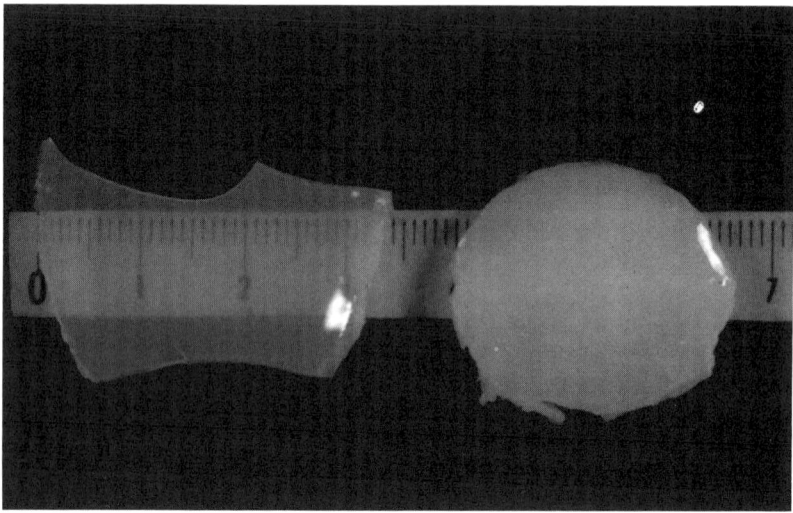

Figure 5.2-12 - Boehmite gels: on left, dried on a glass surface; on right dried over a layer of mercury.

In basic conditions, white gelatinous precipitates sometimes named gels are formed but they do not present the platelike configuration as the acidic gels. Their chemical composition corresponds to a trihydroxide $Al(OH)_3$ which can crystallize in several crystallographic structures depending on the chemical synthesis conditions.

With the aluminosilicate compounds known as clay, the platelike shape of the colloidal particles is well known and similar to that in boehmite $AlO(OH)$ samples.

Iron hydroxide gels with several different crystallographic structures similar to those of aluminum hydroxides are known to exist, although their aging evolution is different [40]. The phases which can crystallize include several monohydroxides $FeO(OH)$ which have a layer structure and trihydroxide structures. Another cation which behaves similarly to Fe and Al is Sc: compounds with both chemical compositions ScO(OH) and Sc(OH)$_3$ are known to form [41].

Apart from those colloidal gels where the solid network has a marked lamellar or fibrous organization, most colloidal gels are composed of a random network of colloidal particles where the proportion of linear linkage between particles is variable. This is the case with zirconia gels made from alkoxides comprising a big alkyl group R or with a high hydrolysis water ratio r_w [19] and with ThO$_2$ gel microspheres made by Yamagishi and Takahashi [42]. GeS$_2$ gels recently synthesized by Stanic et al. [43], by thiolysis with dry H$_2$S of alkoxide solutions in dry toluene, are also composed of submicronic spheres linked to each other in an open random fashion. Such sulfide xerogels can be transparent as illustrated in figure 5.2-13, and have a specific surface area > 400 m^2/g.

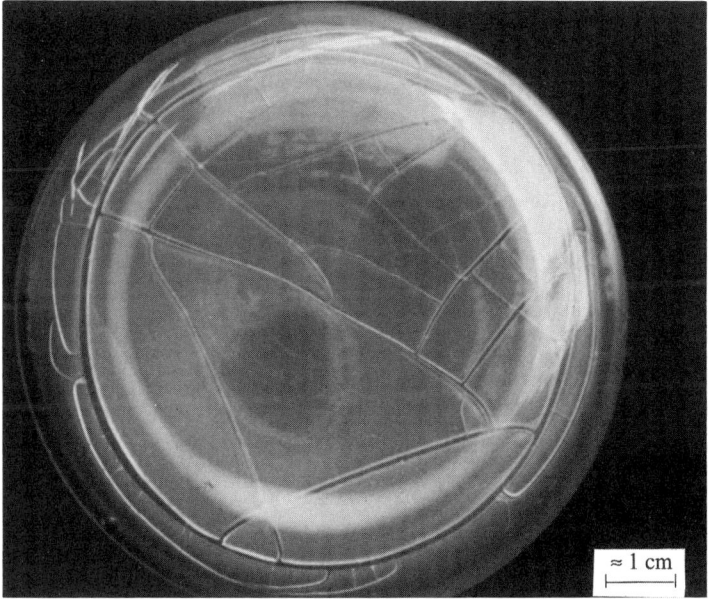

Figure 5.2-13 - GeS$_2$ gel made by thiolysis of alkoxide solutions in Toluene. From Stanic with permission [44].

5.3 - PROPERTIES OF WET GELS

Wet gels are composed of a solid network and a continuous liquid phase. Hence, they have a unique combination of solid and liquid properties.

SOLID PROPERTIES

The solid behavior comes from the existence of elasticity once the gel point is passed during the gelation process. As in all materials, its mechanical properties can be divided in an elastic and a plastic behavior.

Plastic properties

They are important because a tenuous gel network can easily be broken and the liquid matrix in which this network is embedded offers the possibility to easily reconfigure it. However, the network reconfiguration process depends on the nature of bonds which links the colloidal particles or the polymers. In some cases, these bonds can be slowly destroyed under tension, which occurs in competition with the formation of new links after straining. In other cases, the bonds are not easily destroyed but the linear strings of colloidal particle can change configuration by relaxation and new bonds can also form.

For a given gel, the answer to a given strain depends on the strain rate. If the strain rate is too high, an enhanced immediate elastic answer is first observed and it can eventually lead to fracture. If the strain rate is moderate, the kinetic mechanisms responsible for plastic deformation can have time to operate. It is even possible to apply a moderate and constant mechanical stress to a gel which therefore undertakes a constant rate plastic deformation, so that the gel flows and it is possible to speak of its viscosity.

From a mechanical point of view, the difference between a fluid such as a polymeric solution and a solid, such as a gel, can be seen as a difference in the time scale to answer to a stress. For a solution which comprises a polymeric solute with rather linear molecules, entanglement of the polymeric chains is frequent. When submitted to a very sudden stress, the solution behaves as if weak bonds linked the polymers at certain points. However, these links have a very short lifetime which corresponds to the relaxation time of the chains, so that the solution yields under any shear stress even very small, such as due to its own weight. On the other hand, the lifetime of bonds is long in a gel. A permanent network structure is insured and requires a yield stress to be broken. Pieces of gels are able to keep their shape without flowing under their own weight. Flow only occurs above the yield stress, as this can easily be seen in thixotropic gels such as boehmite gels. In the latter case, reconstruction of a gel network after rupture requires an induction time. If the bonds between colloidal particles which are responsible for gel network formation are not to weak, the reconstruction of a small number of them is sufficient to give a marked thixotropic behavior to the gel [3]. These thixotropic gels also usually show a marked hysteresis when submitted to successive heating and cooling so that they can be classified as heat-reversible gels [45].

Elastic properties
The elastic properties of a gel is directly related to the number of active branches N_a of its network, as defined in the previous section. In the Flory-Stockmayer model for instance, it is possible to show that the static elastic shear modulus is:

$$G = \frac{N_a}{V_m} RT \qquad (5.3\text{-}1)$$

where V_m is the volume per mole of polymer chains and N_a was related to the extent of reaction ξ by the equation 5.2-5.
For the oxide gels which present a more or less strong affinity for water, the shear modulus depends on the water content in the gel. This is particularly true for the soluble gels such as boehmite or clays, which must be dried to reach the gel point. For such materials submitted to small deformations, the modulus G can be related to the concentration C of the gel by experimental laws of the type:

$$G \approx C^n \qquad (5.3\text{-}2)$$

in which n has a value from 2 to 7 [46-48].
Kingery and Francl [49] have also defined a "workability" as the product of the yield stress by the maximum deformation and they have shown that the maximum value of this product increases linearly with the surface tension of the liquid which impregnates the material (figure 5.3-1). These mechanical properties are also directly connected to their drying behavior.
Experimentally, as this was indicated for the phenomenon of gelation in chapter 4, it is possible to measure the storage and loss modulus, respectively designated by G' and G", of the complex shear modulus G of a gel submitted to an oscillatory strain of frequency ω. For a material such as a crosslinked gel network which is elastic when submitted to a shear stress below its yield point, G' and G" are practically independent of the oscillation frequency ω, while for a uncrosslinked polymer solution which consists of an entanglement of chains, they are strongly frequency dependent.
Ross-Murphy and McEvoy [36] have designed a model in which crosslinking occurs between monomers with molar mass M and functionality Z, by a dimerization reaction with an equilibrium constant K. The main result was that, for a functionality Z, $\ln(G'K_1)$ could be plotted as a function of $\ln \dfrac{C}{C_o}$, where

$$K_1 = K\frac{(Z-1)(Z-2)}{\xi RT} \qquad (5.3\text{-}2)$$

and
$$C_0 = \frac{M(Z-1)}{KZ(Z-2)^2} \qquad (5.3\text{-}3)$$

and all curves could be superimposed on a single one, by translation. As this graph shows (figure 5.3-2), when $\dfrac{C}{C_0} > 10$, G' tends towards a more simple asymptotic expression

$$G' \sim C^2 \qquad (5.3\text{-}4)$$

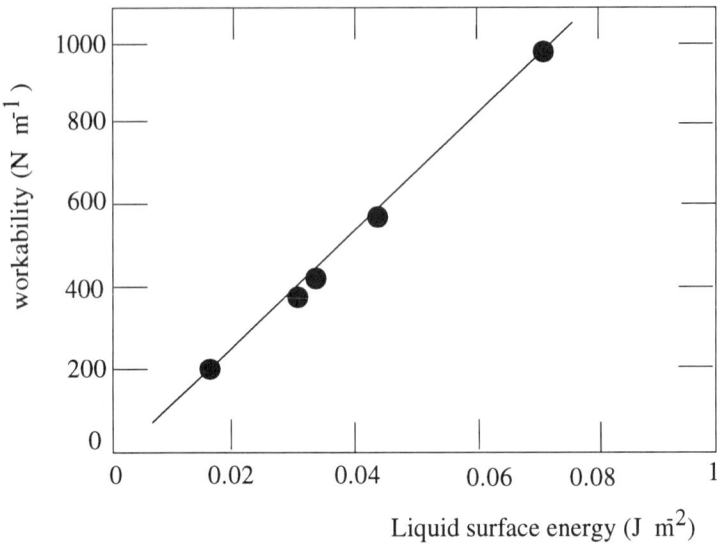

Figure 5.3-1 - Effect of liquid surface tension on the workability of kaolin clay. Adapted from Kingery and Francl [49].

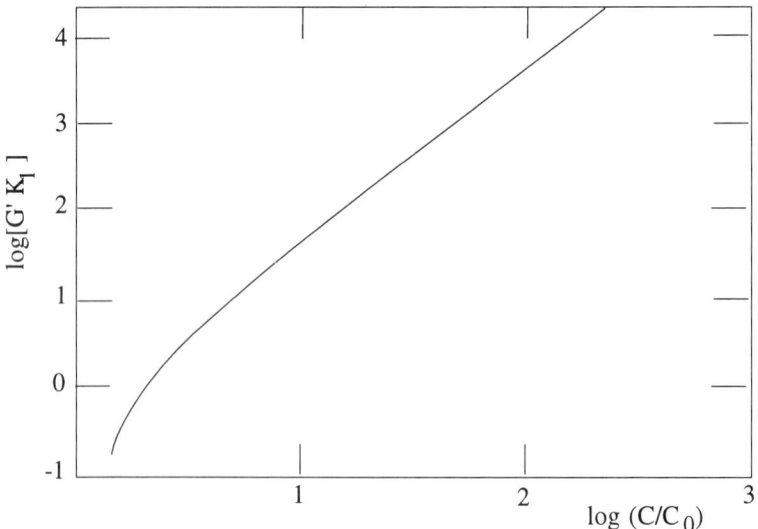

Figure 5.3-2 - Shear modulus data reduced to a master curve for globular proteins gels. Adapted from Clark and Ross-Murphy [50].

Large deformations can also be analyzed according to theories developed for rubber [51], such as with the Mooney-Rivlin equation:

$$\sigma_T = 2(C_1 \varepsilon + C_2)(\varepsilon - \frac{1}{\varepsilon^2})$$ (5.3-5)

where ε is the tensile strain, and σ_T is the true tensile stress and C_1 and C_2 are two constants.

Transport properties in the liquid of a gel
The liquid matrix which impregnates the gel network is important for all transport properties through the liquid medium. However, the gel network has a significant influence, as it can for instance be utilized as a barrier to the diffusion of particles or of big species. For this reason, gels are often used as media to grow all kinds of organic or inorganic monocrystals [3], because polycrystalline aggregation is made difficult by the gel network, while the diffusion of small molecular species is as fast as in a pure liquid and it is not disturbed by convection.

In the case of $Fe(OH)_2$ gels, such a process occurs naturally during aging, by dissolution followed by recrystallization of the solid material. In this material, individual Fe_3O_4 monocrystals grow after some time [52] (figure 5.3-3).

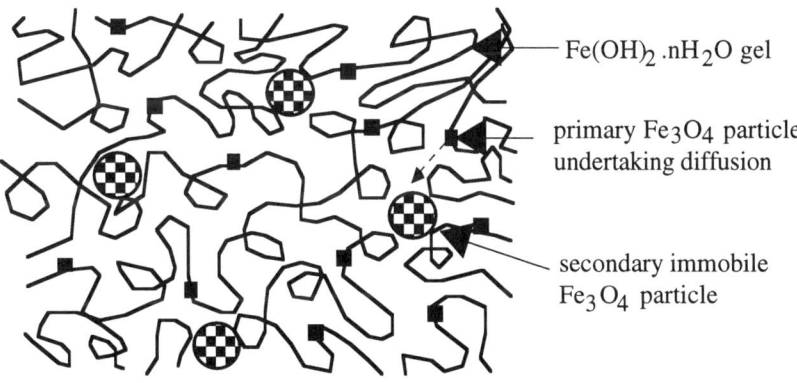

Figure 5.3-3 - Aging evolution of $Fe(OH)_2$ gels.
Adapted from Haruta and Delmon [52].

REVERSIBLE SWELLING AND SHRINKAGE OF WET GELS
The molecular groups which constitute the solid network of a gel can present a more or less strong affinity for molecules in the liquid which impregnates them. This affinity can be strong enough, in some cases, to overcome the mechanical strength of the network without destroying it. That is to say, the gel swells or shrinks and in

some gels such as organic polymeric gels, this change in gel volume is reversible. This behavior has been at the origin for naming such wet gels "xerogel" [3], an appellation which is now restricted to gels dried by evaporation.

Osmotic swelling theory of covalent organic polymeric gels

The most complete theory on the reversible swelling or shrinkage of a gel in its liquid matrix, has been developed by Tanaka [6] for truly covalent polymeric organic gels. Such gels can shrink or swell reversibly inside their liquid by a factor which depends on the conditions and which can be as high as 100. However, the crosslinking remains unchanged during swelling or shrinkage. That is to say the shape of a monolith is changed by a scaling factor.

Tanaka studied the interaction between the solid network and the liquid matrix and he showed that the balance of forces on the gel network was a function of variables which include the temperature, the liquid matrix composition, the pH, and eventually an electrical field. The forces which act on a gel network comprise:

- The network elasticity. To extend a polymer chain it is necessary to apply a tensile force, while to make it contract a compressive force must be applied. Moreover, there is a limit in the extension of a polymer chain due to the number of mers it comprises.

- The Brownian motion. If polymers chains are extended, random thermal agitation will tend to make them shrink. On the other hand if the ends are put together, random agitation will tend to extend the chains.

- The chemical affinity, on one hand of the gel network mers for each other, on the other hand of the gel network mers for the solvent molecules. The resulting interaction is actually equivalent to the Van der Waals interaction in the D.L.V.O theory. It was termed steric interaction in the Flory-Huggins theory of polymer solutions and responsible for the steric stabilization of colloidal sols (Chapter 3). The fact that polymers are crosslinked in a tridimensional gel network does not suppress this interaction: if a gel network comprises molecular groups which have a strong affinity for water, the gel network shows a tendency to swell in an attempt to maximize the number of polymer-solvent contacts. An expression for the equilibrium swelling ratio $q = \dfrac{V}{V_0}$ given by Dusek [53] is:

$$q^{5/3} \sim \frac{V_0}{N_a} \, f(X) \qquad\qquad (5.3\text{-}6)$$

where V_0 is the gel volume in a reference state, X is a parameter which represents the polymer-solvent interaction, f(X) is a thermodynamic function of this interaction, and N_a is the number of active branches defined in a preceding section. This expression indicates that, for a given solvent, the greatest swelling occurs for the most scarcely crosslinked systems (N_a small).

- The coulomb interaction. It depends on possible local electric charges on some molecular groups in a polymer chain and on their electrostatic interaction, as well as on the interaction with the counterions scattered in the solvent.

As a result of these interaction, a polymeric gel can practically adopt one of two possible states: a shrunken state and a swollen state. These two states can coexist, exactly as a liquid and a gas state can coexist in a classical liquid-gas phase transformation. Moreover, a critical point exists above which the gel passes gradually from one state to the other. Below the critical point it is possible to obtain either one of these two states; or the coexistence of both, as in classical liquid-gas state transition. The critical point corresponds to critical conditions where the compressibility of the gel network becomes infinite; this is a true critical point in a thermodynamic sense.

Overall, the state transition of a covalent polymeric gel network can be described within the theory of critical phenomena, and a state diagram can be drawn as in figure 5.3-4.

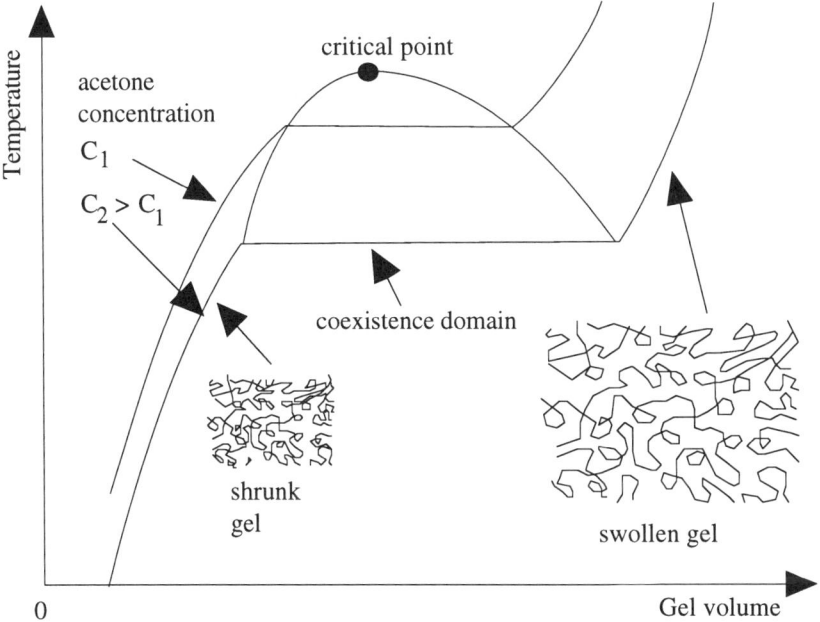

Figure 5.3-4 - State diagram of an organic polymer gel. Adapted from Tanaka [6].

When polymeric gels carry electric charges on some of their mers, the principal results of the D.L.V.O theory applies. That is to say, it is possible to dissolve a non-potential determining electrolyte in the solvent ; the critical concentration of this electrolyte necessary to make the gel shrink (the equivalent of the critical concentration for coagulation C_c) is proportional to z^{-6} where z is the valence of the counterions.

This theory therefore achieves a synthesis of the D.L.V.O. and steric theories and it appears quite similar to the coagulation theory of sols. The states equivalent to a shrunken or swollen gel are respectively a coagulated and a dispersed sol. It suggests the possible coexistence of the dispersed and the coagulated states for a sol, in appropriate conditions.

Swelling of oxide gels

The main difference between the preceding polymeric organic gels and the oxide gels concerns the crosslinking which is responsible for the formation of the gel network. Oxide gels are not really polymeric with a fixed crosslinking but colloidal, with a variable number of contact points between colloidal particles and a variable size of the necks which link the colloidal particles. If these necks are due to strong metaloxane bonds which cannot be destroyed by hydrolysis once they are formed, swelling is impossible but shrinkage remains possible: it is an irreversible and slow process named syneresis and examined in the next section.

On the other hand, if these necks comprise bonds which can easily be hydrolyzed, gelation does not occur spontaneously; it requires drying. In this case, syneresis does not occur in liquids where these bonds are destroyed, but swelling occurs if a piece of gel is placed in an excess of such liquids. However, the crosslinking points can be somewhat modified during successive swelling and drying steps, so that this process is not exactly reversible.

This behavior of oxide gels in aqueous solvents has been known for a long time [54], and has lead to a classification of these gels in two categories with respect to swelling. A first categories comprises gels which undertake syneresis and do not swell, such as SiO_2, TiO_2, ZrO_2, V_2O_5, MnO_2, SnO_2. For silica, swelling and return to a sol state is only possible in very basic hydrothermal conditions, it involves a chemical digestion [55].

The second categories comprises gels in which water must be evaporated to reach the gel point and which can swell in aqueous solvents, such as boehmite $AlO(OH)$, the equivalent $CrO(OH)$ and $FeO(OH)$, and clays such as montmorillonite. All these gels have a layer structure with hydrogen interlayer bonds. In clay-water systems, the interlayer distance can become quite large by adsorption of water, which rejects the interlayer X-ray diffraction line to small diffraction angles [56]. The ease of these gels to swell also depends on their structure. A well crystallized solid comprises hydrolyzable bonds on extended areas, which limits the swelling. On the contrary, with an amorphous material, complete dissolution can become possible.

The phenomenon of swelling of this second categories of oxide gels has been extensively studied and osmosis plays a major role; exactly as in the polymeric organic gels.

Osmosis has been observed for the first time in 1748 by Nollet, in solutions comprising a solute which has a strong chemical affinity for the solvent, for instance water. It has been explicitly detailed in textbooks on colloids and extended to electrolyte solutions where the modified mechanism which is involved leads to the Donnan equilibrium[57]. It has been extended later to clays [58] and to organic gels which present a strong affinity for water [3]. In wet gels, the solid network of the gel plays both the roles of a solute and of a membrane.

The modification of the solvent free energy, due to its chemical affinity for the solute, can be expressed by:

$$\Delta G_{os} = -\pi V_A \tag{5.3-7}$$

where V_A is the partial volume of the solvent in the solution and π is a coefficient which has the dimension of a pressure and named the osmotic pressure. ΔG_{os} is also related to the ratio of the equilibrium vapor pressure p_v of the solvent in the solution, to the vapor pressure of the pure solvent p_o by:

$$\Delta G_{os} = RT \ln \frac{p_v}{p_o} \qquad (5.3-8)$$

Because, the origin of the osmotic free energy term in equation (5.3-7) is due to interactions between a solute and a solvent, it applies to the liquid in a gel as well. However, it must be noted that in spite of the name "osmotic pressure", the osmotic mechanism does not introduce a mechanical hydrostatic pressure in the solvent as long as the dilution of a gel is uniform (Figure 5.3-5a). A hydrostatic pressure gradient will only occur if a dissymmetry in solute concentration (i.e., gel network concentration) is created, such as between a wet gel and an excess of liquid in which the gel is immersed (Figure 5.3-5b).

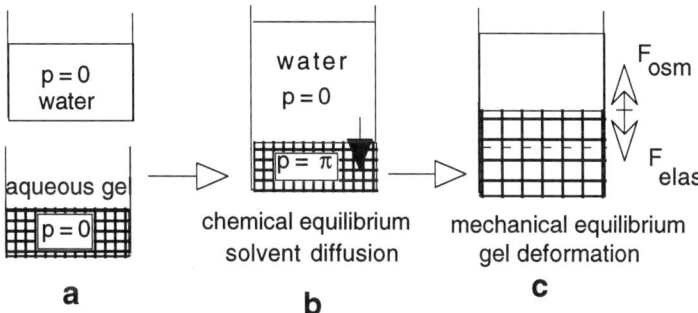

Figure 5.3-5 - The osmotic swelling mechanism: (a) separate gel and solvent; (d) creation of an overpressure inside the gel (e) swelling of the gel.

When a wet gel is immersed in excess liquid, the solvent initially does not have the same free energy in the excess liquid and inside the gel, because the solute (i.e. the gel network) concentration is not the same in the gel and in the excess liquid. The initial difference between the two molar free energies of the solvent can be used to define the osmotic pressure of the solvent inside the gel, by using equation (5.3-7). However, an evolution will occur so that the solvent tends to reach the same molar free energy inside the gel and in the excess liquid. To achieve this, the solvent molecules will start diffusing in or out of the gel, (depending on the gel and the solvent). Let us suppose that some solvent molecules from the excess liquid start to diffuse in the gel network (Figure 5.3-5b), then the liquid inside the gel is placed under compression. Because liquids are roughly incompressible, the in diffusion of a very small number of solvent molecules is sufficient to induce a significant compression ΔP in the liquid inside the gel. At this point, the initial dissymmetry in solute (i.e. gel network) concentration, inside compared to outside the gel, has given rise to a dissymmetry in solvent hydrostatic pressure ΔP [57]. In terms of free energy, a difference of mechanical nature is created between the gel and the excess liquid, until it balances exactly the free energy difference of chemical nature (i.e. the osmotic free energy), once equilibrium is reached.

The process does not stop there. The compression in the liquid inside the gel stresses the gel network, which itself responds by starting to expand (Figure 5.3-5c). That is to say, the gel starts to swell.

The flux of solvent J which is transferred during the diffusion step can be described by a macroscopic flow according to a modified Darcy's law :

$$ J = - K \, (\Delta P - \Delta \pi) \qquad (5.3\text{-}9) $$

where ΔP designates the actual hydrostatic pressure difference between the liquid inside the gel and the excess solvent , $\Delta \pi$ the osmotic pressure difference between the solvent in the excess liquid and inside the gel, and K is a permeability constant.

Equilibrium is reached when the elastic resistance of the gel balances the residual hydrostatic overpressure in its own liquid. In this case equation (5.3-9) gives for J = 0:

$$ \Delta \pi = \Delta P_{eq} = \sigma_{el} \qquad (5.3\text{-}10) $$

where σ_{el} designates the elastic stress in the gel network.

In practice, any swelling magnitude can be achieved in a gel by selecting an appropriate solvent composition. A gradual transition exists, between a limited swelling and an infinite swelling. In practice however, it is necessary to create moderate solvent concentration gradients between inside the gel and outside from it, otherwise a gel will not swell uniformly throughout its thickness, which will introduce stress gradients on the gel network and may eventually fracture it. Swelling can also be an exothermic process, in which case the swelling degree can be modified by modifying the temperature and vice versa.

Donnan equilibrium

The osmotic mechanism can occur even when the gel network does not carry any electric charges. When electrically charged sites are present on the gel network itself, the osmotic property still operates, however it is made more complex by the action of the electrostatic interaction. The component of the osmotic pressure due to the electrostatic character is taken into account in the so-called Donnan equilibrium previously mentioned.

Let us consider an aqueous covalent gel which has swollen to its maximum extent because of its affinity for water, and which comprises positive charges due to the adsorption of H^+ on some specific sites. This gel can be described by some concentration y of polymer molecules GH^+, where G designates the average polymer associated segment with each charged site. If the gel was made with an acid such HCl for instance, an excess concentration y of free Cl^- ions is present inside the pores of the gel, so that the wet gel is globally neutral.

Let us now establish a contact between this gel and an excess HCl solution, through a porous membrane. The membrane will let water molecules pass through it, as well as H^+ and Cl^- ions. At equilibrium, the concentrations of GH^+, free H^+ and Cl^- ions inside the gel will respectively be y, z and y+z, while on the other side of the membrane they will be respectively 0 for GH^+, and x for instance for both H^+ and Cl^-, because the solution outside the gel must also be globally neutral. Chemical equilibrium for the ionization of HCl between the two sides of the membrane gives:

$$x^2 = z(y+z) \qquad (5.3\text{-}11)$$

The validity of such an equation was verified by measuring the pH outside and inside a gel [59]. A consequence of this equilibrium is that a difference in concentration ΔC of the mobile ions (H^+ and Cl^-) exists, between both sides of the membrane. At equilibrium :

$$\Delta C = 2z + y - 2x = 2(x^2+y^2)^{1/2} - 2x \qquad (5.3\text{-}12)$$

This can be described by a free energy term and an osmotic pressure contribution of electrolytic origin, π_{don} , directly proportional to this mobile ions concentration difference. This contribution adds up to the chemical affinity of the gel for water molecules and contributes to gel swelling. Starting from an isoelectric point where y=0, this contribution increases first with y, then it reaches a maximum and decreases again, when x increases. Swelling is increasingly enhanced by anions respectively in the order: SO_4^{2-} ; CH_3COO^- ; Cl^- ; NO_3^-; CNS^- ; I^-, and by cations in the order Mg^+ ; NH_4^+ ; K^+ ; Na^+ ; Li^+.

Osmosis can be used to replace water by another solvent, in the technique known as dialysis. A contraction occurs immediately if an aquagel is immersed into alcohol. This is due to an osmotic withdrawal of water by alcohol, which does not occur if the alcohol concentration outside the gel is increased gradually so as to let time to the molecular species to diffuse without straining too much the gel network beyond its elastic limit.

SYNERESIS

Oxide gels which are not soluble in water (e.g., SiO_2, TiO_2), do not swell but they can shrink. They even do this spontaneously inside their mother liquid right after gelation. In practice, the volume increases first during the process of gelation, and this is due to the expulsion of water molecules by the condensation reactions (figure 5.3-6 b and c). When the gel is formed, it remains in its expanded state in a first step. Then in a second step it begins to shrink within its mother liquor (figure 5.3-6 d), an irreversible phenomenon which is named syneresis.

During syneresis, the shape of the gel is maintained. The shrinkage rate can also be accelerated with reagents such as H_2O_2 such as in the TiO_2 gels made from Ti alkoxides by Yoldas [17]. The collapse of a TiO_2 gel creates new $\equiv$Ti-O-Ti$\equiv$ links by conversion of the terminal alkyl groups. As a result, the equivalent oxide content increases, from 70 to 82% before syneresis, to 96% after syneresis. The supernatant liquid expelled from the gel during syneresis comes from the condensation water due to the formation of these new bonds.

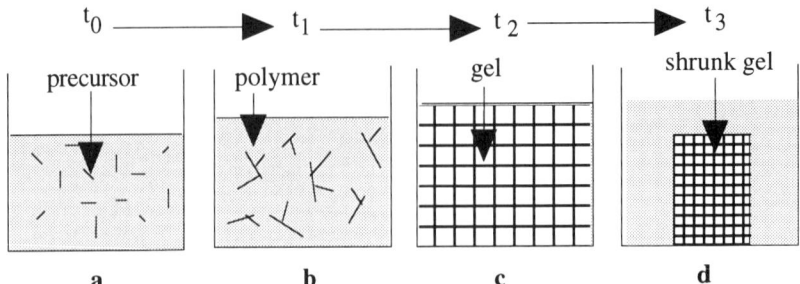

Figure 5.3-6 - Volume change in a SiO$_2$ gel during syneresis.

The kinetics of syneresis were studied for instance in alumina-silica gels by Snel [29], who measured a syneresis rate defined by:

$$r_{sy} = \frac{V_l}{t.V_g}$$

$$(5.3-13)$$

where V_l is the volume of supernatant liquid above the gel, t the time and V_g the gel volume.

The results showed that the extent and initial rate of syneresis depended drastically on the pH where the gel was made (figure 5.3-7). They both increased with the pH, at pH values above the isoelectric point of silica ($\approx$ 2.5). However, the total duration of syneresis was roughly independent of the pH, in a pH range from 7 to 10.5 [60,61], of the order 0.3 10^6 s (figure 5.3-8). The fact that both syneresis and gelation have slower kinetics close to the isoelectric point, suggests a that common principles underlie both phenomena. Concerning the texture evolution in the gel during syneresis, the X-ray diffraction crystallite size increased, the specific surface are decreased as the small pores disappeared, but the colloidal particles density also decreased which can be explained by the closure of an increasing proportion of pores.

It has been known for a long time that syneresis can also be viewed as part of a slow coagulation. For instance when a concentration C of a non-potential determining electrolyte is added, it is possible to define a critical syneresis concentration C_s intermediate between the classical fast coagulation concentration C_c and gelation concentration C_g, introduced in chapters 3 and 4. C_s is such that

$$C_g < C_s < C_c$$

$$(5.3-14)$$

As seen in chapter 4, when $C < C_g$, a sol is stable. For $C_g < C < C_c$, the bonding architecture between particles is such that linear structures are formed (chapter 4). When $C_s < C < C_c$, misalignments in the linear aggregation of particles become more frequent and the particles have a tendency to slide slowly under the tangential component of the Van der Waals attraction [61]. This is precisely the phenomenon of syneresis. Its overall magnitude increases with the gel concentration and its initial kinetics with the temperature [62].

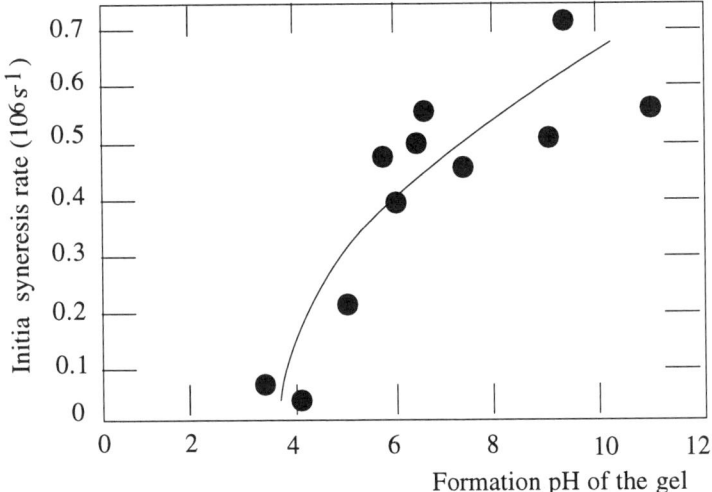

Figure 5.3-7 - Variation of the syneresis rate of an alumino-silicate gel as a function of the pH. Adapted from Snel [29].

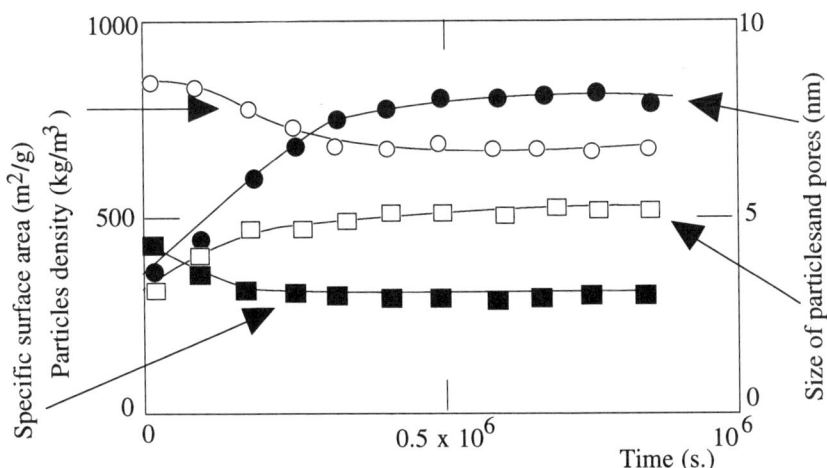

Figure 5.3-8 - Evolution of the main characteristics of an alumino-silicate gel during syneresis. Adapted from Snel [29].

Eventually, syneresis can go up to breaking a gel network so as to transform it to a powder which precipitates, as in MnO_2 gels [63]. At last, for $C > C_c$, the bonding between particles is not linear from the beginning, hence random dense structures form immediately (figure 5.3-9).

AGING WET GELS

A wet gel keeps transforming when it is stored for a long time in its liquid and this evolution is known as aging. Two fundamental types of transformations underlie aging evolution.

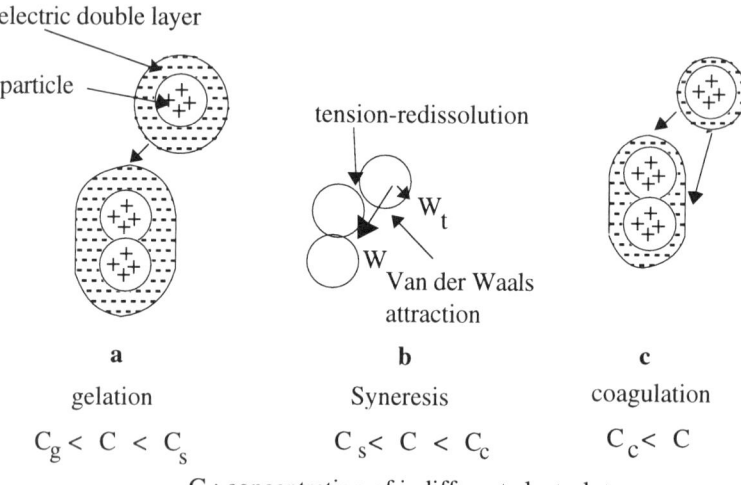

C ; concentration of indifferent electrolyte

Figure 5.3-9 - Unified mechanism of gelation, syneresis and coagulation. Adapted from Pierre and Uhlmann [61].

Chemical evolution during aging

A gel has no reason to be composed of the most stable chemical species when it is formed for the first time, as a result of the hydrolysis-polymerization reactions in liquid medium. A critical supersaturation of intermediate chemical complexes leading to gelation, may well be due intermediate species which are not the most stable. The anions in the solution can themselves have a marked influence as they can be slowly expelled from a gel by re-dissolution.

No universal mechanism can explain the aging evolution of a gel. With Al precursors in acidic conditions for instance, a pseudoboehmite monohydroxide gel network easily forms, which comprises ol bridges and oxo bridges, as shown in chapter 2. However this boehmite is metastable. Aging, especially in basic conditions, helps to transform the monohydroxide $AlO(OH)$ gel to a more stable trihydroxide $Al(OH)_3$ structure [15]. Anything which prevents the deprotonation of hydroxo groups in he monohydroxide, such as the adsorption of glycerol or ethanediol, prevents its aging transformation to bayerite.

A similar aging evolution occurs in the case of tungsten oxide gels [39] which are known to be unstable and crystallize after a few days. They can also be stabilized by foreign colloids such as MoO_3. Even silicic acid gels, which always show an amorphous X-ray diffraction pattern, have a tendency to crystallize after aging in dehydrating conditions [3]. The reverse can also be true: crystalline zirconia immersed in water undergoes a surface hydrolysis modification yielding a thin gel-like coating [64].

Physical evolution during aging
The origin of the physical aging evolution is the same as in syneresis; this is the high specific area of the solid gel network which tends to decrease slowly.
In silica gels at low pH the polymer species which make the gel network are not very soluble, so the network is "frozen in"; no modification occurs during aging [65]. However, at higher pH, colloidal silica gel slowly transform to coarser structures, as indicated by the evolution of water vapor adsorption desorption isotherms of these gels (figure 5.3-10). These isotherms show a displacement of the hysteresis loops towards higher pore sizes. That is to say both the pores and the gel network becomes coarser. The specific surface area of a gel decreases as the aging time increases.

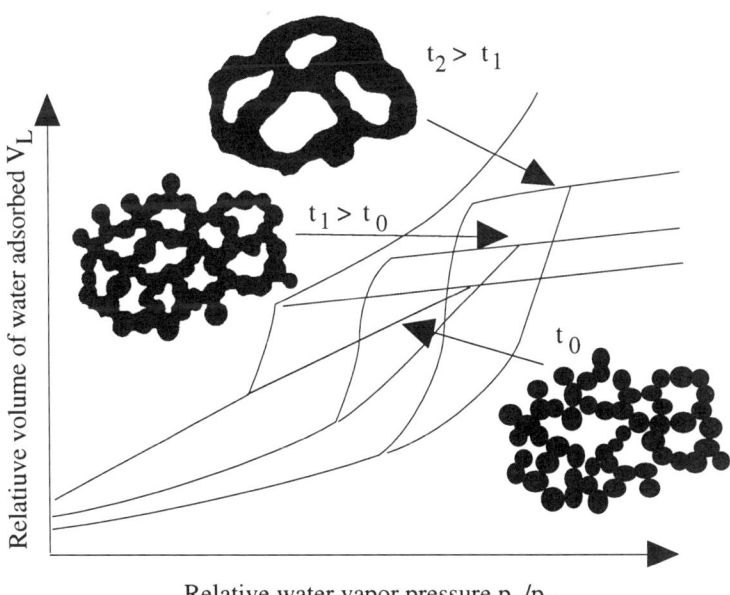

Figure 5.3-10 - Adsorption-desorption isotherms on silica gels after various aging times. Adapted from Van Bemmelen [2], Hermans [3] and Iler [9].
Sometimes, completely different textures develop, as in gelatinous ZrO_2 precipitates made from Zr propoxide in cyclohexane by Kindu and Ganguli [66], where

coalescence of the colloidal particles into rods, with a periodic swelling along the
length of a rod, occurs during aging .

5.4 - DRYING GELS

DRYING RATE BY EVAPORATION

It is known for clay [58], boehmite [67] and in a more controversial fashion for silica
[68], that the drying rate per unit area of gels follows two successive regimes: a
"constant-rate" period, followed by a "falling-rate" period (figure 5.4-1, curves a and
b). The transition between these two regimes occurs at a very sharp point named the
"critical-point" [49] or "leatherhard point" [58] depending on authors.
Two different types of mechanisms have been proposed and summarized by
Hermans, to explain these dual regimes [3].

Capillary mechanism

The capillary mechanism explains well the reproducible adsorption hysteresis curves
of water in silica gels. This mechanism has been summarized as follows by
Hermans (Figure 5.4-2):

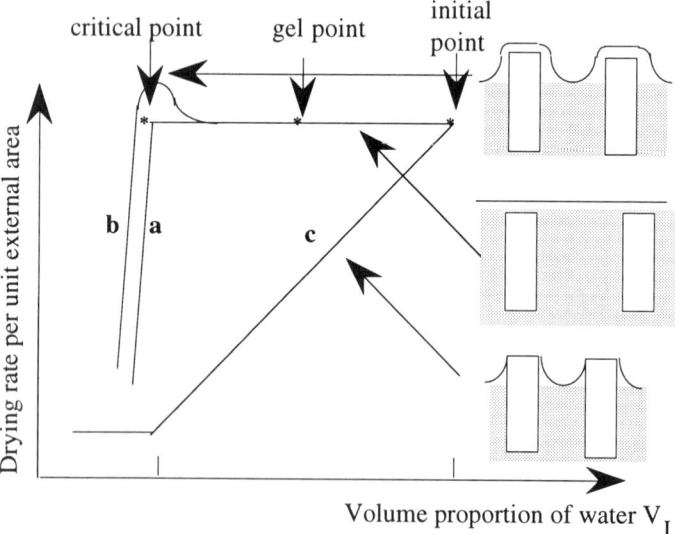

Figure 5.4-1 - Evolution of the drying-rate per external unit area of wet gel as a
function of the liquid volume proportion V_L : (a) and (b) experimental graphs; (c)
graph if the drying rate was proportional to the relative surface of menisci.
- (1) Evaporation creates a liquid vapor meniscus at the exit of pores in the gel

- (2) This induces a hydrostatic tension in the liquid, which is balanced by an axial compression on the solid.
- (3) The latter compression makes the gel shrink.
- (4) As a result of shrinkage, more liquid is fed to the menisci at the exit of the gel pores, where it is evaporated and so on.

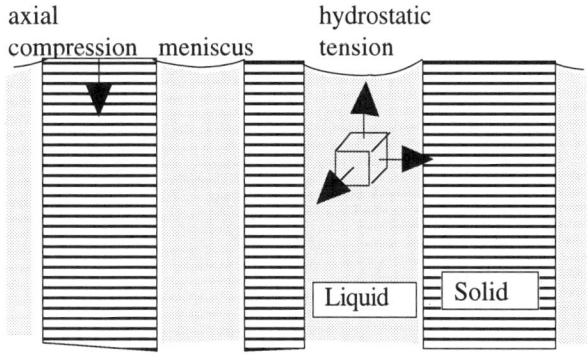

Figure 5.4-2 - Capillary mechanism.

One consequence of capillarity is also the existence of a hysteresis loop on adsorption-desorption isotherms, for instance of silica gels (figure 5.4-3). This comes from the fact that the equilibrium vapor pressure p_v above a meniscus of principal radii r_1 and r_2, is:

$$p_v = p_o \exp(-\gamma(\frac{1}{r_1} + \frac{1}{r_2})V_m RT]$$ (5.4-1)

where p_o is the vapor pressure above a planar surface and V_m is the molar volume of the liquid.

Branch IC in figure 5.4-3a corresponds to the initial drying of the gel. At point C, the meniscus is as in figure 5.4-3b. Further on, it penetrates inside the gel pores which stop shrinking. Along CD, the meniscus is deeper inside the pores (figure 5.4-3c), but this is a spherical meniscus with two radii of curvature equal to the radius of the pore r_{por}. Its curvature is therefore $\frac{2}{r_{por}}$. From D to S, only a thin layer of water remains on the cylindrical walls of the pores. The cylindrical configuration of point S maintains if water is readsorbed in the gel and the curve SDF is followed during re-adsorption. At point F, the gel pores are again full of water. Along FC, during the second desorption cycle, a spherical meniscus is formed again.

Detailed calculations according to the capillary mechanism have been made by Scherer [69]. The equilibrium meniscus radius r_m at any instant, is given by the compressive stress that the solid network of the gel can support. The contact wetting angle at the liquid-solid-vapor interface is undetermined along a sharp solid edge. Higher stresses are required for higher compression states of the solid, which

Gels

requires a higher hydrostatic tension in the liquid. It follows therefore that the menisci are sharper and sharper (smaller radius r_m) when contraction increases. This keeps a gel shrinking until the meniscus radius r_m reaches the pores radius r_{por}. The latter event defines the drying critical-point. Beyond this critical-point, capillarity cannot increase the compressive stresses anymore on the solid network. For instance, for a cylindrical pore of radius r_{por} and cross section:

$$A_{por} = \pi r_{por}^2 \qquad (5.4-2)$$

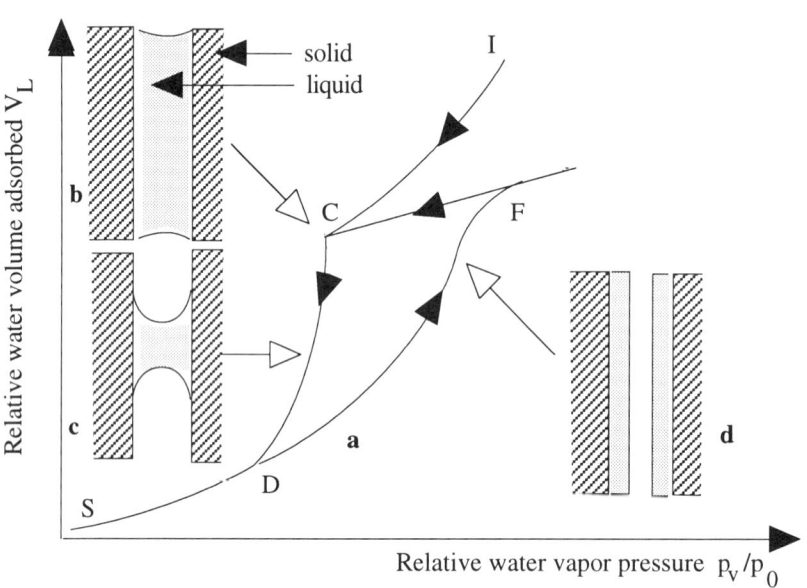

Figure 5.4-3 - Adsorption - desorption isotherms of water vapor in a silica gel. Adapted from Van Bemmelen [2] and Hermans [3].

The hydrostatic tension σ_T in the pore liquid increases when r_{por} decreases, as:

$$\sigma_T = 2\,\gamma\,\frac{\cos\Theta}{r_{por}} \qquad (5.4-3)$$

where Θ is the wetting angle of the liquid over the solid; γ the surface tension in the liquid. For a perfect wetting ($\Theta = 0$):

$$\sigma_T = \frac{2\gamma}{r_{por}} \qquad (5.4-4)$$

However, the tensile force decreases as:

$$F_T = \sigma_T A_{por} = \pi \, \gamma \, r_{por} \qquad (5.4-5)$$

Hence the compressive force on the gel decreases so that the gel contraction stops at some moment. For a higher liquid surface tensions γ, the gel reaches a higher shrinkage at the critical point, which is in agreement with data by Kingery and Francl [49].

The capillary mechanism is not able to explain in a straightforward fashion the existence of a "constant-rate" regime. A mathematical model by Suzuki and Maeda [70] has shown that the rate of drying could be constant when evaporation occurs from menisci, even when these menisci only occupy a minor relative proportion A_L of the gel external surface. This assumes the solvent vapor transport is the slowest step and controls the kinetics of drying. However, an experimental study has shown that the drying rate falls drastically as soon as menisci appear, then it remains constant while the menisci penetrate the inner part of a gel [71].

It is also possible to argue that the heat transfer to vaporize the liquid can control the kinetics of drying. In this case the drying rate should, in a first approximation, be proportional to A_L. As long as a gel is full of liquid, this is a random two phase composite material. In these conditions, it is known [72] that the proportion of each phase (liquid L and solid S), along a cutting line (respective linear proportions L_L, L_S), in a cutting plane (respective area proportions A_L and A_S), or in volume (respective volume proportions V_L and V_S), are independent of the determination mode:

$$L_L = A_L = V_L \qquad (5.4-6)$$
$$L_S = A_S = V_S \qquad (5.4-7)$$

Hence the drying rate per unit area should decrease linearly with the liquid content V_L (figure 5.4-1 curve c). The experimental existence of a "constant-rate" period before the critical-point is therefore generally interpreted as being due to the evaporation of a continuous liquid film, which covers the entire surface of the gel. Obviously the existence of such a film can be related to the affinity of oxides for water, which depends largely on the nature of the oxide. In the case of boehmite and montmorillonite, this affinity is very important. The latter gels are able to swell back in water and the constant rate can account for up to 80% of the gel volume shrinkage .

Osmotic mechanism

The second drying mechanism proposed for gels is osmosis. As discussed in a previous section, the gel solid network plays both the roles of a solute and a porous membrane. The osmotic mechanism corresponding to drying can be described in a manner similar to swelling and illustrated in figure 5.4-4. In particular, a gel can be made to shrink by immersion in another solvent than water.

The osmotic mechanism involves first the creation of a dissymmetry between a gel and a new solvent (Figure 5.4-4b). In second step, water diffuses through the gel network (equivalent to a porous membrane) and creates a hydrostatic tension in the liquid of the gel. This tends to reestablish an identical total free energy, of chemical and mechanical origins, in the gel solvent and in the solvent outside the gel. In a third step, this hydrostatic tension in the gel liquid makes the the gel network shrink (figure 5.4-4c). Consequently, the liquid free energies of chemical origin come close to each other in the liquid inside and outside the gel. Simultaneously the difference in hydrostatic pressure decreases until it reaches an equilibrium value balanced by the compressive stress in the gel network.

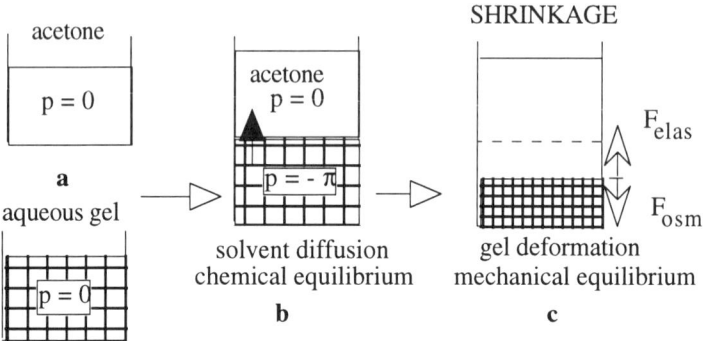

Figure 5.4-4 - The osmotic shrinkage mechanism: (a) separate gel and solvent; (b) creation of an overpressure inside the gel (c) shrinkage of the gel.

Drying and the DLVO theory
To explain both the constant-rate regime, and the behavior at the critical-point, a mixed osmosis-capillarity can be proposed [73]. It takes into account the structuration of monomolecular water layers around a solid particle, which is well known to occur in clays and boehmite. Recent studies by Livage [74] on V_2O_5 gels have shown that the number of water layers of this type can be very important. This structuration is a consequence of both steric and electrostatic interactions. For instance, in the case of the electrostatic interactions between two parallel and planar surfaces according to the DLVO theory (chapter 3), the hydrostatic pressure P inside a pore of colloidal size is not uniform, it is a function of the distance x from the nearest pore surface which is given by (figure 5.4-5a):

$$P_x = P_M + \frac{\varepsilon_o \varepsilon}{2} \left(\frac{d\Psi}{dx}\right)^2 \qquad (5.4-8)$$

where M is the mid-distance point inside the liquid and Ψ is the electrical potential . During drying, when a meniscus is created at the end of a pore, the liquid is placed under hydrostatic tension ($P_M < 0$) in the middle of the pore, so that the liquid-air meniscus just above the mid distance point is concave. However the hydrostatic tension P_x increases algebraically (is negative and decreases in magnitude) as the distance x from the nearest particle decreases. At a close distance x, it can even become a compression so that the meniscus profile becomes convex and insures a continuity with a liquid film which covers the solid. An inflexion point exists in the meniscus profile, which corresponds to a null hydrostatic tension in the liquid.

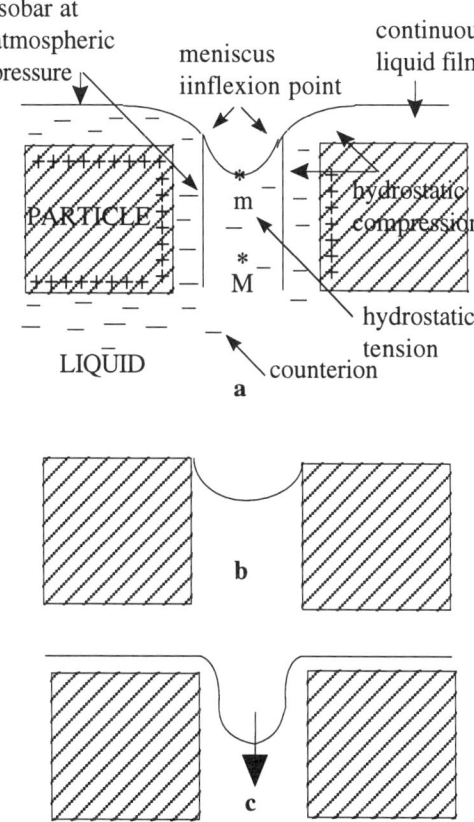

Figure 5.4-5 - Modification of the capillary mechanism by electrostatic osmosis according to the DLVO theory. Adapted from Pierre [73].

Actually, the capillary and osmotic theories constitute two approximations of the above colloidal view. The traditional capillary theory applies well when the pores have a diameter well above the colloidal dimensions, it approximates a solid-liquid interface by a sharp transition described by a surface tension γ_{SL}. The osmotic theory applies well to molecular solutions or truly polymeric gels: the solid-liquid interface loses all significance and the solvent structuration is replaced by an average effect in the solvent.

The structuration at the liquid-air interface can be neglected so that this interface can be described by a surface tension γ_w. If p_{hyd}^m designates the hydrostatic tension in the middle of the meniscus, the meniscus curvature is maximum at this point and given by:

$$r_m = \frac{\gamma_w}{\frac{m;}{p_{hyd}}} \qquad (5.4-9)$$

This radius of curvature is smaller, hence the curvature is more marked as γ_w is lower.

During drying, the continuous solvent film above the solid becomes less thick, while the maximum curvature of the meniscus increases. This evolution can end in two fashions:

- The film disappears first so that drying and shrinkage of the gel keep going with a decreasing rate (figure 5.4-5b).

- The meniscus radius reaches the pore radius and penetrates inside the gel, so that shrinkage and drying of the gel stop suddenly. This exactly corresponds to a behavior such at the drying critical point (figure 5.4-5c).

The influence of the liquid surface tension γ_w on the critical point has been well studied. If the chemical affinity of the liquid for the solid is maintained constant, for instance by keeping water as the main liquid, it is possible to modify the water surface tension by adding a small amount of surfactant. This was done by Kingery and Francl in the kaolinite-water system [49]. Their results are in agreement with the predictions of the mixed model, concerning the decrease in the water content of the gel after the critical point when the surface tension increases (figure 5.4-6).

Instead of modifying the surface tension γ_w and keeping the chemical affinity constant, the opposite can be done. A comparison has been made between the different drying behaviors of boehmite sols hydrolyzed in the same conditions, then diluted in excess solvents of a different nature [5]. The solvents comprised formamide, ethylene glycol and water in which various surfactants were added so as to compare the drying behavior with formamide and ethylene glycol for a same surface tension. The results have shown that the drying behavior changed drastically from water to formamide or ethylene glycol. With formamide, the critical-point roughly coincided with the gel point and it was reached for a total liquid volume much larger than with a water-surfactant liquid having the same surface tension. Also, the volume at gel point increased linearly with the formamide initial volume. A similar behavior of different magnitude was observed with ethylene glycol. These results can be only be explained by a different chemical affinity of boehmite for each solvent.

During the constant-rate regime, two parameters influence in an opposite direction the observed evaporation rate. One is the surface area of the evaporation front, per unit cross-section area of drying gel. Because of roughness on the evaporation surface due to nascent menisci, the surface area of this evaporation profile increases slowly during drying and hence it tends to increase the drying rate. On the contrary, the water vapor pressure decreases both over the nascent menisci and above the adsorbed layer, which tends to decrease the drying rate. Actually, these effects become measurable only when the liquid surface tension is high and close to the drying critical point. In these conditions, water evaporates from a film which follows more closely the particles contour and the increase in the evaporation surface becomes non-negligible. Data of Kingery and Francl [49] showed that the drying rate per unit cross-section area of gel increased first significantly, before falling down suddenly at the critical point (figure 5.4-2, curve b). That is to say, the increase in drying surface area outweighed the opposite decrease in water vapor pressure.

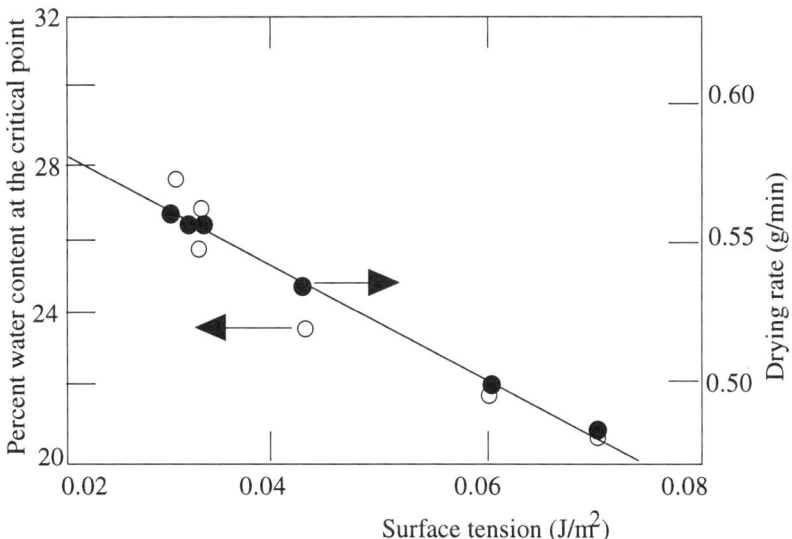

Figure 5.4-6 - Effect of surface tension on the drying critical point and maximum constant drying rate of kaolin clay. Adapted from Kingery and Francl [49].

STRESSES DEVELOPED IN A GEL DURING DRYING BY EVAPORATION

Practically, the procedure which must be selected to dry a gel depends on the material and on its degree of crystallinity [75]. To obtain a monolith, it is necessary to minimize the internal stresses due, on one hand to differential volume change inside a gel , on the other hand to stresses of capillary origin.

Internal stresses due to differential volume change have been investigated by Cooper [76] and by Scherer [69]. Cooper assumed that drying gels were submitted to a plastic flow behavior similar to that in glasses. This approach showed that a gel slab of thickness 2w, dried by evaporation from its two flat sides with an evaporation rate j, is submitted at time t to the tensile stress $\sigma(w,t)$ which inceases with w. An approximate relation often numerically valid is :

$$\frac{\sigma(w,t)(1-v)}{E(w,t)} \approx - \frac{j \ w}{9 \ D_w} \qquad (5.4\text{-}10)$$

where E(w,t) is the Young's modulus of the slab and ν its Poisson's ratio; which both depend on the average local liquid content. This formula shows that to avoid fracturing a gel monolith, the drying rate j must be decreased in inverse proportion to the monolith thickness. This explains that the most appropriate application of gels is for coatings.

The relationship between stresses developed during drying and the presence of capillary pores in a gel has been investigated by Zarzycki and al. [77] and by Scherer [69]. According to equation (5.4-3), a larger pore radius decreases the capillary stresses and makes it easier to synthesize an unfractured gel monolith, as shown by Shoup for SiO_2 [78] and Yoldas for Al_2O_3 [79]. The size and pore distribution in a gel can be modified in particular by choosing a different chemical synthesis procedure, such as modifying the hydrolysis water ratio r_w [80], or the nature of a solvent [81]. Hydrolysis with a high water proportion produces a gel with a strong tridimensional crosslinking, which is less prone to cracking [82]. With a low water proportion, a gel tends to crack in a powder as this is often the case with mixed alkoxides chemistry [75].

To decrease the effect of capillary stresses, for a given gel structure, various solvent extraction techniques can be used, in particular: freeze-drying; drying in a humid atmosphere; heating in a microwave oven [83]; adding a surfactant; slow drying; and supercritical drying. The three latter techniques have been compared by Zarzycki ad al. [77] on silica gels. The most efficient to make uncracked monoliths is supercritical drying which is examined in the next section.

To make uncracked silica gels monoliths with a diameter of 65 mm and ≈ 10 mm thick by slow evaporation, the time which is necessary is of the order of one month [84]. During this process, the weight loss is ≈ 50% and the radial shrinkage ≈ 25% [65]. The shrinkage is less important in a humid atmosphere [85]. Drying in a controlled humidity atmosphere (up to 100% relative humidity) makes it possible to attenuate cracking both during and after drying, as in the ThO_2 gel spheres made by Yamagishi and Takahashi [42]. Similarly, when the main solvent is not water but for instance methanol, control over cracking and turbidity can be gained by controlling the solvent vapor pressure [86].

Gel monoliths can also crack because of a restraint of the gel by adherence to the container. This effect can be limited by using materials containers in Teflon®, or by performing gelation over a layer of mercury.

SUPERCRITICAL DRYING

Supercritical drying, also named hypercritical drying, can be viewed as an application in an extreme case, of the drying laws presented in the previous paragraphs. It corresponds to drying in a case when the surface tension of the liquid cancels out (γ_w =0). In this case the equation (5.4-9) gives a meniscus radius r_m=0. This indicates that the meniscus radius reaches at once the radius of the pore where it penetrates. No constant rate regime can establish, all pores are emptied without any contraction of the gel, which is in agreement with experimental observations.

The supercritical drying of silica gels was first investigated by Kistler in 1932 [87]. This technique consists of heating a gel at temperatures and pressures exceeding the critical point of the solvent. Then, the supercritical solvent can slowly be evacuated by flushing the autoclave with dry argon. Supercritical drying has been applied to

SiO_2 as well as to Al_2O_3 and TiO_2 by Teichner et al. [88], to binary systems (SiO_2-B_2O_3; SiO_2-P_2O_5, NiO-Al_2O_3, Fe_2O_3-Al_2O_3) and ternary systems (SiO_2-B_2O_3-P_2O_5) [21].

Several heating schedules can be used , such as the one illustrated in figure 5.4-7. The autoclave is first partly filled with the solvent. Then it is closed and heated: the solvent evaporates and its pressure increases altogether with its temperature. If an appropriate quantity of solvent was placed in the autoclave, the pressure goes over the solvent critical point value p_c and the temperature over its critical point temperature T_c, before the gel begins to loose its liquid by evaporation. The supercritical solvent can then be evacuated as a gas, very slowly to not decrease the autoclave temperature below T_c.

This technique can be implemented with diverse fluids, of which a partial list with their critical parameters is given in Table 5.4-1.

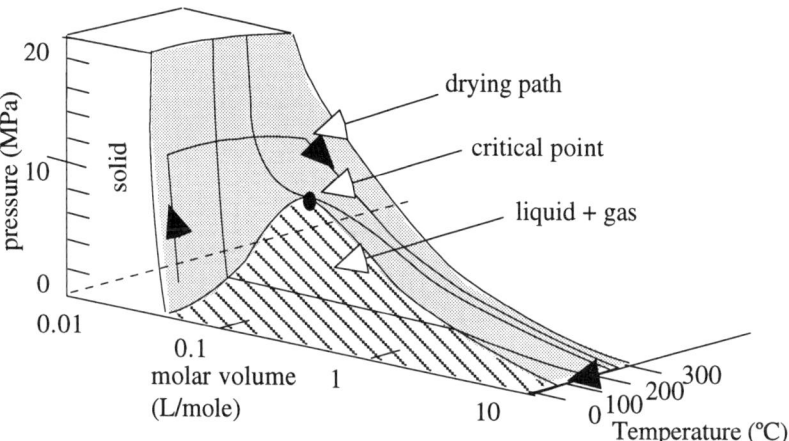

Figure 5.4-7 - Possible path for hypercritical drying of aqueous gels. Adapted from Fricke [8].

As this table shows, supercritical drying is difficult with water because of the high critical pressure and temperature. It is more feasible with methanol and ethanol, and easy with carbon dioxide [90]. However, carbon dioxide does not dissolve water and is not an outstanding solvent for methanol and ethanol: it is a much better solvent for acetone, so that aqueous liquids in a wet gel must first be exchanged for acetone by dialysis, before supercritical drying in CO_2. It produces relatively large dry silica aerogel monoliths [8] with 84 to 91% porosity and a specific area which can reach up to 800 m^2/g.

Table 5.4-1 - Critical point parameters of common fluids.
Adapted from D.W. Matson and R.D. Smith [89]

Fluid	Formula	T_c (°C)	P_c (MPa)
Water	H_2O	374.1	22.04
Carbon dioxide	CO_2	31.0	7.37
Freon® 23	CHF_3	25.9	4.82
Freon® 116	CF_3-CF_3	19.7	2.97
Ammonia	NH_3	132.5	11.27
Acetone	$(CH_3)_2O$	235.0	4.66
Benzene	C_6H_6	288.9	4.89
Nitrous oxide	N_2O	36.4	7.24
Methanol	CH_3OH	239.4	8.09
Ethanol	C_2H_5OH	243.0	6.38
1-Propanol	C_3H_7OH	263.5	5.17
1-Butanol	C_4H_9OH	289.7	4.39

5.5 - PROPERTIES OF DRY GELS

STRUCTURE AND TEXTURE OF DRY GELS

As the wet gels, dry gels are often amorphous according to X-ray diffraction patterns [23]. For instance, in the case of silica, this amorphous state differs from the glassy state in that not all oxygen atoms are bridging (Si-O-Si), a larger number of these oxygen atoms terminate in hydroxyl groups (Si-OH), or in alkoxy groups (Si-OR) where R can be any alkyl group.

Nonetheless, some modifications occur in the gel solid network structure, during drying. For instance in silica, the tri-siloxanes in the surface of wet gels are absent and replaced by cyclic tetra-siloxanes [11].

Gels dried by evaporation, that is to say xerogels, often present a lower surface area than the wet gels, although it remains quite high. Gels dried by supercritical drying, that is to say aerogels, usually have a higher specific surface area than xerogels.

The high specific surface area provides many interesting properties to gels. First, gels present a higher chemical reactivity than conventional powders or porous solids, so that they have important catalytic properties which will are reviewed in chapter 9 [88]. Also, their high chemical reactivity is responsible for the special phase transformation and sintering behavior which are respectively reviewed in chapters 7 and 8. Secondly, because they are very porous materials, dry gels constitute an original class of solid materials gifted with very special physical properties

PHYSICAL PROPERTIES OF DRY GELS

In spite of their high porosity, aerogels such as silica aerogels cane remain quite transparent. This is related to their texture which is composed of solid colloidal particles and pores which have a size smaller than the visible light wavelengths. Because of their texture, they are also excellent thermal insulators with a thermal conductivity λ from 0.01 to 0.02 W/m K.

If we consider the contributions to thermal conductivity, through the solid λ_s, through the gas λ_g, and by radiation λ_{rad}, the solid contribution λ_s scales up with the apparent density ρ_s according to:

$$\lambda_s \sim \rho_s^\alpha \qquad (5.5\text{-}1)$$

where $\alpha=1.6$ for bulk samples [91] and $\alpha=0.9$ for thin films [92] (figure 5.5-1).

The absolute value of λ_s ranges from 10^{-3} w/m K for $\rho_s=80$ kg/m^3 to 1.4 w/m K for $\rho_s=270$ kg/m^3. The extrapolated value for a completely dense solid is one order of magnitude smaller than for glassy silica.

This low thermal conductivity through the solid network goes altogether with a very low sound velocity transmission v_s, which is < 100 m/s ; aerogels are excellent sound insulators.

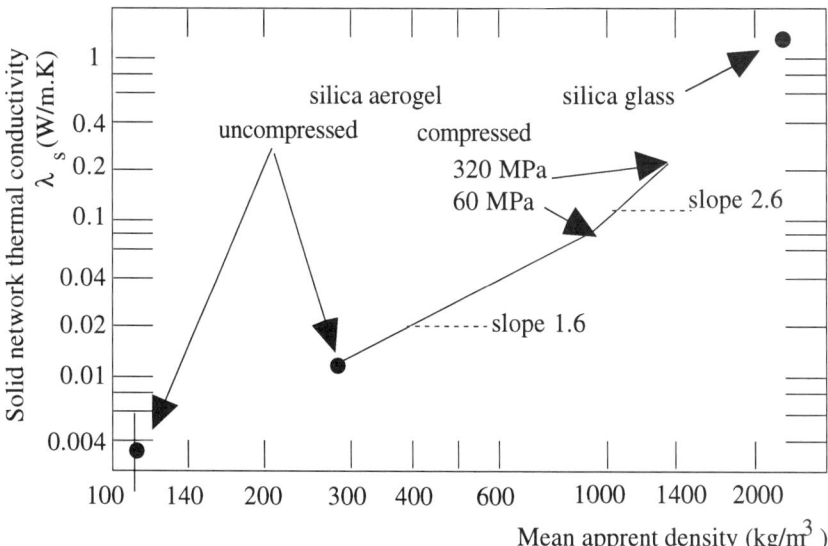

Figure 5.5-1 - Contribution of solid phase conduction to the thermal conductivity of silica aerogels. Adapted from Fricke [8].

The gas contribution to thermal conductivity can be written:

$$\lambda_g \sim \frac{1}{3} \lambda \ C_v \ \rho_g \ v_g \qquad (5.5-2)$$

where ρ_g is the gas density, C_v is the specific heat per unit mass of gas at constant volume, v_g is the gas molecules average velocity and λ the mean free path between two collisions of a gas molecule. This mean free path depends on the gas pressure P_g, except when the latter falls below a critical pressure P_{cr}, in which case λ is given by the average gel pore size. Hence the thermal loss factor $k = \dfrac{\lambda_g}{e_g}$ where e_g is the gel thickness, is first a constant at low gas pressure P_g, then it rises sharply with Pg (figure 5.5-2).

The radiation component λ_{rad} of the thermal conductivity is negligible in the visible light wavelengths. Light absorption occurs in the infra-red, for light wavelengths λ_L between 7 and 30 μm. This radiation conduction is displaced towards shorter wavelengths λ_L and its magnitude rises sharply as the temperature increases.

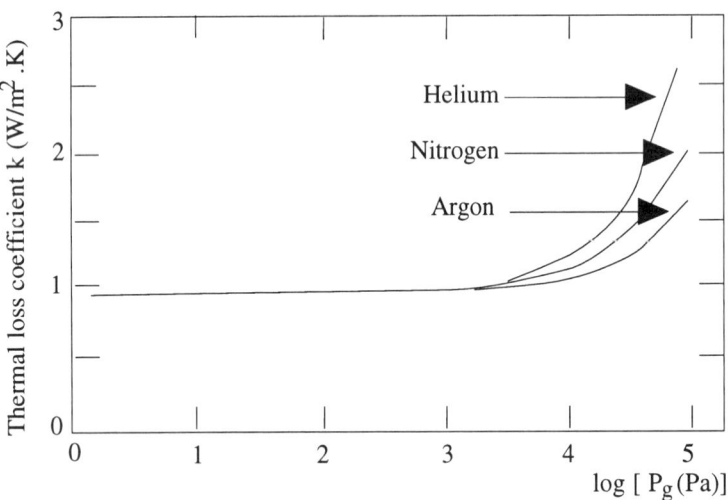

Figure 5.5-2 - Contribution of gas phase to the thermal conductivity of silica aerogels. Adapted from Fricke [8].

The mechanical properties of silica aerogels are characterized by a Young's modulus with a low value, in the range of 10^6 to 10^7 Pa (figure 5.5-3).

The material can be compressed between fingers. It makes a metallic noise which sounds like an aluminum foil. Its modulus increases with its apparent density as:

$$E_y \sim \rho_a^{\gamma} \qquad (5.5-3)$$

where $\gamma \approx 3.7$.

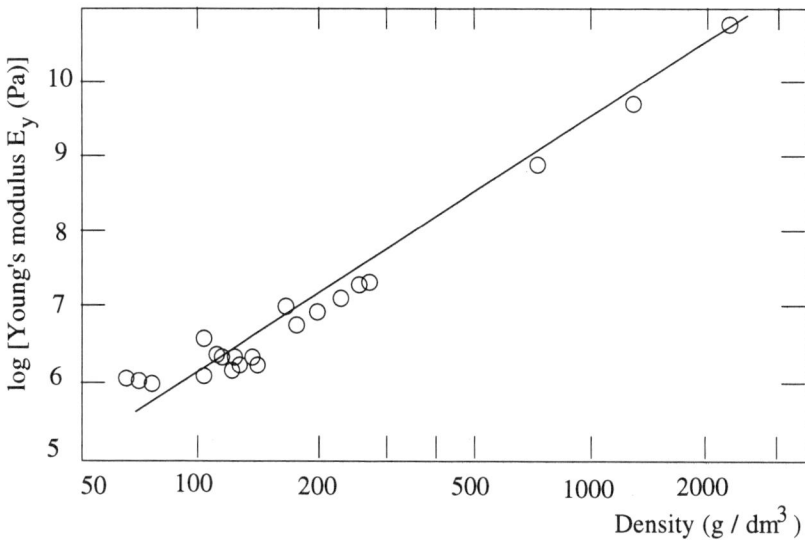

Figure 5.5-3 - Young's modulus of silica aerogels. After Fricke [8].

5.6 - REFERENCES

1 - Von Nägeli C., "Pflanzenphysiologische Untersuchungen", F. Schulthess, Zurich, 1858.

2 - Van Bemmelen J.M., Z. Anorg. Chem. 18 (1898) 98. Reference quoted by Hermans P.H., "Reversible systems", in "Colloid Science", vol II, Ed. H.R.Kruyt, Elsevier, New-York, 1952," Gels" p 486 et 522.

3 - Hermans P.H., "Reversible systems", in "Colloid Science", vol II, Ed. H.R.Kruyt, Elsevier, New-York, 1952," Gels" p 483-650; "Thermodynamics of Long chain Molecules", p 49-91.

4 - Flory P.J., Disc. Faraday Soc., 57 (1974) 7-18.

5 - Pierre A.C., Uhlmann D.R., Mater. Res. Soc. Sympsium 73 (1986) 481-487.

6 - Tanaka T., "Gels", Scientific American , 244 No. 1 (1981) 124-138.

7 - Gerngross O., Herrmann K., Kolloid Z., 60 (1932) 276-288.

8 - Fricke J., J. Non-Crystalline Solids, 100 (1988) 169-173.

9 - Iler R.K., 'The Chemistry of Silica", Wiley, New-York (1979).

10 - Yoldas B.E., J. of Non-Crystalline Solids 51 (1982) 105-121.

11 - Brinker C.J., Bunker B.C., Tallant D.R., Ward K.J., J. de Chimie Physique, 83 (1986) 851-858.

12 - Warren B.E., Biscoe J., J. Am. Ceram. Soc., 21 (1938) 49-54.

13 - Livage J., Lemerle J., Ann. Rev.Mater. Sci. 12 (1982) 103-122.

14 - Yamaguchi O., Tomihisa D., Kawabata H., Shimizu K., J. Am. Ceram. Soc 70 (1987) C94-C96.

15 - Marboe E.C., Bentur S.,   Silicate Industriels 26 (1961) 389-399.

16 - Pierre A.C., Uhlmann D.R., J. of Non Cryst. Solids, 82 (1986) 271-276.

17 - Yoldas B.E., J. of Mat. Sciences 21 (1986) 1087-1092.
18 - Komarneni S., Roy R., Breval E., J. Am. Ceram. Soc., 68 (1985) C41-C42.
19 - Yoldas B.E., J. of Mat. Sciences 21 (1986) 1080-1086.
20 - Debsikdar J.C., J. of Non Crystalline Solids 86 (1986b) 231-240.
21 - Woignier T., Phalippou J., Zarzycki J., J. of Non-Crystalline Solids 63 (1984) 117-130.
22 - Colomban P., L'industrie Céramique 792-3 (1985) 186-196.
23 - Dislich H., J. of Non-Crystalline Solids 57 (1983) 371-388.
24 - Hayashi T., Saito H., J. Mater. Sci. 15 (1980) 1971-1977.
25 - Mukherjee S.P., Mat. Res. Soc. Symp. Proc. 9 (1982) 321-331.
26 - Ritter J.J., Roth R.S., Blendell J.E., J. Am. Ceram. Soc. 69 (1986) 155-162.
27 - Blum J.B., Gurkovich S.R., J. of Mater. Sciences 20 (1985) 4479-4483.
28 - Melling P.J., Am. Ceram. Soc. Bull., 63 (1984) 1427-1429.
29 - Snel R., Applied Catalysis 11 (1984) 271-280.
30 - Debsikdar J.C., Advanced Ceramic Materials 1 (1986) 93-98.
31 - Tewari P.H., Hunt A.J., Lieber J.G., Lofftus K., "Microstructural Properties of Transparent Silica Aerogels", in "Aerogels", Ed. Fricke J., Springer Proc. in Physics, vol.6, Springer Verlag Heidelberg, New-York, (1986) 142-147.
32 - Lampert C.M., Mazur J.H., "Microstructure of Silica Based Aerogel Using High resolution transmission electron microscope", in "Aerogels", Ed. Fricke J., Springer Proc. in Physics, Springer Verlag Heidelberg, New-York, 6 (1986) 154-159.
33 - Hunt A.J., Berdahl P., Mat. Res. Soc. Symposium Proceedings, 32 (1985) 275-280.
34 - Fricke J., Caps R., "Aerogels- A fascinating Class of porous Solids", in "Ultrastructure Processing of Ceramics, glasses and Composites", Eds. Mackenzie J.D., Ulrich D.R., John Wiley & Sons, (1988) 613-622.
35 - Weitz D.A., Huang J.S., "Self-Similar Structures and the Kinetics of Aggregation of Gold Colloids", in "Kinetics of Aggregation and Gelation",Eds. Family F. et Landau D., Elsevier Science Publishers B.V. (1984) 19-28.
36 - Ross-Murphy S.B., McEvoy H., British Polymer Journal, 18 (1986) 2 7.
37 - Music S., Ljubesic N., Colloid Polym. Sci. 258 (1980) 194-195.
38 - Heller W., Wojtowicz W., Watson J.H.L., J. Chem. Phys. 16 (1948) 998-999.
39 - Lemerle J., Lefebvre J., Can. J. Chem. 55 (1977) 3758-3762.
40 - Parks G.A., Chem. Rev., 65 (1965) 177-198.
41 - Milligan W. O., McAtee J. L., J. Phys. Chem. 60 (1956) 273-277.
42 - Yamagishi S., Takahashi K., J. of Nuclear Materials 144 (1987) 244-251.
43 - Stanic V., Etsell T.H., Pierre A.C. and Mikula R.J., J. Materials Chemistry, 7 (1997) 105-107.
44 - Stanic V., PhD Thesis, Department of Chemical and Materials Engineering, The University of Alberta, 10 February 1997.
45 - Hardy C. J., Roy. Soc. Proc. 66 (1900) I: 95-109, II: 110-125.
46 - Poole H.J., Trans. farad. Soc., 21 (1925) 114-142.
47 - Walter A.T., J. Polym. Sci., 13 (1954) 207-228.
48 - Ellis H.S., Ring S.G., Carbohydr. Polymers, 5 (1985) 201-203.
49 - Kingery W.D., Francl J., J. Am. Ceram. Soc., 37 (1954) 596-602.
50 - Clark A.H., Ross-Murphy S.B., Br. Polym. J. 17 (1985) 164-168.
51 - Flory P.J., Br. Polym. J., 17 (1985) 96-102.
52 - Haruta M. Delmon B., J. Chimie Physique, 83 (1986) 859-868.
53 - Dusek K., J. Polym. Sci. (symp.) C42 (1973) 701-712.
54 - Prakash S., Dhar N.R., J. Ind. Chem. Soc., 6 (1929) I, 391-409; II, 587-598.
55 - Trail H.S., "Method of Preparing Silica Sols", U.S. Patent 2,572,578 oct 23, 1951.

56 - Van Der Gaast S.J., Jansen J.H.F., Simonton T.C., Clay and clay minerals, 33 (1985) 471-472.

57 - Hiemenz P.C., "Principles of Colloid and Surface Chemistry", Marcel Dekker, New-York (1977).

58 - Ford R.W., "Drying"; Institute of ceramics, textbook series; MacLaren and Sons, London, England (1964).

59 - Procter H.R., Wilson J.A., J. Chem. Soc. London 109 (1916) 307-319.

60 - Plank C.J., J. Colloid Sci., 2 (1947) 413-427.

61 - Pierre A.C., Uhlmann D.R., Mater. Res. Soc. Symp. 121 (1988) 207-212.

62 - Prakash S., Dhar N.R., J. Ind. Chem.Soc. 7 (1930) 417-432.

63 - Witzemann E.J., J. Amer. Chem. Soc.I- 37 (1915) 1080-1091, II- 39 (1917) 25-33.

64 - Blesa M.A., Maroto A.J.G., Passagio S.I., Figliolia N.E., Rigotti G., J. of Mater. Sciences 20 (1985) 4601-4609.

65 - Brinker C.J., Keefer D.W., Schaefer D.W., Ashley C.S., J. of Non-Crystalline Solids 48 (1982) 47-64.

66 - Kindu D., Ganguli D., J. of Materials sciences Letters 5 (1986) 293-295.

67 - Pierre A.C., Uhlmann D.R., J. Amer. Ceram. Soc. 70 (1987) 28-32.

68 - Hench L.L., Wilson M.J.R., J. of Non-Crystalline solids, 121 (1990) 234-243.

69 - Scherer G.W., J. of Non Cryst. Solids, 109 (1989) 171-182.

70 - Suzuki M., Maeda S., J. Chem. Eng. Japan 1 (1968) 26-31.

71 - Castro D., Ring T.A., Haggerty J.S. , Adv. Ceram. Mat., 3 (1988) 162-166.

72 - Kingery W.D., Bowen H.K., Uhlmann D.R., "Introduction to Ceramics", Wiley, New-York (1975); p 526.

73 - Pierre A.C., J. Can Ceram Soc., 59 (1990) 52-59.

74 - Livage J., Mat. Res. Soc. Symposium Proceedings, 121 (1988) 167-177.

75 - Zelinski B.J.J., Uhlmann D.R., J. Phys. Chem. Solids 45 (1984) 1069-1090.

76 - Cooper A.R., "Quantitative Theory of Cracking and warping During the Drying of Clay Bodies", in "Ceramic Processing Before Firing", Ed. Onoda G.Y.Jr., et Hench L.L. Wiley, New-York (1978) 261-276.

77 - Zarzycki J., Prassas M., Phalippou J., J. of Mater. Sciences 17 (1982) 3371-3379.

78 - Shoup R.D., "Controlled pore silica bodies gelled from silica sol-alkali silicate mixtures", in "Colloid and Interface Science", Ed. Kerker M., vol.III, Academic Press, New-York (1976) 63-69.

79 - Yoldas B.E., J. Mater. Sci.10 (1975) 1856-1860.

80 - Astier M., Sing K.S.W., J. Chem. technol. Biotechnol., 30 (1980) 691-698.

81 - Cormack B., Freeman J.J., Sing K.S.W., J. Chem. Technol. Biotechnol. 30 (1980) 367-373.

82 - Sakka S, "Gel method for making glass", in "Treatise on Materials Science and Technology", Ed. Tomazawa M. et Doremus R.H. , Vol 22 (1982) 129-167.

83 - Higuchi K., Naka S. Hirano S.S., Advanced Ceramic Materials 1 (1986) 104-107.

84 - Klein L.C., Garvey G.J., J. of Non-crystalline solids 48 (1982) 97-104.

85 - Lanutti J.J., Clark D.E., Mat. Res. Soc. Symp. Proceed. 32 (1984) 369-381.

86 - Yamane M., Aso S., Sakaino T., J. Mater. Sci. 13 (1978) 865-870.

87 - Kistler S.S., J. Phys. Chem. 36 (1932) 52-64.

88 - Teichner S., Nicolaon G.A., Vicarini M.A. and Gardes G.E.E, Adv. colloid Interface Sci; 5 (1976) 245-273

89 - Matson D.W. and Smith R.D., J. Am. Ceram Soc, 72 (1989) 871

90 - Tewari P.H., Hunt A.J., Lofftus K.D., "Advances in Production of Transparent Silica aerogels for Window Glazings", in "Aerogels, Ed. Fricke J., Springer Proc. in Physics, Vol.6, springer Verlag Heidelberg, New-York, (1986) 31-37.

91 - Nilsson O., Fransson A., Sandberg O., "Thermal Properties of silica Aerogel", in "Aerogels", Ed. Fricke J., Springer Proc. in Physics, Vol.6, Springer Verlag Heidelberg, 6 (1986) 121-126.
92 - Heinemann U., Hümmer E., Büttner D., Caps R., Fricke J., High Temp. High Pressures 18, (1986) 517-526.

NEW TYPES
OF SOL-GEL DERIVED MATERIALS

6.1 - INTRODUCTION

Since about 1992, two types of completely new materials have appeared and sol-gel is the obliged route to synthesize them. These sorts of materials do not exist in nature. They are the hybrid organic-inorganic materials and the ordered mesoporous materials with a very high specific area made from liquid crystal templates. These new types of materials also open very broad and promising horizons and a tremendous amount of research is devoted to them. Reviewed papers on these fields have begun to appear, by Sanchez and Ribot [1] for the hybrid organic-inorganic materials and by Corma for the ordered mesoporous ones [2]. Being given their importance, they deserve a special chapter.

6.2 - HYBRID ORGANIC INORGANIC MATERIALS [1]

DEFINITION OF HYBRID MATERIALS
New types of materials can be made by combining organic and inorganic components. Such materials are termed hybrid materials, and the organic and inorganic components can interpenetrate each other on a scale from the sub-micronic range down to the nanometer level. They can be divided in two classes :
- Hybrids which consist of organic molecules, oligomers or low molecular weight polymers embedded in an inorganic matrix to which they are held by weak hydrogen or Van der Waals bonds.
- Hybrids in which the organic and inorganic components are bonded to each other by strong covalent or partially covalent bonds.

HYBRIDS WITH WEAK ORGANIC INORGANIC BONDS
These hybrid materials can be either amorphous or ordered, and they can be divided them in two sub-classes

Amorphous weakly bonded hybrids
They can be made by

- hydrolysis-condensation of alkoxides inside soluble organic polymers
- mixing alkoxides and organic compounds in a common solvent
- Impregnating a porous oxide gel with organic compounds

Important hybrids made by the first technique comprise organic dyes embedded in an amorphous sol-gel matrix, mostly silica or aluminosilicate and to some extent in transition metal oxide matrices [3]. In these sol-gel matrices, dyes remain thermally stable contrary to what occurs in organic matrices. This is due to an interaction between the dye molecules and the sol-gel matrix, possibly by hydrogen bonds with the M-OH groups of the gel or chemical reactions of esterification or complexation between these groups and the dye molecules. In favorable cases, this interaction is stronger than between the dye molecules only. However it depends on the nature of the gel network. For instance, colorless spiropyran embedded in sol-gel SiO_2 made from $Si(OEt)_4$ transforms to a colored form of merocyanine when irradiated (photochromism), according to the reaction :

$$(6.2-1)$$

The colored dye is stable in the dark but it is bleached when it is submitted to UV radiation, hence it shows reverse photochromism. On the other hand, sol-gel SiO_2 made from $EtSi(OEt)_3$, only shows the direct photochromism transformation, due to the formation of micelles involving the spirospyran by interaction with the hydrophobic methyl groups present on the SiO_2 walls [4,5].

This technique can also yield hybrids composed of polymers in an inorganic amorphous sol-gel matrix, with weak bonding between the two components. Sol-gel oxides have M-OH hydroxyl groups which are Brönsted acid sites, while amide carbonyl groups present in some polymers are strong acidic proton acceptors. Hence, hydrogen bonds interaction between such polymers and the sol-gel oxide can be achieved. Homogeneously dispersed hybrids made in this way include silica-POZO [6], silica-PDMAAm [7] and silica-PVP composites [8], in which POZO, PVP and PDMAAm respectively designate the polyamides poly(2-methyl-2-oxazoline), poly(N-vinyl pyrrolidone), and poly(N,N-dimethylacrylamide). A schematic representation of the PVP-silica hybrid structure is shown in Fig. 6.2-1.

Other hybrids of this type where the homogeneity is not so good include PDMS-SiO_2 [9], Poly(n-butylmethacrylate)-titanium oxide [10], PMMA-silica [11] and polyphosphazene metal oxo where the metal can be Si, Ti, Al and Zr [12], and PDMS stands for poly-dimethyl siloxane and PMMA for poly methylmethacrylate.

At last, it is possible to use this technique in the opposite way, to incorporate inorganic particles in a polymeric gel. This can be achieved by adding acrylamide monomers and a radical transfer agent (N,N,N',N' - tetra1ethylethylenediamine) to a solution of metals cations complexed by citric acid. A polyacrylamide gel is formed which traps the complexed cations [13].

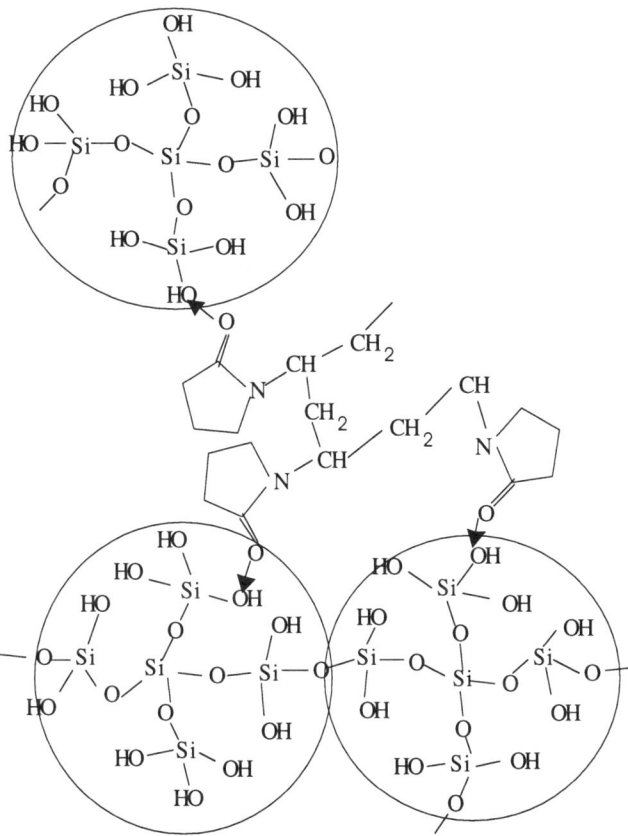

Figure 6.2-1 - Schematic representation of PVP-silica hybrid material. After [6].

The second synthesis technique consists of generating a hybrid material by simultaneous gelation of the organic and inorganic components so that they can form interpenetrating three dimensional networks. This is not easy because organic phases often precipitate when they are polymerized in water-alcohol solvents, so that heterogeneous materials are obtained. However, success could be reached by selecting cyclic alkenyl monomers which polymerize by ring opening metathesis or unsaturated alcohols which polymerize by a radical mechanism, as the organic component. In this case the inorganic component must be a modified silicon alkoxide made by reaction of silicon alkoxides with either cyclic alkenols or unsaturated alcohols (Figure 6.2-2) [14]. If the hydrolysis ratio r_w exactly corresponds to stoichiometry, the full solution gels without requiring any drying and without any shrinkage.

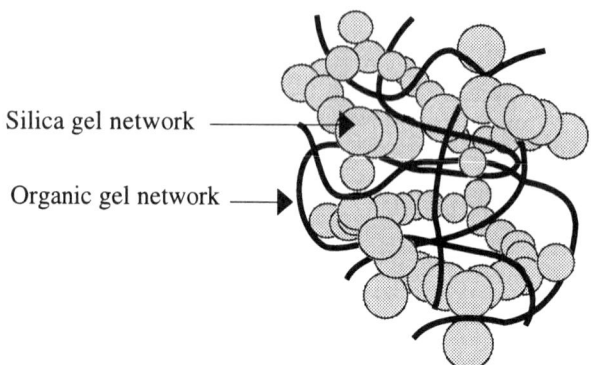

Silica gel network

Organic gel network

Figure 6.2-2 - Hybrids with interpenetrating SiO_2 and organic gel networks.

In particular ormosils gels made by cocondensation of TEOS and silanol terminated-PDMS (polydimethylsiloxane) were made to interpenetrate with organic polymers such as polypyrrole, polyaniline (PA) and polyparaphenylene (PPV) [15]

The third synthesis technique is useful when a commune solvent to the organic and inorganic components cannot be found. In this case, the inorganic gel can first be made, then it is impregnated with the organic monomer which is made to gel. This has been applied to make PMMA-silica and silica-PMMA-Perylene hybrids [16], where perylene is a dye initially dissolved in MMA. Silica gels termed class VI gels with a large pore size, up to ≈ 9 nm and a very narrow size distribution, developed by Hench et al. [17] are very useful for this application .

Ordered weakly bonded hybrids
Ordered hybrids can be made by intercalation of organic compounds in ordered inorganic hosts, such clay silicates, metal phosphates, layered metal oxides, halides or chalcogenides., on which reviewed paper have been published [18].
As an example, V_2O_5 gels made from $VO(OPr^n)_3$ in n-propanol with a high hydrolysis ratio r_w (> 100) comprise flat ribbons and can be dried to layered xerogels [19]. Organic molecules can be intercalated in between these layers and they take a particular orientation. Alkyl amines $C_nH_{2n+1}NH_2$ with short chains (n < 6) remain parallel to the V_2O_5 ribbons after intercalation, while long chains (n > 12) are perpendicular to these layers (Figure 6.2-3).
Vanadium has redox properties, so that it can induce an oxidative polymerization of the organic components, in particular of conducting polymers such as polyaniline, polypyrrole or polythiophene.

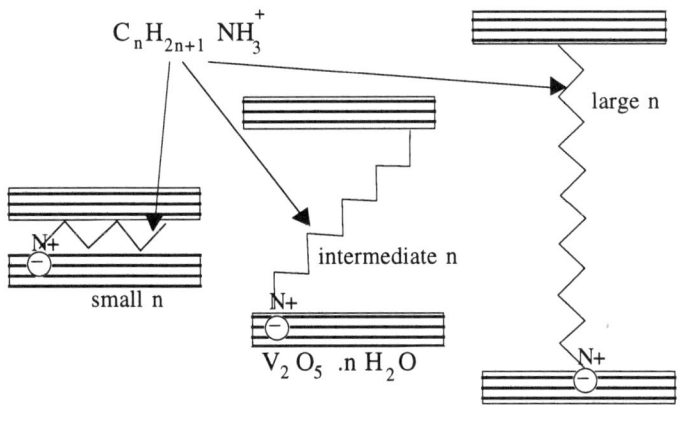

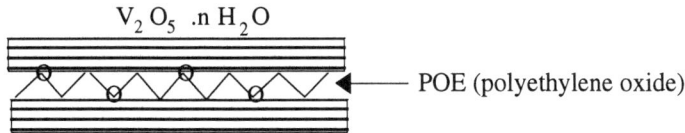

Figure 6.2-3 - Intercalation of organics in V_2O_5 xerogels of Alkylammonium $C_nH_{2n+1}NH_3^+$ with different n [20] and polyethylene oxide POE [18]. Adapted from [1].

STRONGLY BONDED ORGANIC-INORGANIC HYBRIDS

In these hybrid materials, the bonding between the organic and inorganic components is of the covalent or partly covalent type. In some cases, such hybrids can be formed in a sequential manner, for instance when an inorganic gel network of oxopolymers is first formed. If these oxopolymers comprise organic functionalities, these functionalities may be able to crosslink with an organic network in a second step. In other cases both networks can also form simultaneously from monomers comprising organic and inorganic functionalities.

The strong organic -inorganic bonds can come from organic compounds such as the polysaccharides [21], cellulosic materials [22] and vegetable oil derivatives [23]. More recently, organic macromonomers with reactive functionalities towards inorganic precursors, or inorganic clusters with organic functionalities have been studied.

Hybrid networks made from organic macromonomers with reactive functionalities

Such organic macromonomers were indicated in section 2.6 and their combination was mostly studied with silicon and titanium alkoxides, titanium acetylacetonates and triethanolamine titanium chelates as the inorganic precursor. The properties of such hybrid materials and their homogeneity depends on many parameters, including the nature of each component, their relative proportion, their functionalities, the hydrolysis ratio, the solvent, the temperature, catalysts.

With functionalized PDMS and PTMO, phase separation between the organic and the inorganic components generally occurs on a micrometer scale, and the interface between the two components plays an important role to improve the mechanical properties of the composite. When oxopolymers are embedded in the organic matrix, the hybrid material is termed a CERAMER [24,25]. With the silicates, no phase separation was observed.

With the polyimides, phase separation between the organic and inorganic components is frequent. However, success could be achieved by prebinding Ti(OEt)$_4$ or Si(OEt)$_4$ to the carboxylic sites of polyamic acid according to the reaction in (6.2-2) [26]. Imidization occurs at 300°C and the freed water hydrolyzes the alkoxide or can react with an ethoxysilyl functionalized polyimide. Up to 70% silica can be incorporated by reaction of TEOS with the ethoxysilyl functionalized polyimide macromonomer shown in Fig.2.6-1 (Chapter 2), in solution in dimethylacetamide. Phase separation between silica and polyimide occurs but the size of silica particles can be monitored down to the nanometer range.

(6.2-2)

Hybrid networks made from functionalized inorganic complexes

Apart from the siloxane clusters terminated by polymerizable organic ligands, mentioned in section 2.6, this domain is presently limited.

Trifunctional alkoxysilanes R'Si(OR)$_3$ mainly produce inorganic networks carrying organic groups. Polyfunctional alkoxysilanes such as those in figure.2.6-2 are also studied. On the other hand bifunctional alkoxysilanes R'R''Si(OR)$_2$ where R' and R'' are methyl or phenyl groups do not normally produce three-dimensional networks. Copolymerization of diethoxydimethyl silane with alkoxides of Ti, Zr, Al and V was studied by Babonneau et al. who showed that mixed M-O-Si bonds were established [27,28]. The coordination number of an alkoxide such as Ti(OR)$_4$ is unsaturated. It is electrophilic and undergoes fast alcoholysis with R'OH solvents and with hydroxyl-terminated short OH-PDMS (polydimethoxysilane) chains so as to form $Ti(OR)_{4-x-y}(OR')_x(OPDMS)_y$ species. Overall, $Ti(OR)_4$ catalyzes the condensation of siloxane units all along the PDMS chains. The products obtained can be used to make thick uncracked coating [29].

Other oxo clusters such as $Ti_6O_4(OEt)_8(OMc)$ and $[Zr_{10}(\mu_4\text{-}O)_2(\mu_3\text{-}O)_4(\mu_3\text{-}OH)_4(\mu_2\text{-}OPr^n)_8(OPr^n)_{10}$ where the complexing ligands are located at the periphery of the clusters [30,31] are very sensitive to hydrolytic cleavage below a certain size. Their polymerization must be performed in an organic solvent.

Bigger colloidal clusters of this type are more stable, especially when they are capped with the organic ligands. In this way methacrylate modified TiO_2 particles could be dispersed in an aqueous solution of sodium dodecyl sulfate, a surfactant organized in micelles. The TiO_2 particles occupied the core of the micelles and after polymerization of the methylmethacrylate dispersed inside the organic micelle walls, a core shell hybrid material was obtained (Fig 6.2-4).

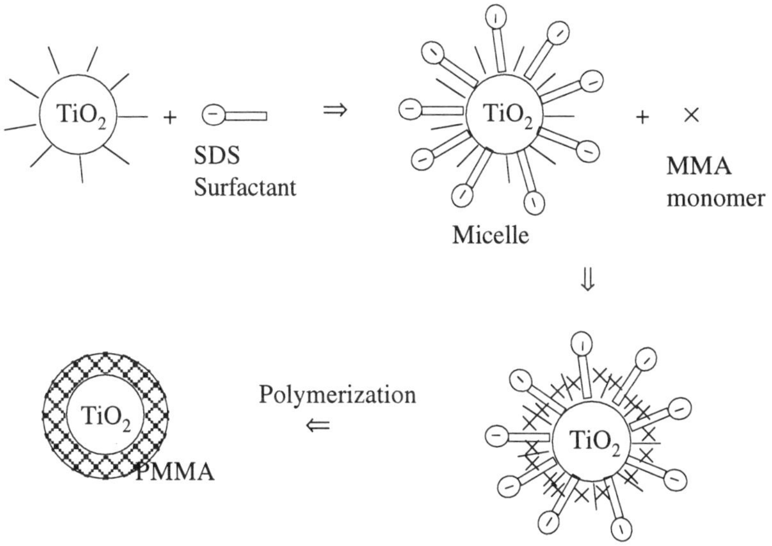

Figure 6.2-4 - Emulsion polymerization at the surface of TiO_2 particles. Adapted from Sanchez and In [32].

Other TiO_2 based hybrids stable towards hydrolysis could be made by complexation of $Ti(OR)_4$ with cinnamic acid ($C_6H_5\text{-}CH=CH\text{-}COOH$) and methacrylamidosalycilate (MASA), or by chelation with acetoacetoxyethylmethacrylate (AAEM) [32].

6.3 - ORDERED MESOPOROUS MATERIALS MADE WITH SURFACTANTS

SOLUTE ADSORPTION AT A LIQUID INTERFACE

Gibbs adsorption isotherm [33]

Let us consider an interface between a solution and another fluid medium, such as air. The solution is composed of a solvent and a solute respectively designated by the labels 1 and 2. Generally, the solvent and solute concentrations change progressively across a thin layer between the solution and the other fluid. For instance if each component concentration is measured along an axis x perpendicular to the interface between the two medium, as illustrated in figure 6.3-1, the solvent concentration C_{1x} goes progressively from its average solution concentration C_1 in the solution, far from its interface with air, to a value = 0 at an abscissa x_E. A similar gradation in the solute concentration C_2 exists.

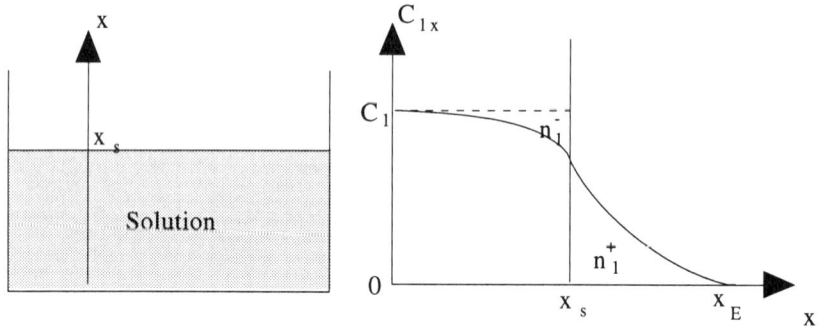

Figure 6.3-1 - Interface between two media and solvent Gibbs surface excess.

It is desirable to replace this actual progressive composition interface by a more simple representation which consists of a well defined surface at abscissa $x = x_s$, such that the solvent concentration is $C_{1x} = C_1$ for $x < x_s$ and $C_1 = 0$ for $x > x_s$. When doing this, we neglect a missing solvent mole number n_1^- for $x < x_s$ and a surplus solvent mole number n_1^+ for $x > x_s$. This can be accounted for by attributing an excess solvent mole number n_1^s to the surface at $x = x_s$, defined by :

$$n_1^s = n_1^+ - n_1^- \qquad (6.3-1)$$

Let A designate the area in m^2 of the surface at $x = x_s$. The solvent Gibbs surface excess is defined as

$$\Gamma_1 = \frac{n_1^s}{A} \tag{6.3-2}$$

recalling Gibbs-Duhem relatiosnhip

$$0 = SdT - VdP + A\,d\gamma + n_1\,d\mu_1 + n_2\,d\mu_2 \tag{6.3-3}$$

where S is the entropy, T the temperature, A the surface area and γ the surface tension, n_1 the number of moles of the solvent 1 and μ_1 its chemical potential, n_2 the number of moles of the solute and μ_2 its chemical potential. One sees that at constant T and P , for the interface defined by a surface at $x = x_s$:

$$0 = A\,d\gamma + n_1^s\,d\mu_1 + n_2^s\,d\mu_2 \tag{6.3-4}$$

which transforms to the so called Gibbs adsorption equation :

$$d\gamma = -\Gamma_1\,d\mu_1 - \Gamma_2\,d\mu_2 \tag{6.3-5}$$

It is convenient to chose the interface position x_s such that $\Gamma_1 = 0$; In this case

$$d\gamma = -\Gamma_2\,d\mu_2 \tag{6.3-6}$$

$$\text{As } \mu_2 = \mu_2^0 + RT\,Ln\,a_2 \tag{6.3-7}$$

where a_2 is the chemical activity of the solute, equation (6.3-6) transforms to

$$d\gamma = -\Gamma_2\,RT\,\frac{da_2}{a_2} \tag{6.3-8}$$

For a dilute solution where C_2 is the average solute concentration

$$d\gamma = -\Gamma_2\,RT\,\frac{dC_2}{C_2} \tag{6.3-9}$$

Solute classification [33]
Equation (6.3-9) indicates that the surface tension γ of a solution is modified by addition of a solute. Hence γ can be written

$$\gamma = \gamma_0 - \pi \tag{6.3-10}$$

where γ_0 is the surface tension of the pure solvent and π is a modification due to the solvent and to the concentration profile of the solute at the solution interface with the surrounding medium, termed the film pressure. However this is not a real pressure; π has the dimension of $\dfrac{\text{Force}}{\text{Length}}$. The real film pressure is

$$P = \frac{\pi}{\tau} \qquad (6.3\text{-}11)$$

where τ is the interface thickness.

The film pressure π depends on the solute concentration C_2 and for a small concentration it can be written

$$\pi = mC_2 \qquad (6.3\text{-}12)$$

The solutes can be classified in three classes depending on their effect on the surface tension γ of the solution. As illustrated in figure 6.3-2; m is slightly negative for the electrolytes. That is to say the solution surface tension slightly increases with C_2,. On the other hand, for many organic solutes m > 0, the solution surface tension slightly decreases as C_2 increases. At last for some organic solute termed surface active or surfactants and composed of molecules known as amphipathic, γ decreases fastly as C_2 increases, until a critical solution concentration termed the critical micellar concentration or c.m.c.

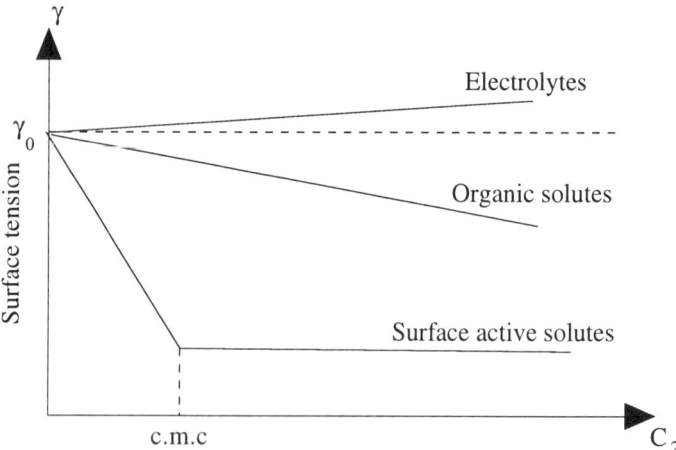

Figure 6.3-2 - Effect of different solutes on the surface tension of a solution.

By derivation from (6.3-12) and comparison with equation (6.3-9), as long as m is a constant

$$m = \frac{d\pi}{dC_2} = -\frac{d\gamma}{dC_2} = \frac{\Gamma_2\, RT}{C_2} \qquad (6.3\text{-}13)$$

If m is replaced by its expression taken from equation (6.3-12) , this gives :

$$\frac{\pi}{C_2} = \frac{\Gamma_2\,RT}{C_2} \qquad\qquad (6.3\text{-}14)$$

That is to say $\qquad\qquad \pi = \Gamma_2\ RT \qquad\qquad (6.3\text{-}15)$

or $\qquad\qquad \pi A = n_2^s\ RT \qquad\qquad (6.3\text{-}16)$

The latter equation looks like the ideal gas equation.   It indicates that at low concentration, the solute in the solution interface with air behaves as a perfect bidimensional gas.

SURFACTANTS [34]

General structure of surfactant molecules
They are composed of organic molecules not much soluble in a solvent and which comprise two parts with a different polarity as illustrated in figure 6.3-3 :
- An organic polymer tail which is nonpolar, hence hydrophobic (i.e. insoluble in water) and lipohilic (i.e., soluble in organic solvents)
- A smaller head part which is polar, hence hydrophilic (i.e., soluble in water).
Because of this general structure, such molecules tend to easily adsorb at air/aqueous solutions or hydrocarbon/aqueous solutions interfaces, in such a way that the hydrophilic head is turned towards the aqueous solution .

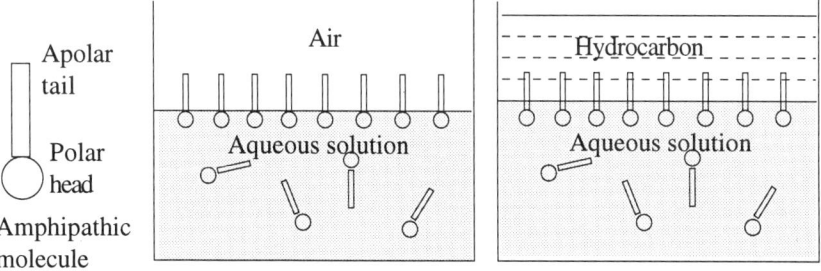

Figure 6.3-3 - Surfactant molecules and their orientaion at aqueous medium/air and aqueous medium/liquid hydrocarbon interfaces

Surfactant molecules can be classified in four families known as the anionic, cationic, nonionic and amphoteric surfactants.

Anionic surfactants
They are the most extensively used in industry. They include:

- Soaps for which the hydrophobic tail is an alkyl chain comprising 11 to 21 carbon atoms and the hydrophilic head is a sodium or potassium salt of an organic acid.
- Sulfonated compounds with general formula $R\text{-}SO_3^- Na^+$ mostly used in detergents
- Sulfated compounds with general formula $R\text{-}OSO_3^- Na^+$ such as the well known normal sodium dodecyl sulfate $nC_{12}H_{25}\text{-}OSO_3^- Na^+$

Cationic surfactants
They are used as components in softening liquids for cloths and in the textile industry. Their general formula is

$$
\begin{array}{c}
CH_3 \\
| \\
CH_3\!-\!\overset{+}{N}\!-\!R_1 \quad X^- \\
| \\
R_2
\end{array}
\qquad (6.3\text{-}17)
$$

Positively charged
tail

such as the cetyl trimethyl ammonium bromide (CTAB) $C_{16}H_{33}N(CH_3)_3^+ Br^-$

Nonionic surfactants
They comprise no ions when in solution. Their hydrophilic head is a polar group such as:
- Ether : R-O-R
- Alcohol R-OH
- Carbonyl : $R\text{-}\underset{\underset{O}{\|}}{C}\text{-}R$

- Amine R-NH-R
An example of nonionic surfactant is the "20 stearyl ether" or "Brij 78" $C_{18}H_{37}\text{-}(CH_2\text{-}CH_2\text{-}O)_{20}\text{-}H$ where the ethoxy chain $(CH_2\text{-}CH_2\text{-}O)_{20}\text{-}H$ chain is polar and hydrophilic. These nonionic surfactants are compatibles with most ions, including Ca^{2+} in hard water. Their properties change progressively with the length of the ethoxy chain for instance. The Pluronic which have the formula below can be solids or liquids depending on the values of the a and b numbers.

$$
H\text{ - }(O\text{ - }CH_2\text{ - }CH_2)_a\text{ - }(O\text{ - }\underset{\underset{CH_3}{|}}{CH}\text{ - }CH_2)_b\text{ - }(O\text{ - }CH_2\text{ - }CH_2)_a\text{ - }OH
$$

Polyoxyethylene   Polyoxypropylene   Hydrophilic
Hydrophilic   Hydrophobic

$$(6.3\text{-}18)$$

Amphoteric surfactants

Some of their properties are similar to the nonionic surfactants while other properties are similar to the ionic surfactants. Exemples of such amphipathic molecules are the betaïnes

$$R\!-\!\underset{\underset{\displaystyle CH_3}{\displaystyle |}}{\overset{\overset{\displaystyle CH_3}{\displaystyle |}}{N^+}}\!-\!CH_2\!-\!COO^- \tag{6.3-19}$$

and the lecithines or phospholipides

$$
\begin{array}{l}
CH_2\!-\!O\!-\!CO\!-\!R \\
| \\
CH\!-\!O\!-\!CO\!-\!R \\
| \\
CH_2\!-\!O\!-\!\underset{\underset{\displaystyle O^-}{\displaystyle |}}{PO}\!-\!O\!-\!CH_2\!-\!CH_2\!-\!N^+(CH_3)_3
\end{array} \tag{6.3-20}
$$

Hydrophilic - Lipophilic balance (HLB) of a surfactant [35]

The HLB is an empiric number which characterizes the relative weight of the hydrophilic part realtive to the complete amphipathic molecule.
For a ionic surfactant it is given by

$$HLB = 7 + \Sigma_{al\ groups}(group\ number) \tag{6.3-21}$$

A few group numbers are provided in Table 6.3-1.

Table 6.3-1 - Number to use in the calculation of the HLB for ion surfactants. Adapted from Berthod [34].

Hydrophilic group	Number	Lipophilic group	Number
$-SO_4^-\ Na^+$	38.7	$-\overset{|}{\underset{|}{C}}-$	- 0.475
$-COO^-Na^+$	19.1		
$-SO_3^-\ Na^+$	11	$-O-\underset{\underset{\displaystyle CH_3}{\displaystyle |}}{CH}-CH_2-$	- 0.16
$-OH$	1.9		

By example, for SDS : $C_{12}H_{25} - SO_4^-\ Na^+$

$$HLB = \quad 7 \; - \; 12 \; x \; 0.475 \; + 38.7 \; = \; 40 \tag{6.3-22}$$

For a nonionic surfactant, this number is given by

$$HLB = 20 \; \frac{H}{H+L} \qquad (6.3\text{-}23)$$

where H is the molecular mass of the hydrophilic part and L that of the lipophilic part.

The HLB numbers are interesting because their value can be correlated with the main properties of a surfactant such as its solubility in water, its interfacial adsorption enrgy and its c.m.c.

BEHAVIOR OF SURFACTANTS AS A FUNCTION OF THEIR CONCENTRATION [33,34,36-39]

Behavior below the c.m.c

For a concentration of surfactant which is $C_2 <$ c.m.c, equation (6.3-16) shows that the excess surfactant at the surface of the solution behaves as a bidimensional gas. This equation can be transformed to

$$\frac{\pi A}{n_2^s \, N_A} = \frac{R}{N_A} \; T = k_b T \qquad (6.3\text{-}24)$$

where N_A is Avogadro's number (6.02×10^{23}) and k_b is Boltzmann's constant. In this case

$$\sigma = \frac{A}{n_2^s \, N_A} \qquad (6.3\text{-}25)$$

is the average free area around a surfactant molecule in the solution surface film. Equation (6.3-25) can then simply be written

$$\pi \, \sigma = k_b \, T \qquad (6.3\text{-}26)$$

Behavior above the c.m.c

As C_2 increases, π increases and σ decreases. At some moment, σ reaches a value σ_0 which is the minimum area occupied by a surfactant molecule in the interface (figure 6.3-4). This corresponds to the c.m.c. in figure 6.3-2. After this point the Surfactant Gibbs surface excess by reference to the solution remains a constant and $\frac{d\gamma}{dC_2} = 0$.

The surfactant film is now a compact and almost uncompressible bidimensionnal liquid-like film. All supplementay surfactant added to the solution will not participate in the surface film, nor in the solution as individual molecules. It will phase separate and build colloidal aggregates termed micelles, dispersed in the solution. The formation of a micelle from m amphipathic molecules S can be represented by the chemical reaction [34]

$$m \, S \quad \Leftrightarrow \quad S_m \qquad (6.3\text{-}27)$$
$$\text{micelle}$$

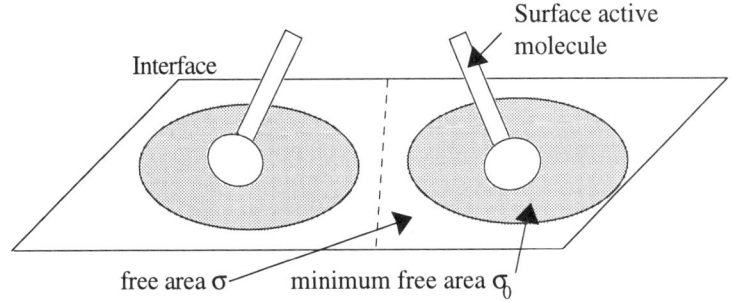

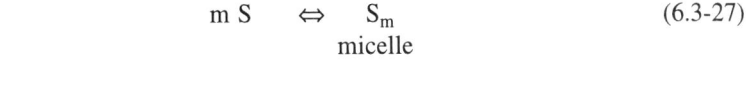

Figure 6.3-4 - Gas-like surfactant film

Micelle structure [34,36-39]
Several micelle structures are possible and they depend on the surfactant concentration C_2.
For a concentration just above the c.m.c, micelles are spherical and their structure is illustrated figure 6.3-5. A spherical micelle looks like a typical colloidal particle with an electrical double layer, in the case of an ionic surfactant, which can be described by the Gouy-Chapman model.
As C_2 keeps increasing, micelles with a cylindrical rod shape are formed   (figure 6.3-6). The radius of such micelles is typically from 1 to 3 nm, it can comprise several thousands of surfactant molecules. For higher surfactant concentrations, these rods take on an ordered parallel hexagonal packing orientation (figure 6.3-7), which is characteristic of some liquid-crystals.
For still higher concentrations, the shape changes again and consists of lamellar micelles (figure 6.3-8). At last if the surfactant concentration is larger than that of water, inverse micelles with a hydrophobic tail turned outside the micelles are formed (figure 6.3-9).

Factors influencing the c.m.c.
The nature of a surfactant molecule has an important effect on the c.m.c, in particular the number of carbon atoms N_c of the hydrophobic chain. For a homologous series of surfactants [40]:

$$\log_{10} \text{c.m.c} \; = \; a \; + b \, N_c \qquad (6.3\text{-}28)$$

where a and b are some positive constants.
The nature of the hydrophilic head itself is not so important, apart from the charge of the associated ions.

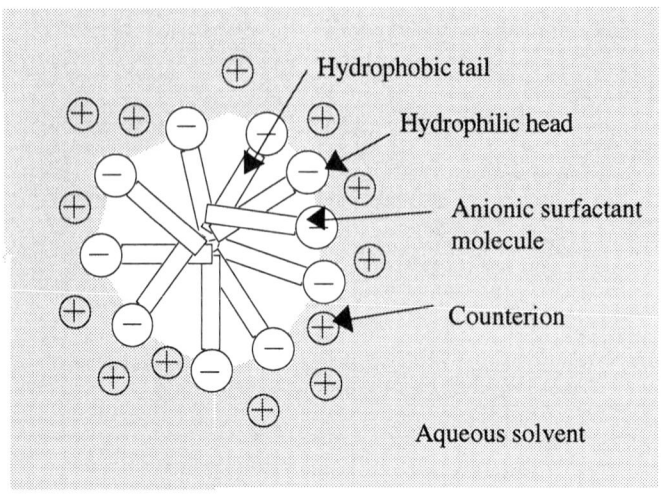

Figure 6.3-5 - Spherical micelle.

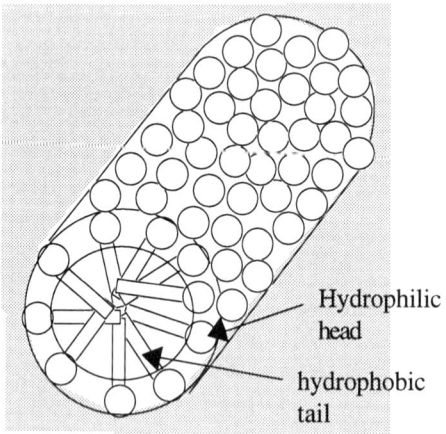

Figure 6.3-6 - Cylindrical micelle

In the case of anionic surfactants the c.m.c decreases as the formal charge z+ of the cationic counterions increases [41]. Hence, the c.m.c of divalent counterions such as Ca^{2+}, Mg^{2+}, Pb^{2+} and Zn^{2+} is lower than that of the monovalent cations Na^+ or K^+. For cationic surfactants, the valence of the anionic counterions is important; the c.m.c is lower for SO_4^{2-} than for NO_3^-, Br^-, Cl^- and F^-. This result is the same as the one already presented for solid particle colloids, in chapter 3. Similarly, the c.m.c.

decreases as I increases in the case of ionic surfactants. When a concentration C_e of non-potential determing electrolyte is added to a surfactant solution, the c.m.c follows an empirical relationship of the type [34]

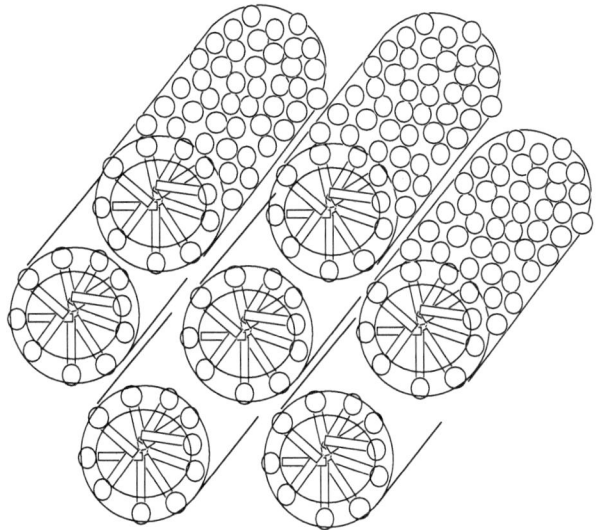

Figure 6.3-7 - Hexagonal packing of cylindrical micelles

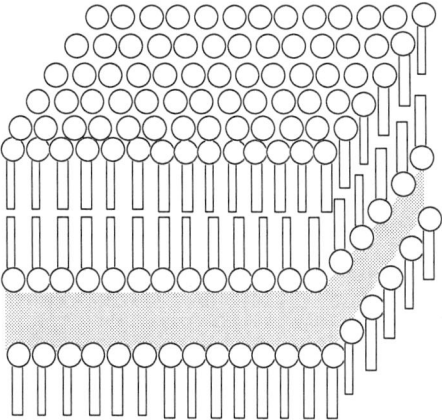

Figure 6.3-8 - Lamellar micelles

$$Ln (c.m.c.) = - a Ln (c.m.c. + C_e) + b \qquad (6.3\text{-}29)$$

where a and b are positive constants depending on the surfactant and the electrolyte. Electrolytes favor the formation of cylindrical micelles.

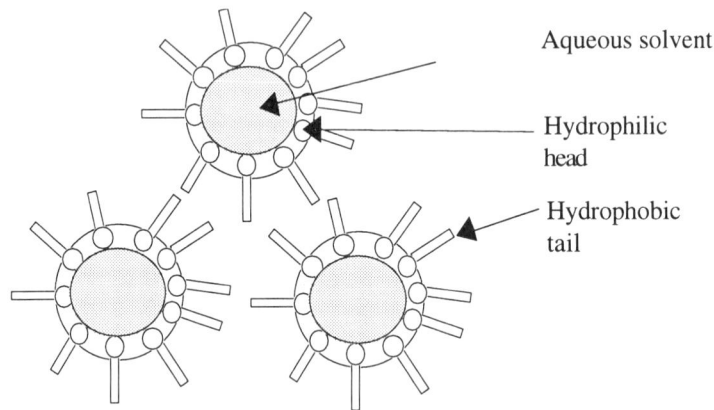

Aqueous solvent

Hydrophilic head

Hydrophobic tail

Figure 6.3-9 - Inverse micelles

The solubility of an ionic surfactant rapidly increases as the temperature T increases [42]. The main temperature characteristics of an ionic surfactant solution is its Krafft point K. This is a particular temperature which depends on the surfactant. For T < K, the surfactant solubility is low and due to monomers. On the other hand for T > K, solubility occurs by the formation of micelles, unless the temperature is too high and the micelles are destroyed. The size of cylindrical micelles decreases as the temperature increases, an effect which is very moderate on spherical miceles. The qualitative effect of temperature is summarized in the graph of figure 6.3-10. The counterion valence has a signifiacnt influence. For instance $K(Ca^{2+}) > K(Na^+)$, which explains that soap is precipitated in hard water.

Nonionic surfactants are not characterized by a Kraft point, but by a cloud temperature which depends on the surfactant concentration. When a solution is heated above the cloud temperature, it phase separates in two solutions with a different micellar concentration. The c.m.c of nonionic surfactants is much lower than that of ionic ones, and it is very moderately affected by electrolytes. It decreases as T increases. The ionic strength I of an electrolyte solution has little influence in the case of nonionic surfactants.

SOLUBILISATION OF COMPOUNDS BY MICELLES [34, 43]

The solubility of many compounds in water can be improved with the help of micelles, a technique which is quite uesful in the pharmaceutical and detergent industries.

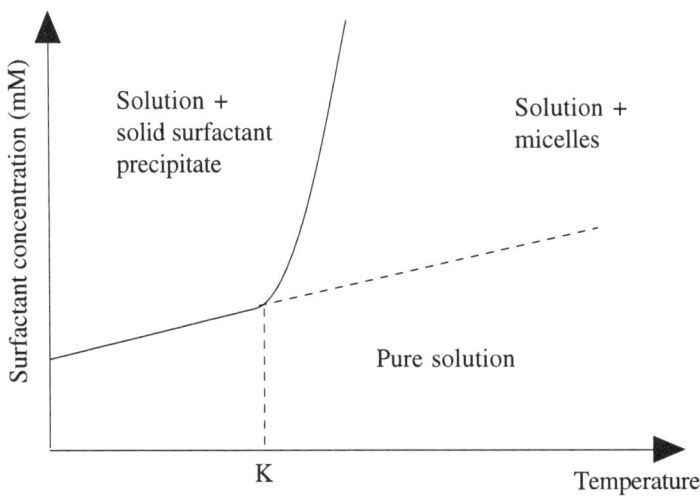

Figure 6.3-10 - Structure of a surfactant solution as a function of Temperature and surfactant concentration. Adapted from Berthod [34]

Micellar solution.
Nonpolar organic compounds such as hydrocarbons, toluene and cyclohexane can be solubilized with the help of micelles. These molecules are dispersed in the heart of micelles made by the hydrophobic tails of the surfactant molecules (figure 6.3-11a). Hence, their solubility increases with the size of the micelles and with the length of the surfactant hydrophobic tails. A solubility limit exists beyond which an emulsion is formed as this is discussed further on.
On the other hand, polar compounds often participate in the structure of micelles and form mixed micelles, as this occurs in particular with alcohol molecules (figure 6.3-11b).

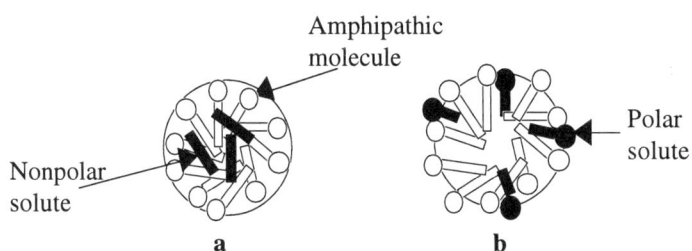

Figure 6.3-11 - Solubilisation of (a) nonpolar and (b) polar compounds by micelles. Adapted from Mukerjee [44].

Microemulsions
An emulsion is a two-phases liquid mixture, which forms each time two liquids are only partially soluble. Just after mechanical dispersion of the two liquids by agitation, the least abundant liquid phase is dispersed as fine droplets inside the most abundant liquid phase. Normally an emulsion is not stable, it readily tranforms to a simple two-phases layered separation, either by flocculation followed by coalescence, or by flotation followed by coalescence.
On the other hand in a microemulsion, the dispersed microdroplets are kinetically stabilized because they are embedded inside surfactant micelles (figure 6.3-12). An emulsion can be of the O/W (oil in water) type when the oil is inside the micelle, or of the W/O (water in oil) inverse micelle type when water is inside the inverse micelle (figure 6.3-13). The size of the droplets can go from 10 nm to 0.2 μm. In the first case an emulsion is transparent while in the second case it bluish and shows the same Tyndall effect as sols.

State diagrams of ternary solution systems made with surfactants
The state of a ternary system composed of water (compound W), an organic liquid (compound O) and a surfactant (compound T) can be represented in an equilateral triangle diagram as illustrated in figure 6.3-14. In such diagrams the percentage of each compound can be read by drawing lines parallel to the sides of the triangle. For instance in figure 6.3-14, the composition of system M is 55% W, 32% O and 13% T. In such state diagrams, it is possible to indicate whether spherical micelles (O/W , W/O or both types), cylindrical micelles, or surfactant precipitation occurs, as in figure 6.3-15. Another classification by Winsor [45] considers the number and nature of phases formed (figure 6.3-16). A WINSOR I system comprises a W/O emulsion phase and an oil solution on top of it. A WINSOR II system comprises a water solution and an O/W emulsion on top of it. In a WINSOR III system we have three phases: a water solution, a O/W emulsion and an oil solution. In a WINSOR IV system, only an emulsion phase exists.

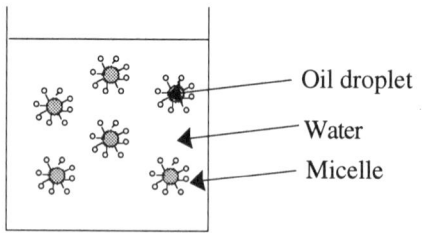

Oil droplet
Water
Micelle

Figure 6.3- 12 - Micellar emulsion

APPLICATION OF SURFACTANTS TO SYNTHESIZE ORDERED MESOPOROUS MATERIALS [2]
Micelles and particularly hexagonal packing of cylindrical micelles, cubic packing of spherical micelles or planar packing of lamellar micelles, can be used as templates to make sol-gel oxide materials with a regular pore texture, termed molecular sieves.

These new types of materials were the first time investigated by the Mobil Oil company under the generic name M41S [48]. The porous channels can have a diameter of 1.5 to 10 nm and be organized in a hexagonal, cubic or a lamellar array, in which cases the corresponding mesoporous materials are respectively designated under the names MCM-41 , MCM-48 and MCM-50 [49]. Their specific surface area is frequently very high, > 700 m^2/g.

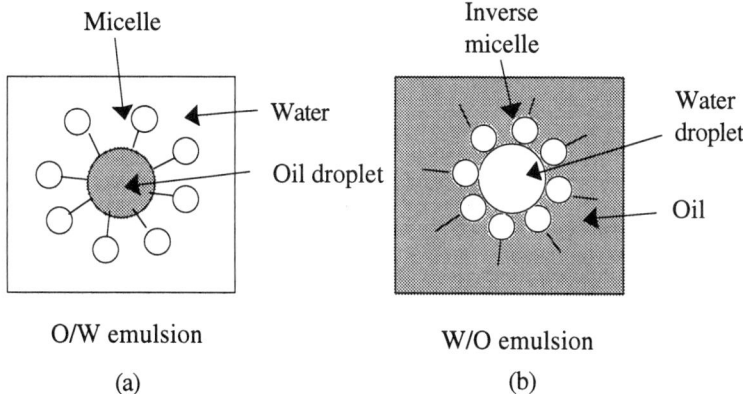

Figure 6.3-13 - Oil-water (O/W) and Water-oil (W/O) emulsions

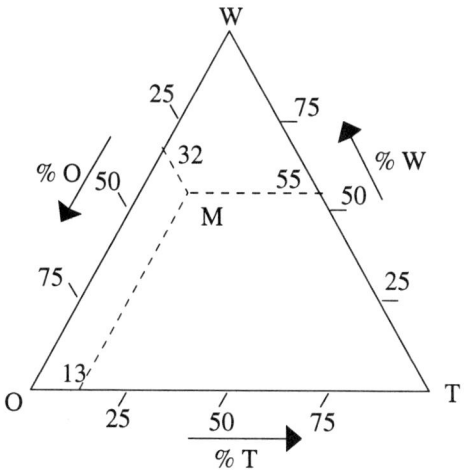

Figure 6.3-14 - Ternary state diagram representation

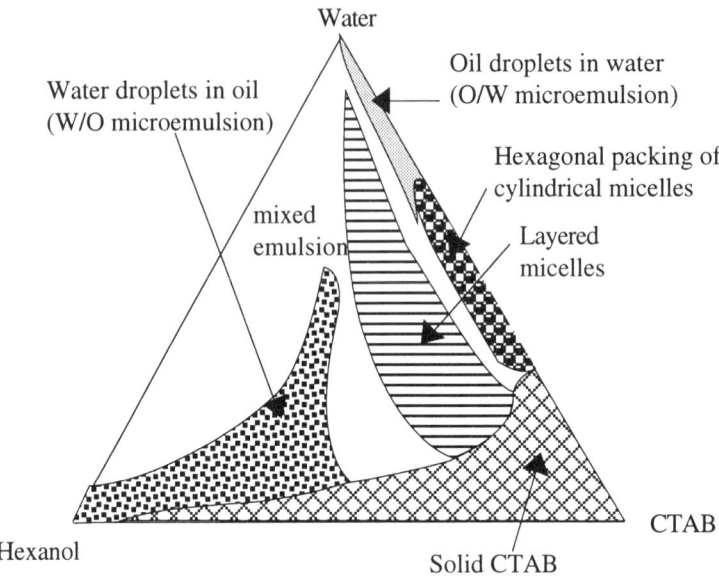

Figure 6.3-15 - Ternary diagram in the system Water - Hexanol - Cetyltrimethylammonium Bromide(CTAB). Adapted from Ekwall et al. [46]

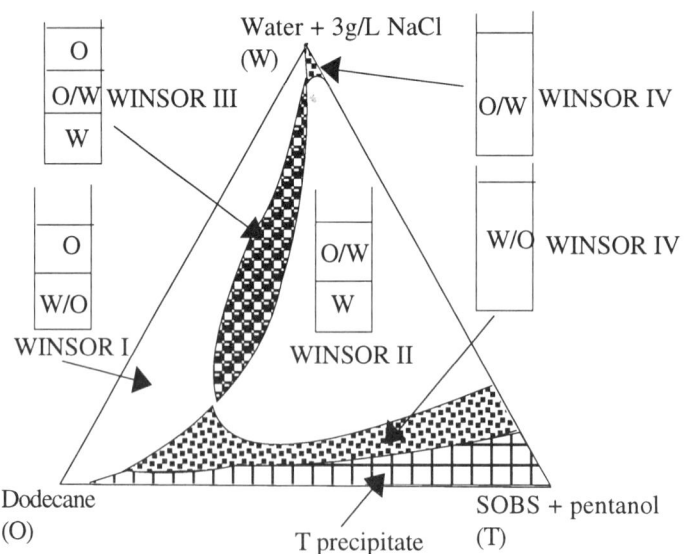

Figure 6.3-16 - Winsor classification in the ternary system Water + NaCl (3g/L) - Dodecane - paraoctylbenzene sodium sulfonate (SOBS) + pentanol. Adapted from Bellocq et al. [47]

Fabrication procedure

According to Beck et al. [50], for a hexagonal pore array for instance, the process first requires to make micellar rods with a surfactant, next to organize the micelles in a hexagonal array and then to add an inorganic precursor solution in a polar solvent which surrounds the micelles on their hydrophilic side. After hydrolysis, a regular array of hollow oxide cylinders is formed, which contains the hydrophobic tail of the micelle inside. These organic heart must finally be eliminated, either by washing in an appropriate solvent or by calcination, which is not an easy operation as this may destroy the oxide porous array.

However, according to several researchers, ordering of the cylindrical micelles before adding the inorganic precursor is not necessary. Silicate, especially, participate in the ordering of the hexagonal array [51] (Figure 6.3-17).

Multidentate binding of cationic surfactant to silicate oligomers favors the silicate polymerization in the interface with the surfactant and these silicate layers themselves induce an ordering of the micelles by the interplay of the classical electrostatic and steric interactions in colloidal materials [52,53] . This polymerization can be done at low temperature as in any sol-gel process, or at a higher temperature by a hydrothermal process, which makes more crystalline stronger oxide channels and facilitates the elimination of the surfactant template. Microwave heating of gels has also be applied with success to make MCM-41 hexagonal molecular sieves [54]

The electrostatic interactions play a major role in the synthesis procedure of mesophase derived ordered mesoporous materials. Cationic surfactants have a positively charged hydrophilic head S^+ which can directly combine with anionic inorganic complexes C^-, according to the reaction

$$S^+ + C^- \Rightarrow S^+ C^- \qquad (6.3\text{-}30)$$

They can also combine with cationic inorganic complexes by the intermediate of negative ion X^- which can be the counterions of the surfactant itself.

$$S^+ + X^- + C^+ \Rightarrow S^+ X^- C^+ \qquad (6.3\text{-}31)$$

In a similar way, anionic surfactants have a negatively charged head S^- which can directly combine with cationic inorganic complexes C^+, according to the reaction

$$S^- + C^+ \Rightarrow S^- C^+ \qquad (6.3\text{-}32)$$

They can also combine with anionic inorganic complexes by the intermediate of positive ions A^+

$$S^- + A^+ + C^- \Rightarrow S^- A^+ C^- \qquad (6.3\text{-}33)$$

A few mesoporous materials made by these techniques according to Corma [2] are gathered in Table 6.3-2. It is also possible to use micelles made by nonionic surfactants such as polyethylene glycol [55] and ethylene glycol hexadecyl [56]. The inorganic component is then linked by hydrogen bonds.

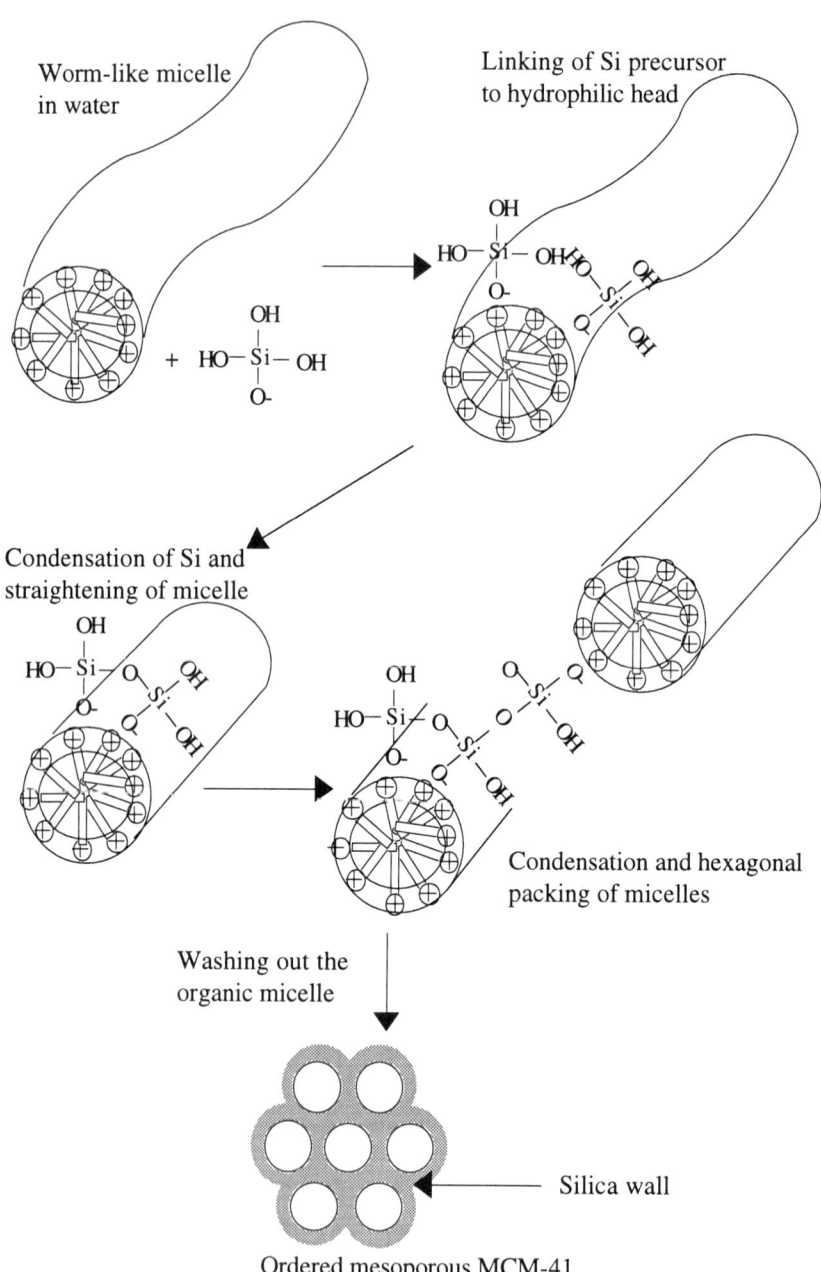

Ordered mesoporous MCM-41

Figure 6.3-17 - Formation mechanism of a MCM 41 . Adapted from Corma [2]

Table 6.3-2 - Ordered mesoporous materials made with surfactant templates. Adapted from Corma [2]

Surfactant type	Inorganic species	Linking type	Examples
Cationic S+	Anionic C-	S+ C-	- Aluminosilicate M41S - Aluminosilicate MCM-48 - Antimony oxide - Tungsten oxide (pH < 7)
Cationic S+	Cationic C+	S+ X- C+	- Silica (pH < 2) - Zinc phosphate (pH < 3)
Anionic S-	Cationic C+	S- C+	- Iron oxide - Lead oxide - Aluminum oxide
Anionic S-	Anionic C-	S-A+ C-	- Zinc oxide (pH > 12.5)

Other related synthesis techniques

It is possible to make a tridimensional but disordered network of worm-like channels with a uniform pore diameter, when combining a surfactant with a hydrothermal treatment, as this was done for sodium silicate and ethylenediaminetetraacetic acid tetrasodium salt [57]. Random branching of the cylindrical channels provide a tridimensional pore network texture very useful for catalytic applications.

The inside diameter of cylindrical channels can be increased, up to ≈ 10 nm, by making microemulsions with an organic liquid, such as 1,3,5-trimethylbenzene (TMB) [50] or alkanes [58] (Figure 6.3-18)

Another original method to synthesize materials with a regular pore texture from silica, consisted in intercalating a surfactant in Kanemite, a layered polysilicate comprising tetrahedral flexible silica sheet. The surfactant folded the silica sheet and made micelles with silica walls coming from the kanemite layers. In this case, a mesoporous silica form termed FSM16, with a tridimensional regular and ordered pore network was obtained [59].

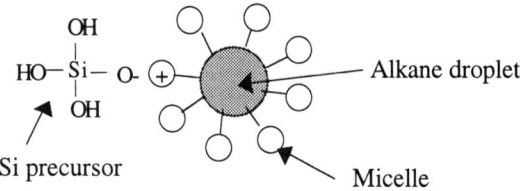

Figure 6.3-18 - Use of microemulsions to increase the diameter of micelles.

Materials composition

Mesoporous oxide materials made from ordered micellar templates have been studied with other oxide compositions than pure silica. In particular MCM-41 and

MCM-48 with walls containing Si and Al have been synthesized. In the best cases, Al could be incorporated in a tetra coordination, [52,60], but the results depended enormously on the synthesis conditions, including the type of precursors used, such as sodium silicate or Al alkoxide. Ti could be incorporated in MCM-41 and MCM-48 structures in tetrahedral coordination, by using a cationic surfactant [61, 62] or a nonionic surfactant [63]. Other metals which could be incorporated in such structures comprise vanadium in MCM-41 [62] and manganese in hexagonal, cubic and planar structures [64], unfortunately Mn atoms migrate out of the walls during calcination.

Pure W, Fe or Pb oxides micellar structures have been studied, but the oxide walls collapsed when eliminating the surfactant [65]. With pure Al_2O_3, lamellar [66] and hexagonal mesophases [67,68] could be made and the mesoporous texture did not collapse when using nonionic surfactants (polyethylene oxide). Cylindrical channels with a regular diameter could be obtained in some cases, but they became worm-like during calcination at 600°C so that the ordered packing was lost [68]. Hudson and Knowles made pure ZrO_2 mesoporous materials with a cationic quaternary ammonium surfactant, but the pores had no ordered packing organization [69]. On the other hand, a hexagonal mesophase was formed with an amphoteric surfactant, the cocamidopropylbetaine (CAPB) [70] or a nonionic amine [71]. With the nonionic surfactant, the mesophase structure changed from lamellar to hexagonal when adding the hydrolysis water. A good acid catalyst could also be obtained when using lauryl sulfate as a surfactant. With TiO_2, a hexagonal mesophase was made with alkyl phosphate surfactants but the structure collapsed when removing the surfactant [72].

6.4 - REFERENCES

1 - Sanchez C., Ribot F., New J. Chem., 18 (1994) 1007-1047.
2 - Corma A., Chem. Rev. 97 (1997) 2373-2419.
3 - Avnir D., Levy D., Reisfeld R., J. Phys. Chem., 88 (1984) 5956
4 - Levy D., Avnir D., J. Phys. chem., 92 (1988) 734.
5 - Levy D., Einhorn S., Avnir D., J. Non-cryst. solids 113 (1989) 137.
6 - Saegusa T., Chujo Y., Polymer prep., 1 (1989) 39
7 - Kure S., Matsuki H., Jordan R., Chujo Y., Saegusa T., Polymer, Prep. Jpn. 39 (1990) 1684
8 - Toki M., Chow T.Y. Ohnaka T., Samura H., Saegusa T., Polymer. Bull., 29 (1992) 653
9 - Ning Y.-P., Mark J.E., Polymer. bull., 12 (1984) 407
10 - Mauritz K.A., J. Appl. polym. Sci., 40 (1990) 1401
11 - Yoldas B.E., J. Mater. Sci. 14 (1979) 1843-1849.
12 - Landry C.J.T., Coltrain B.K., Wesson J.A., Zambulyadis N., Polymer, 33 (1992) 1496
13 - Douy A., Odier P., Mat. Res. Bull., 24 (1989) 1119
14 - Novak B.M., Davies C., Macromolecules 24 91991) 5481
15 - Nishida F., Dunn B., Knobbe E.T., Fuqua P.D., Kaner R.B., Mattes B.R., Mat. Res. Soc. Symp. Proc. 180 (1990) 747

16 - Reisfeld R., Brusilovsky D., Eyal M., Miron E., Bursheim Z., Ivri J., Chem. Phys. Lett.m 160 (1989) 43

17 - Hench L.L., West J.K., Zhu B.F., Ochoa R., in Sol-Gel Optics I, Mackenzie J.D., Ulrich D.R., Eds., Proc. SPIE 1328, SPIE, Washington D.C., 1990, p.230

18 - Ruiz-Hitchky E., Adv. Mater. 5 (1993) 334

19 - Gharbi N., Sanchez C., Livage J., Lemerle J., Nejem L., Lefebvre J., Inorg. Chem., 21 (1982) 2758

20 - Adler D., Henisch H.K., Mott N.F., Rev. Mod. Phys. 50 (1978) 209-220.

21 - Kramer J., Prud'homme R.K., J. Colloid and Interface Sci., 118 (1987) 294

22 - Monchâtre G., L'Actualité Chimique, January 1984, 11

23 - Kiselev V.S., Ermolaeva T.A., J. Appl. Chem. USSR 31 (1958) 100.

24 - Wilkes G.L., Orler B., Huang H.H., Polymer Prep., 26 (1985) 300

25 - Sur G.S., Mark J.E., Eur. polym. J., 21 (1985) 1051

26 - Nandi M., Conklin J.A., Salvati L. Jr;, Sen A., Chem. Mater., 3 (1991) 201.

27 - Sanchez C., Alonso B., Chapusot F., Ribot F., Audebert P., J. Sol-Gel sci. Technol. 2 (1994) 161

28 - Diré S., Babonneau F., Carturan G., Livage J., J. Non-Cryst. Solids, 147/148 (1992) 62.

29 - Diré S., Babonneau F., Sanchez C., Livage J., J. Mater. Chem. 2 (1992) 239.

30 - Schubert U., Arpac E., Glaubitt W., Helmerich A., Chau C., Chem. mater., 3 (1992) 291

31 - Sanchez C., In M., Toledano P., Griesmar P., Mat. Res. Soc. Symp. Proc. 271 (1992) 669

32 - Sanchez C., In M., J. Non Cryst. solids, 147/148 (1992) 1.

33 - Hiemenz P.C., "Principles of Colloid and Surface Science", Marcel Dekker, New-York (1976).

34 - Berthod A., J. Chim Phys. (Fr.), 80 (1983) 407.

35 - Griffin W.C., J. Soc. Cosmetic Chemists, 5 (1954) 249

36 - Perron R., 'Physicochimie des composés amphiphiles", Editions du CNRS, Paris (1979)

37 - Mittal K.L., (Ed.), "Micellization, Solubilization and Microemulsions", 2 vol., Plenum Press, New-York (1977)

38 - Mittal K.L., (Ed.), "Solution Chemistry of Surfactants", 2 vol., Plenum Press, New-York (1979)

39 - Mittal K.L., Fendler E.J., (Eds.), "Solution Behavior of Surfactants", 2 vol., Plenum Press, New-York (1982)

40 - Israelachvili J.N, Mitchell D.J., Ninham B.W., J. Chem. Soc. Far. trans. II, 72 (1976) 1525.

41 - Lindman B., Wennerstrom H., in "Topics in Current Chemistry", Boschke F.L., Ed., Springer-Verlag, Heidelberg, 87 (1980) 1

42 - Krafft F., Wiglow H., Chem. berichte, 28 (1895) 2566

43 - Robb I.D., (Ed.) "Microemulsions", Plenum Press, New-York (1982).

44 - Mukerjee P., in "Solution Chemistry of Surfactants", Mittal K.L., Ed., Plenum Press, New-York, vol. 1 (1979) p. 153.

45 - Winsor P.A., "Solvent Properties of Amphiphilic Molecules", Butterworths, London (1954).

46 - Ekwall P., Mandell L., Fontell K., Mol. Cryst. Liquid, 8 (1969) 157

47 - Bellocq A.M., Bourdon D., Lemanceau B., J. Colloid Interface Sci., 79 (1981) 419

48 - Kresge C.T., Leonowicz M.E., Roth W.J., Vartulli J.C., Beck J.S., Nature 359 (1992) 710

49 - Vartulli J.C., Schmitt K.D., Kresge C.T., Roth W.J., Leonowicz M.E., McCullen S.B., Hellring S.D., Beck J.S., Schlenker J.L, Olson D.H., Sheppard E.W., Stud. Surf. Sci. Catal. 84 (1994) 53.

50 - Beck J.S., Vartulli J.C., Roth W.J., Leonowicz M.E., Kresge C.T., Schmitt K.D., Chu C.T-W., Olson D.H., Sheppard E.W., McCullen S.B., Higgins J.B., Schlenker J.L,J. Am. Chem. Soc., 114 [1992] 10834.

51 - Beck J.S., Vartulli J.C., Kennedy G.J., Kresge C.T., Roth W.J., Schramm S.E., Chem. Mater., 6 (1994) 1816.

52 - Chen C.Y., Burkett S.L., Li H.X., Davis M.E., Microporous Mater. 2 (1993) 27.

53 - Monnier A., Schüth F., Huo Q., Kumar D., Margolese D., Maxwell R.S., Stucky G.D., Krishnamurty M., Petroff P., Firouzi A., Janicke M., Chmelka B.F., Science 261 (1993) 1299.

54 - Arafat A., Janson J.C., Ebaid A.R., van Bekkum H., Zeolites 13 (1993) 162.

55 - Bagshaw S.A., Prouzet S.A., Pinnavaia T.J., Science 269 (1995) 1242.

56 - Attard G.S., Glyde J.C., Goltner C.G., nature 378 (1995) 366.

57 - Ryoo R., Kim J.M., Ko C.H., Shin C.H., J. Phys. Chem., 100 (1996) 17718.

58 - Ulagappan N., Rao C.N.R., J. Chem. Soc. Chem. Commun., (1996) 2759.

59 - Inagaki S., Fukushima Y., Kuroda K., Stud. Surf. Sci. Catal. 84 (1994) 125.

60 - Schmidt R., Junggreen H., stocker M., ellestad O.H., J. Chem. Soc. Chem. Commun., (1996) 875.

61 - Corma A., Navarro M.T., Pérez-Pariente J., J. Chem. Soc. Chem. Commun., (1994) 147.

62 - Zhang W., Pinnavaia T>, J. Catal., Lett., 38 (1996) 261.

63 - Tanev P.T., Chibwe M., Pinnavaia T.J., Nature 368 (194) 321.

64 - Zhao D., Goldfarb D., J. Chem. Soc. Chem. Commun., (1995) 875.

65 - Ciesla U., Demuth D., Leon R., Petroff P., Stucky G., Unger K., Schüth F., J. Chem. soc. Chem. Commun., (1994) 1387

66 - Huo Q., Margolese D.I., Ciesla U., feng P., Sieger P., Leon R., petroff P., Schüth F., Stucky G.D., Nature 368 (1994) 317

67 - Yada M., machida M., kijima T., J. Chcm. Soc. chem. Commun., (1996) 769.

68 - Bagshaw, S.A., Pinavaia T.J., Angew. chem. Int. ed. Engl., 35 (1996) 1102.

69 - Hudson M.J., Knowles J.A., J. mater. Chem., 6 (1996) 89.

70 - Kim A., bruinsma P., chen Y., Wang L.Q., Liu J., Chem. Soc. Chem. Commun., (1997) 161.

71 - Ulagappan N., Battaram N., raju U.N., Rao C.N.R., J. Chem. Soc. Chem. Commun., (1996) 2243.

72 - Antonelli D.M., Ying J.Y., Angew. Chem. Int. Ed. Engl., 34 (1995) 2014.

PHASE TRANSFORMATIONS

7.1 - INTRODUCTION

Once a gel is dried, it is often heated at temperatures much higher than for drying, so that transformation to more stable phases can occur. In many cases, this first involves a chemical evolution, each time a gel does not have the same chemical composition as the desired technical ceramic. Simultaneously, the matter composing the solid network undertakes a crystallographic reorganization and the pore texture changes. Such transformations proceed along kinetic routes which do not usually produce the most stable thermodynamic phases, but metastable ones, during the first steps. In a temperature range from 100°C to 1000°C which is termed intermediate temperature range in the present chapter, thermal treatments often produce either a crystalline transition phase or a glassy phase. These metastable phases eventually transform to a stable thermodynamic phase at high temperatures, which in the present book designates temperatures above 1000°C.

The main steps addressed in this chapter follow closely the chronological evolution of gels during thermal treatment. The intermediate temperature kinetic processes are examined first, in particular those transformations known as topotactic. A special section is devoted to the formation of glassy phases. Stable thermodynamic phases are frequently formed by nucleation and growth from a transition phase, which is the subject of further section. The formation of non-oxide ceramics by high temperature chemical reactions from oxide gels, is examined last.

7.2 - CHEMICAL TRANSFORMATIONS AT INTERMEDIATE TEMPERATURES

In a temperature range which depends on the gel being considered, but which typically goes from 100°C to 300°C, the crystallographic structure of the solid network remains close to what it was in the dry gel state. For instance, titania gels made from alkoxides by Komarneni et al. [1] presented a structure which remained amorphous at 300°C. It was only above 300°C and when observed in a TEM, that very fine anatase particles could be observed. Their size was comprised between 10 and 15 nm at 400°C.

During this first stage, the main transformation of a gel actually consists in a mass loss [2] as shown in figure 7.2-1a, on thermogravimetric data curves. This mass loss keeps occurring with a decreasing magnitude as the temperature increases above 400°C, throughout the intermediate temperature range. As an example dry silica gels made from TEOS typically lose 25 % of their mass during heating up to 900°C [3].

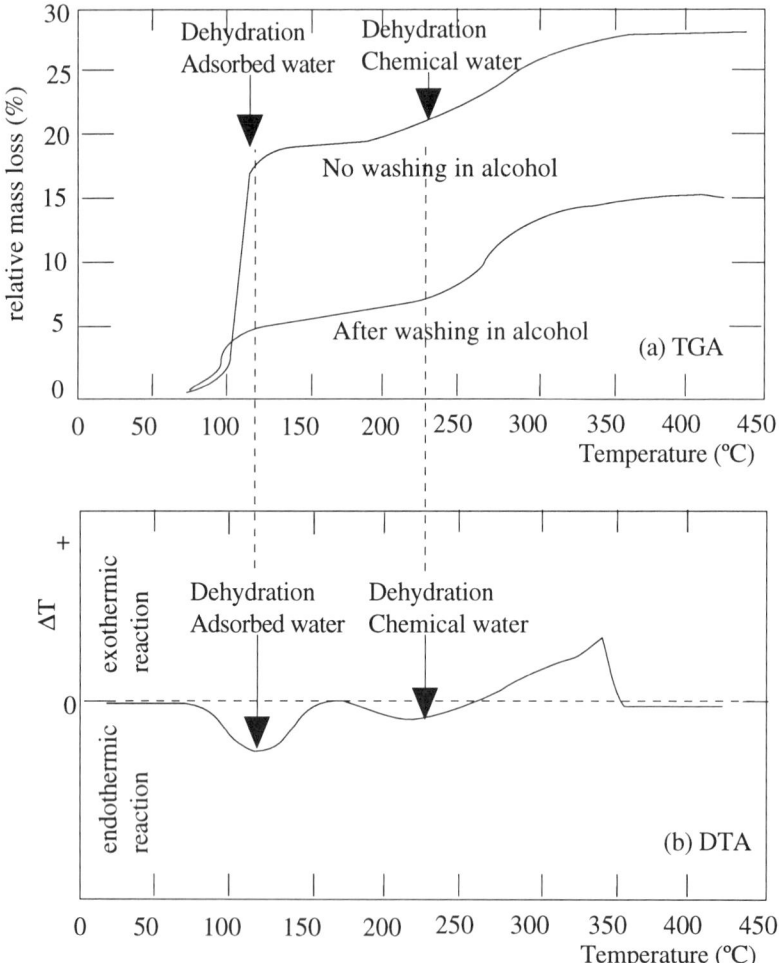

Figure 7.2-1 - Thermogravimetric analysis (TGA) and differential thermal analysis (DTA) of Alkoxy-derived PLZT (lead zirconate titanate) ceramic . Adapted from Mazdiyasni [4].

GEL DEHYDRATION

One of the biggest contributions to the mass loss comes from the departure of the residual water. The latter can be divided in adsorbed water and structural water, also called "chemical water". These two types of dehydration are characterized by endothermic peaks when a material is examined by a differential thermal analysis (DTA), as illustrated in figure 7.2-1b.

These two types of dehydration occur at temperatures which largely depend on the synthesis conditions of a gel. For instance a WO_3 gel prepared by hydrolysis of W ethoxide can behave differently depending on whether it has been prepared at 20°C or 80°C [2]. The gel type obtained at the lower temperature only contains adsorbed water and dehydrates at 200°C. On the other hand, the higher temperature gel type keeps dehydrating up to 310°C because this gel corresponds to the chemical formula $WO_3.H_2O$ and its structural water leaves. The removal of adsorbed water is always easier than that of chemical water. For instance, adsorbed water dehydrates between 25 and 140°C in alumina gels, while the departure of the structural water necessary to form the first transition alumina, requires a temperature of the order of 400°C [5]. Moreover, the relative proportion of adsorbed water to structural water depends on the initial hydrolysis conditions under which a gel was made.

Dehydration of adsorbed water.

To some extent, the dehydration of adsorbed water can be considered as a continuation of the drying process examined in chapter 5. If the process is not properly monitored, it can also break a sample. However, the origin of cracking is not always the same as during drying. This is due to the fact that the porous texture begins to change and in particular some pores can close and prevent the desorbed water vapor to escape. To obtain uncracked ceramic samples, it is necessary to eliminate first the water adsorbed on the micropores walls before these pores can close. This can be done by heating a sample under vacuum, a technique which made it possible to synthesize $SiO_2-P_2O_5$, SiO_2-TiO_2 and $SiO_2-B_2O_3$ glasses with less than 10 ppm of residual water [6]. If the residual water is not eliminated before pore closure, a gel may not only fracture but it may also bloat [7].

The desorption of the chemical species adsorbed onto the pore walls is not limited only to water, it also concerns all types of solvents.

Dehydration of structural water

The structural water comprises OH and H_2O groups which belong to the crystallographic structure of the gel solid network phase. Therefore, their departure already corresponds to a chemical phase transformation of a gel and this induces a gel volume contraction. Moreover, the chemical reactions corresponding to the departure of structural water are also responsible for transformation of a gel network.

Example: Chemical dehydration of boehmite

The departure of OH groups from a structure is well documented in the transformation of boehmite gels to transition alumina (figure 7.2-2) [8]. In a first stage, these gels produce a transition alumina phase known as γ alumina, according to the reaction:

$$2[\gamma\,AlO(OH)] \Rightarrow \gamma\,Al_2O_3 + H_2O \qquad (7.2\text{-}1)$$

Boehmite is composed of octahedral oxygen layers. The bonding between two layers consists of hydrogen bonds, by the intermediate of hydrogen atoms located between the terminal oxygen atoms of two boehmite layers. The chemical water of this monohydroxide is made by combination of one terminal (OH) group on top of a crystallographic boehmite layer, with one H atom from a neighbor (OH) group .

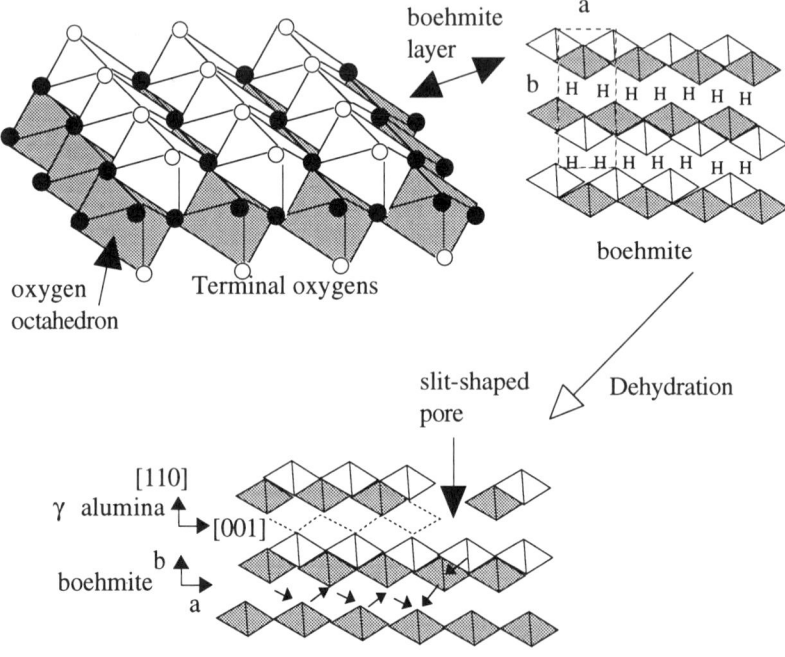

Figure 7.2-2 - Dehydration and topotactic transformation of boehmite to transition aluminas. Adapted from Yang et al. [9].

The chemical water dehydration of boehmite gel is similar to the chemical water dehydration of crystalline boehmite which was studied by Wilson and Stacey [10]. These authors showed that a boehmite monocrystal remained a monocrystal during the formation of transition γ-alumina. Moreover, even in a boehmite monocrystal without pores, some pores are created by departure of the chemical water. The newly formed pores are ordered in an array of slits perpendicular to the (020) crystallographic planes of the former boehmite. Their initial width is ≈ 1.8 nm and their spacing ≈ 4 nm. The main difference in a gel, is that the boehmite microcrystallites have a size ≈ 3nm, that is to say not very different from the distance between slit shaped pores formed in monocrystalline boehmite. Hence, an ordered array of these thin slit shape pores cannot be observed in sol-gel γ-alumina. In any case, micropores are formed by departure of chemical water from boehmite gel and this is not a typical transformation of gels as it occurs in any low temperature phases such as most hydroxides. Similar behaviors were observed in other materials with a crystallographic layered structure such as $MoO_3.H_2O$ [11].

OTHER CHEMICAL TRANSFORMATIONS

Residual organics

Another important contribution to the chemical evolution of a gel during its transformation at intermediate temperatures, concerns the residual organic groups which come either from an organic solvent or from the precursors [12,13]. These residual OR can either hydrolyze at intermediate temperatures and be spontaneously eliminated as organic molecules such as alcohol, or they can remain in the pores of the gel network. In the latter case, they can be decomposed inside a gel and leave a carbon residue which gives a black color to the material. This occurs in particular with alkoxide mixtures hydrolyzed in under-stoichiometric conditions, that is to say by water vapor in a humid atmosphere, or with a water proportion r_w lower than required for complete hydrolysis, for instance $r_w < 2$ in the fabrication of silica fibers [14].

This phenomenon is not limited to silica but extends to other oxides such as TiO_2 [15]. In this oxide, the decomposition of organics is a strongly exothermic reaction which leads to ignition near 400°C. It also leads to the production of water, but the strong exothermic effect drastically contrasts with endothermic dehydration reactions usually observed between 130 and 160 °C [16]. If a TiO_2 gel is washed with NH_4OH before heat treatment, the exothermic effect occurs at a lower temperature (355°C) and ammonia NH_3 is produced by the decomposition reactions, instead of water [17]. In sol-gel titanate powders made by Mazdiyasni [4], the carbonization temperature of residual organics also occurs near 400°C. In the latter materials, the mass decreases first by ≈ 15 % between 70 and 110°C by dehydration, then it decreases by ≈ 5% by departure of alcohol if the material has been washed with isopropanol (figure 7.2-1). In lanthanum doped barium hexaferrites made by Higuchi et al. [18], the endothermic dehydration occurs up to 180°C while the exothermic oxidation of the residual organics occurs near 280°C. In ZrO_2 gels made from Zr propoxide in cyclohexane, the residual organics volatilize at $T \approx$ 170°C [19].

Exactly as with water, the decomposition of residual organics can be responsible for the formation of micropores. The departure of all these ligands can be followed with various techniques such as infra-red and Raman spectroscopy [20,21].

Residual anions

In addition to residual organics, the metallic salts also decompose. For instance, α-Zr phosphate desintegrates to give ZrP_2O_7 by heating in air at 300°C [22]. In particular, nitrate salts complexed with organic groups or by ammonia often lead to explosive reactions.

All these chemical reactions have in common with each other to induce a first structural modification to the solid network of a gel, which consists mainly in the formation of micropores.

CHEMICAL TRANSFORMATIONS OF NON-OXIDE GELS
Chemical reactions similar to those in oxide gels, occur in non-oxide gels.

Silazanes
Concerning the silazanes, which are compounds of a metal with nitrogen and hydrogen, heat treatment under N_2 liberates NH_3 below 400°C and complex compounds of the type $HNCH_3(SiH_2NCH_3)_xH$ above 400°C [23]. The nitride Si_3N_4 forms according to a relatively complex and not well understood reaction [24],

According to Von Glemser and Naumann [25], the decomposition of $Si(NH)_2$ made by reaction of $SiCl_4$ with liquid NH_3, follows the succession of transformations:

$$6[Si(NH)_2]_x \quad \underset{\Rightarrow}{\overset{400°C}{}} \quad 2[Si_3(NH)_3N_2]_x + 2x\ NH_3 \quad \underset{\Rightarrow}{\overset{650°C}{}}$$

$$3[Si_2(NH)N_2]_x + 3x\ NH_3 \quad \underset{\Rightarrow}{\overset{1250°C}{}} \quad 2x\ \alpha\text{-}Si_3N_4 + 4x\ NH_3 \quad (7.2\text{-}2)$$

With the exception of $\alpha\text{-}Si_3N_4$, all intermediate products are polymeric, amorphous and very hygroscopic; any contact with water must be avoided.

A global reaction which summarizes the transformation of silazanes is [24] :

$$4(H_2SiNH) \Rightarrow 6H_2 + Si + Si_3N_4. \qquad (7.2\text{-}3)$$

Silanes
In the synthesis of SiC from polymeric gels, CH_4 is the gas which leaves during heat treatment [26]. The pyrolysis of polysilanes proceeds according to a two-step transformation [27] :

$$\Rightarrow \quad \beta\ SiC + CH_4 + N_2 \qquad (7.2\text{-}4)$$

GEL NETWORK TRANSFORMATION

Network consolidation
The kinetic transformations at intermediate temperatures do not only consist in the formation of micropores by chemical decomposition. Some concurrent phenomena tend to reduce these micropores.

One of these phenomena consists in a relaxation of the polymeric units which compose a gel network. This results in a continuous shrinkage of the solid material and permits more residual OH to come in contact with each other and to establish new metaloxane M-O-M bonds by condensation reactions. The network is therefore reinforced and densification begins according to a process related to viscous flow, from which it differs by the important role played by the chemical reactions.

Chemical condensation reactions between residual OH groups are important preliminary steps in the evolution of a gel towards a glassy or a crystalline phase [28]. In a first step, they are responsible for a consolidation of the gel network with new active bonds forming.

The structure of a gel can also slowly transform by dissolution-recrystallization, in the aqueous medium coming from its own dehydration water, adsorbed as well as chemical. Actually, heating a gel in the water coming from its own chemical water favors a faster crystallization. In the case of boron, this dehydration water even favors a reversal of the condensation-polymerization reactions and it withdraws part of the boron initially incorporated in the gel network. For instance, studies in the Al_2O_3-SiO_2-B_2O_3-Na_2O system have shown that $\equiv$Si-O-Si$\equiv$ and $\equiv$Si-O-B= bonds became predominant while =B-O-B= disappeared progressively, simultaneously with a decrease of the residual water content, during heat treatment [29].

Pore texture evolution
Overall, the specific surface area of sol-gel materials decreases during heat treatment at intermediate temperatures. In the lowest temperature range, which depends on the material, this decrease is usually moderate [30] and it accelerates as the temperature increases. Simultaneously, the grain size only grows moderately, so that the crystalline topology of the gel is roughly maintained. The new phases which are appear are formed by transformations known as topotactic transformations and described in the next section.

The pore transformations are well known in the case of transition alumina derived from boehmite. In an alumina grain obtained from monocrystalline boehmite, we previously indicated that the pores are slit shaped and crystallographically well oriented. They slowly merge to form planar hexagonal shallow pores with a hexagon diameter $\approx$ 6 nm in δ-alumina , but they remain oriented parallel to the former (010) boehmite planes. During heat treatments at higher temperature ($\approx$ 900°C), these shallow pores slowly merge into bigger faceted hexagonal planar pores, with a diagonal length several tens of nm long, in θ-alumina (figure 7.2-3). Simultaneously, these pores progressively close [10] (figure 7.2-4).

In the transition aluminas made from boehmite gels, the slit shaped pores initially formed during the departure of chemical water, are not organized in a parallel array because of the random aspect of the gel network. However, merging into shallow extended pores does occur when transforming to θ-alumina.

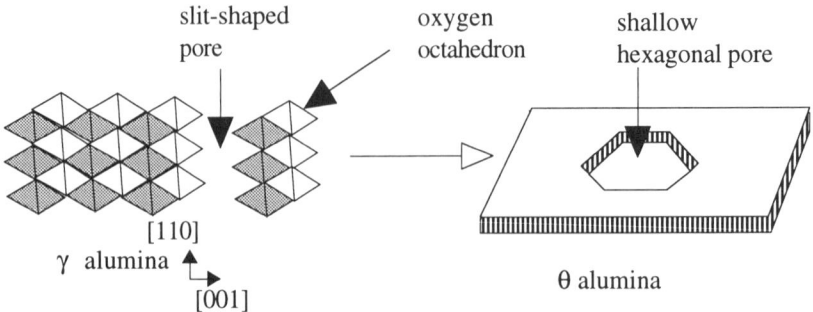

Figure 7.2-3 - Transformation of the pores during the topotactic transformation of γ-alumina to θ-alumina. Adapted from Yang et al. [9].

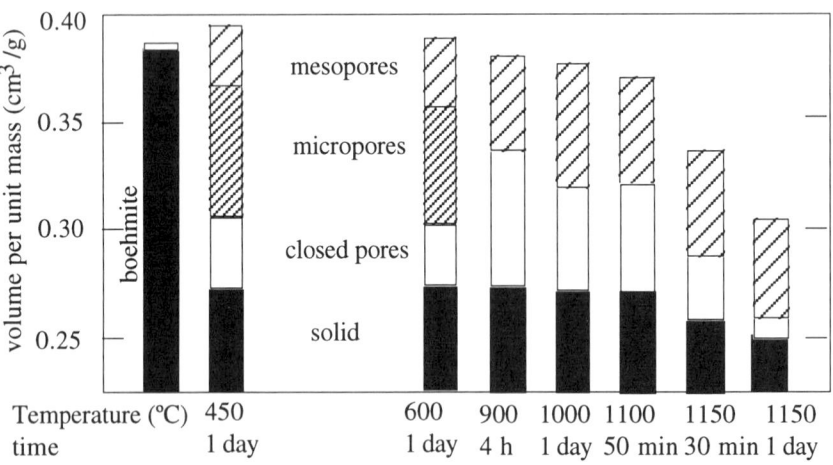

Figure 7.2-4 - Pore distribution in transition aluminas derived from boehmite. After Wilson and Stacey [10].

7.3 - TOPOTACTIC CRYSTALLIZATION

CHARACTERISTICS OF TOPOTACTIC TRANSFORMATIONS

At intermediate temperatures, that is to say in a temperature range which depends on the material but which does not exceed 1000°C, the formation of new phases does not always occur by nucleation and growth, for kinetics reasons. This is particularly true for the most stable thermodynamic phase, when nucleation requires a motion of both the cations and the anions. This impossibility is due to the big size of anions such as oxygen anions which require a high activation energy to move. Such an energy of thermal nature is not available because the temperature is too low.

New crystalline phases can still form, but they do so in maintaining the same relative crystallographic positions of the anions as in the initial phase, that is to say in the initial gel when a material is synthesized by sol-gel process [2]. Such phenomena are not limited to gels however; they are quite common for the transformation at intermediate temperatures of all low temperature stable or metastable phases, such as the hydroxides.

Because they do not affect the relative near neighbor packing of the anions, these transformations are named topotatctic transformations. Roughly, they keep the initial crystalline organization of the initial phase in terms of grains; a monocrystal remains a monocrystal.

In many cases, the new phases which form by topotactic transformation are not the most stable thermodynamic phases. They must be more stable than the initial phase, at a given temperature, but they can be metastable. This explains that a succession of topotactic transformations can be observed as the temperature is raised. More stable phases progressively appear, until the most stable thermodynamic phase is definitely formed. The intermediate phases are termed transition phases. Of course, the sequences of topotactic transformations and intermediate phases formed depends on the initial gel structure, hence on the chemical recipe used in sol-gel processing. The situation for alumina for instance was summarized by Gitzen [31] and a variety of transition aluminas, including rare transition phases difficult to reproduce, have been reported.

It may even occur that the anionic crystallographic organization in a gel is close to that in the most stable phase. In this case the stable crystallographic phase is directly obtained by topotactic transformation such as in the transformation of diaspore α-AlO(OH) to α-Al$_2$O$_3$.

A common characteristics of topotactic transformations is that they are progressive. Hence no sharp transition between the successive phases can be noticed, such as a well defined transition temperature. They are better described in "Transformation, Time, Temperature" diagrams known as TTT diagrams, where the time necessary to achieve a given percentage of phase transformation, at a given temperature, is reported on a graph. An example is given in the next section, for the transition aluminas.

EXAMPLE: THE FORMATION OF TRANSITION ALUMINAS

TTT Diagram of the most common transition aluminas

In the case of alumina, the γ-alumina formed by dehydration of boehmite transforms successively, during heat treatments at increasing temperatures, to other crystalline transition aluminas known as δ-Al_2O_3 at approximately 600°C, and θ-Al_2O_3 at 900°C [8]. These are the most well known transition phases but other more exotic ones have been reported. As mentioned previously, the temperatures reported for these phase transformations provide a broad indication of the temperature range where they usually occur. Because of their kinetic nature, these topotactic transformations are best described in TTT diagrams and such as that established for boehmite monocrystals , that is to say not sol-gel made boehmite, by Wilson and Mac Connell [32] and reported in figure 7.3-1.

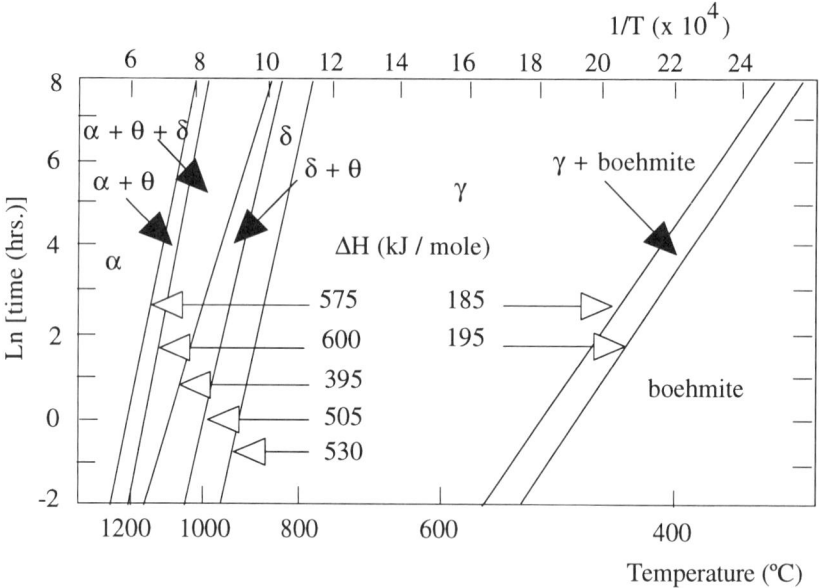

Figure 7.3-1 - Arrhenius plot of the TTT diagram of transition aluminas formed from boehmite. The present phases were determined by X-ray diffraction. Adapted from Wilson and Mac Connell [32].

Mechansim of the topotactic transformations in transition aluminas

The oxygen anions in the transition aluminas as well as in boehmite, have a packing order which is related to the face centered cubic (f.c.c) structure, while the oxygen packing in the stable alumina phase, α-alumina, is hexagonal closed packed (h.c.p). Hence, the thermodynamically stable alumina only form at high temperature, when enough thermal energy is available to make the reconstruction of the oxygen packing, possible. On the other hand, the transition aluminas which appear after dehydration of boehmite are derived from this monohydroxide by topotactic transformation. This consists in a progressive deformation of the oxygen packing, where only the Al cations can diffuse from interstitial sites to other interstitial sites. Crystallographically, γ-Al_2O_3 presents a defect spinel structure with a tetragonal distortion, where vacancies are predominantly distributed on the tetrahedral cation sites. In a first step, the evolution from γ to δ alumina consists in a random diffusion of the Al cations so that the cation vacancies tend to be distributed at random on all the interstitial sites, octahedral and tetrahedral [32]. As a result, the unit cell $\frac{c}{a}$ ratio tends to come closer to the value of 1, which is the value for a cubic material. That is to say, the initial tetragonal distortion of the γ phase attenuates; a increases and c decreases. However, the cubic structure is never achieved, because a concurrent redistribution of the vacancies on the octahedral sites only tends to increase again the tetragonal character of the transition alumina. In the end, a tripling of the unit cell occurs along the c axis when δ-alumina is formed. Actually, there is also no clear cut distinction between γ and δ-Al_2O_3 as a progressive transition occurs. At any intermediate stage during heat treatment, the oxide cannot really be identified as being γ or δ-alumina. The X-ray diffraction lines can be assigned by reference to the two idealized structures of γ-Al_2O_3 and δ-Al_2O_3, but this is only for commodity reasons.

The transition from the δ to the θ phase also consists in a motion of the Al cations only, but this is typically an order-disorder one. The cation vacancies, which are distributed at random on the octahedral interstitial sites in γ-alumina, diffuse so as to adopt an order distribution in these octahedral interstices. When this evolution is terminated, θ-alumina is formed. However, at any intermediate stage during heat treatment, the oxide cannot really be identified as being δ or θ-alumina.

During both the γ to δ and the δ to θ alumina transformations, the relative position of the oxygen anions with respect to each other are not affected. Nonetheless, because the Al cations move and as discussed previously, the pore texture changes progressively, from a fine slit shape to an extended shallow hexagonal one. In the transition aluminas made from boehmite gels, such shallow extended pores still forms simultaneously with θ-alumina.

DIVERSITY OF TRANSITION PHASES

Importance of the sol-gel chemistry

The preceding example for alumina provides an explanation to the fact that, for each initial gel structure, the topotactic transformations can follow a different path. As a result, depending on the sol-gel chemical process chosen, which includes the nature of the precursor and their hydrolysis-polymerization conditions, different metastable transition phases can be formed.

These considerations are quite general for all sol-gel materials. For instance, depending on the hydrolysis temperature, it is possible distinguish a high temperature WO_3 gel form, from a lowest temperature form. The first one is crystalline and transforms to a cubic phase between 350°C and 500°C, while the second one converts to an orthorombic phase [2].

With TiO_2, two different crystalline forms can be directly generated from the gel: rutile and anatase; the phase which is obtained depends on the sol-gel chemical recipe. A high dilution favors the crystallization of anatase at $T \approx 150°C$ just after dehydration [15,17]. The rutile form was obtained by Sakka [14] in TiO_2 gels which crystallized at 200°C, and also by Barringer and Bowen [16]. This crystalline variety transformed to anatase at 800°C.

Example of zirconia

In the case of zirconia, the crystalline tetragonal $_tZrO_2$ phase is known to be thermodynamically stable only above 1100°C. However it can be directly obtained at 300°C by crystallization of amorphous gels [30] and it recrystallizes at 400°C to the monoclinic phase which is stable at this temperature. As mentioned in a preceding chapter, zirconia gels do not consist of a hydroxide but of a hydrated oxide $ZrO_2.xH_2O$, in which the OH groups occupy terminal sites. The transformation of this gel to $_tZrO_2$ is so exothermic that it produces in a glow phenomenon [33-35]. The reason for the formation of this phase instead of $_mZrO_2$, is controversial. However, it is probably a consequence of the initial gel structure which evolves by topotactic transformation. Dissolution-recrystallization in the water liberated by dehydration, according to a hydrothermal process, would lead directly to the stable phase. The $t \Rightarrow m$ transformation is itself known to be of kinetic nature. This is a cooperative and sudden packing reorganization of the same type as martensitic transformations. For such a transformation to occur, one often considers that the grain size must be large enough [36]. However, it is probable that anion impurities [37] or more generally the molecular complexes configuration in the sol-gel chemistry, play an important role [38,39].

Topotactic transformations provide an explanation why different studies on a same compound lead to the production of different crystalline phases. In general, a different chemical recipe was followed in each study, so that the initial gel molecular structure was different and it transformed along a different topotactic path.

For instance, zirconia gels synthesized by Debsikdar [40] remained amorphous up to 300°C. Tetragonal crystallites with a size ≈ 3.5 nm begun to appear at $T \approx 350°C$, and were fully observed at 400°C where only the tetragonal phase was present. At higher temperatures, the grain size stabilized at $\approx 11.5nm$. The monoclinic phase began to appear at 500°C but remained mixed with the tetragonal one up to 600°C, which indicates that the transformatoin was not martensitic in this case. The stable thermodynamic phase diagram was only recovered at 800°C, where only the monoclinic phase was present. During further thermal treatment, heating as well as cooling, the material then obeyed the usual martensitic process. Similar results were found with ZrO_2 gels made from Zr propoxide in cyclohexane by Kindu and Ganguli [19]. On the other hand, the following succession of transformations, where a cubic phase first formed, was found by Yoldas [41]:

$$ZrO_2 \text{ gel} \Rightarrow {}_cZrO_2 \quad \Rightarrow \quad {}_mZrO_2 \quad \Rightarrow \quad {}_tZrO_2 \qquad (7.3\text{-}1)$$
$$500\text{-}1000°C \qquad 1300\text{-}1400°C$$

where $_cZrO_2$ designates a cubic ZrO_2 phase. The variability in the formation of the tetragonal phase is therefore very large, and the pH during the sol-gel preparation can be crucial as a report indicates that tetragonal zirconia could be directly crystallized from a gel in the two pH ranges from 3 to 4 and 13 to 14 [42].

Case of silica

Silica appears as an exception amongst the oxide gels as it does not crystallize until a high temperature. It first gives a glass at temperatures which are termed intermediate in this book.

Glass formation actually is a special topotactic transformation which is examined in the next section. The ease with which an element can form a glass network seems to be approximately the same by sol-gel as by liquid quenching. It is limited to a few elements known as "glass formers", including Si, Ge and B. This is true even with non-oxide ceramics such as Si_3N_4. The silazane compounds of the type $Si(NH)_2$ such as made from silicon precursors in anhydrous liquid ammonia [43,44], can be heated up to 1200°C in a suitable container without decomposing, and when pyrolyzed they give an amorphous Si_3N_4 powder.

Multicomponent oxides

To synthesize a multicomponent oxides, several different cations must be mixed. The temperature where the crystallization first occurs depends on the composition. For instance, gels in the system $Cr_2O_3.Fe_2O_3$ can begin to crystallize in a temperature range from 395°C to 500°C depending on the composition [45]. The lowest cryztallization temperature also depends on the chemical process.

Lower transformation temperatures are provided by double precursors, such as a double hydroxide $2Mg(OH)_2.Al(OH)_3$ [45] or a double alkoxide of Mg and Al [46] which crystallize to spinel at 400°C. The phase $LiAl_5O_8$ can only be crystallized from a double alkoxide of Li and Al, it is not feasible by conventional processing [47]. In these cases, the anionic packing in the gel is already very close to that of this crystalline phase and the distribution of the cations is as in the crystalline phase so that they do not have to diffuse. Hence, the complex crystalline phase can directly form by topotactic transformation.

The variety of phases which can be made is larger than with the monocomponent oxides. Very intricate metastable phase diagrams can often be established, where the transformation temperatures are approximate, especially when a phase forms by topotactic transformation. An example of such a phase diagram concerning the system {Ba;Ti;O} is shown in figure 7.3-2. It comprises several metastable phases which can be synthesized by topotactic transformation from gels made in different conditions. The stability of a phase is of kinetic nature, it is enhanced when a high energy barrier must be overcome to nucleate the stable thermodynamic phase. This is in particular the case of $BaTi_5O_{11}$ and $BaTi_2O_5$. The orthorhombic phase Ba_2TiO_4 made by sol-gel is metastable because of its small crystallite size which makes its disappearance by martensitic transformation difficult [48]. The temperature where these crystalline phases appear correspond to temperature ranges where the different cations can begin to diffuse significantly, while the oxygens are relatively immobile. The compound with a molar ratio Ba/Ti = 1/3 is already crystallized in the gel. The compounds with Ba/Ti molar ratios of 2/1, 1/2, and 1/5 crystallize at T ≈ 700°C.

Those with molar ratios 4/13 and 1/4 crystallize respectively at 780°C and 900°C and those with Ba/Ti molar ratios of 6/17 and 2/9 at T ≈ 1100°C .

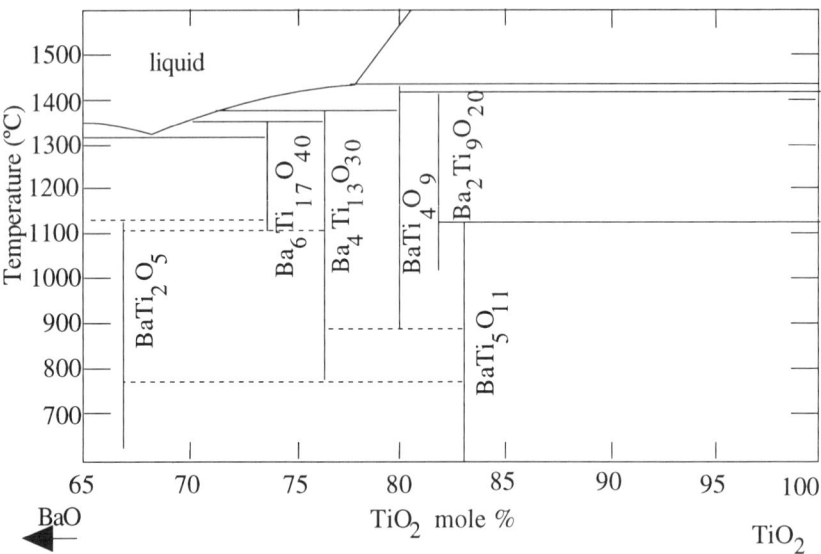

Figure 7.3-2 - Phase diagrams of the compounds obtained in the BaO-TiO_2 system by mixed alkoxide hydrolysis. Adapted from Ritter et al. [48].

The synthesis of several other titanates have been investigated by the sol-gel technique. Some have been directly produced from gel monoliths, such as $PbTiO_3$ made from a double alkoxide of Pb and Ti [49]. The crystallization of this lead titanate occurred at 480°C and it shrunk by up to 80% at 600°C.

For multicomponent as well as for single component oxides, a consequence of crystallization by topotactic transformations is that the crystallite size, which is small in the gels, grows only moderately in the crystalline phases. Hence, the specific surface area remains high. For instance, in the ferrites $Ba_{1-x}La_xFe_{12}O_{19}$ prepared in a liquid medium, the topotactic crystallization is due to a distribution of Fe^{3+} both in tetrahedral and octahedral oxygen interstices, as in transition aluminas [18]. The crystallites grow moderately, from an average size ≈ 10 nm in the gel to ≈ 80 nm at 800°C. Growth accelerates at higher temperatures, after crystallization by nucleation and growth, which is an exothermic process. The crystallites reach an average size ≈ 500 nm at 1000°C. The transition phases often make it possible to achieve exceptional physical properties which are consequences of a small grain size. The ferrites mentioned previously are "superparamagnetic" at 800°C and their coercive strength is 400 kA m^{-1} for x=0.2 .

For many complex oxides often produced as powders, the phase transformation mechanisms have not been studied with as much details as in alumina, hence they remain poorly known [30]. However, it seems realistic to admit that their transformation mechanisms are of a topotactic nature, as this was proposed for HfO_2-Y_2O_3 powder [50], based on the fact that its specific surface area decreased from 210 m^2/g in the initial state to 130 m^2/g at 400°C and then remained stable up to 750°C. In this material, the structure evolution was attributed to a process of alignment of faulted lattice portions, similar to what happens in transition aluminas. The variety of intermediate phases which are feasible by sol- gel is not limited to alkoxides, it concerns all precursors such as the metal salts. For instance, thermal treatments below 1000°C of carbonate solid solutions made from mixed nitrate solutions, made it possible to synthesize the following phases [51]:
- $Mn_{1-x}M_xO$ monoxides with the rock salt structure
 - Perovskite structures with compositions Ca_2FeCoO_5, $Ca_2Co_2O_5$ and $Ca_3Fe_2MnO_8$, in which oxygen vacancies are organized in superstructures.
- $Ca_2Fe_{2-x}Mn_xO_{6-y}$ compounds where the transition metals coexist in three different coordination polyhedra (octahedra, tetrahedra and square pyramidal). One of these compound, $Ca_2Mn_2O_5$ could only be prepared by sol-gel processing.
The compounds $LaCoO_3$ and $LaNiO_3$, made from the corresponding hydroxides, must also be mentioned. The first one transforms by topotactic transformation to a $La_2Co_2O_5$ phase where the oxygen vacancies constitute a superstructure. Also, $Mo_{1-x}W_xO_3$ compounds can be made from aqueous ammonia mixed solutions of MoO_3 and CoO_3. The topotactic transformation which occurs during dehydration is similar to that in boehmite. By reduction in H_2 at 750°C, these mixed oxides give Mo-W solid solutions.

7.4 - GLASS FORMATION

STRUCTURAL CHARACTERISTIC DIFFERENCES BETWEEN A POLYMERIC GEL AND A GLASS.

Gels and the traditional definition of glasses
Glasses are metastable materials. The traditional method to make them is by melting a material and quenching the melt sufficiently fast to not let enough time for the nucleation and growth of crystals to occur. In this way, the random topology of the liquid is maintained when a solid, named a glass forms, so that it is possible to consider that a glass is traditionally obtained by topotactic transformation from a liquid.
Since the existence of glasses is related to crystallization kinetics, the observation of an amorphous X-ray pattern is not sufficient to define whether a solid is a glass.

Other techniques of kinetic type must be used, as a material can be amorphous without being a glass. The two techniques used to experimentally define a glass are:
- measuring the viscosity when cooling a melt. When the viscosity reaches a value of 10^{12} Pl (Poiseuille or Nms^{-2}) and the material is still X-ray amorphous, this is a glass. The temperature where this viscosity is reached is termed the glass transition temperature T_g.
- measuring the transformation heat involved during successive heating and cooling by differential thermal analysis (DTA), when starting from a solid material not made by melt quenching. A glass transition is characterized at the temperature T_g by the existence of a small endothermic inflexion as illustrated in figure 7.4-1. If an amorphous solid material produced by other techniques than melt quenching does not show his DTA characteristics, this is not a glass.
In particular, it is only possible to say that a gel has transformed to a glass when such observations are satisfied.

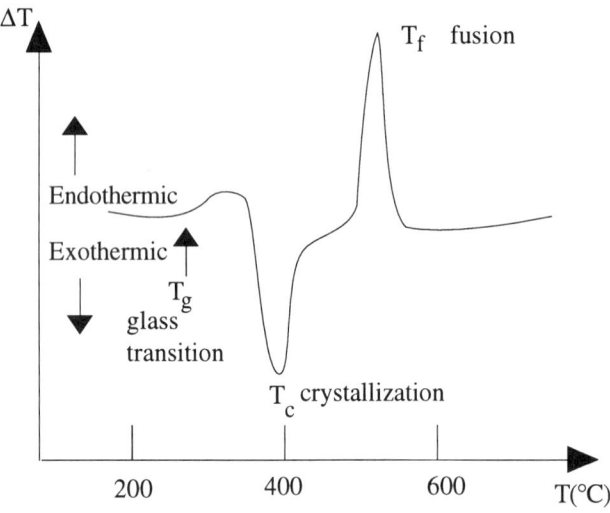

Figure 7.4-1 - Differential thermal analysis of a mixed Sb_2O_3-$PbCl_2$ glass. Adapted from Dubois [52].

X-ray diffraction
The transformation of a gel to a glass can be described by reference to the X-ray diffraction patterns in figure 7.4-2, which concern gels of composition SiO_2-10%TiO_2 [21].
A large diffuse band is present at all temperatures, near a diffraction angle $2\theta \approx 20$ degrees (Cu Kα radiation). The maximum intensity of this diffused band is slightly displaced towards smaller angles during heat treatment but it keeps approximately the same width. The permanency of this broad band, which is similar to the band

observed in glasses made by melt quenching, indicates that the same lack of long range order prevails in the Si-O-Si lattice of a gel and of a glass.

The main difference between a polymeric gel and a glass is the marked central diffusion peak which is observable below 700°C in figure 7.4-2. It is less noticeable at 800°C and it is absent at 900°C. The presence of this central diffusion peak is due to the presence of micro and mesopores which are abundant in a gel network, while such pores are absent in a glass.

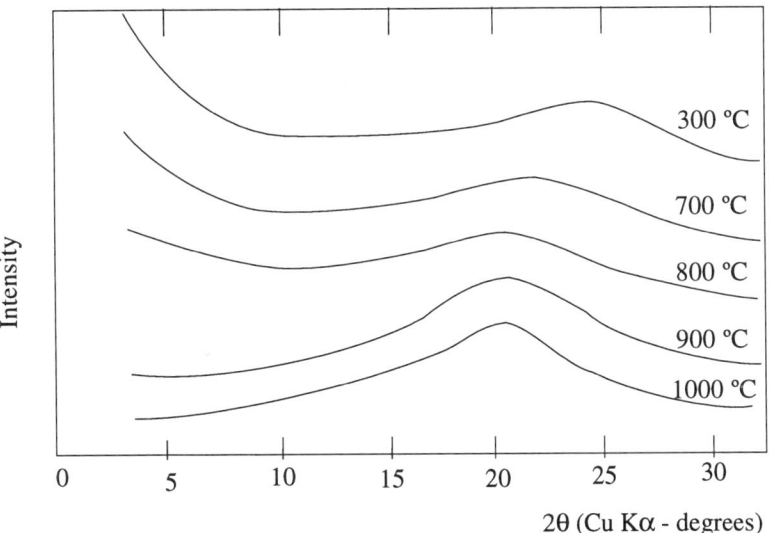

Figure 7.4-2 - X-ray diffraction patterns of 10% TiO$_2$ silica gels, after heat treatment at increasing temperatures. The samples have been taken out of the furnace right after the furnace temperature reached the set point. Adapted from Sakka [14].

Hence, the X-ray diffraction patterns indicate that the material in figure 7.4-1 is still a gel at 700°C, while it has transformed to a glass at 900°C.

GLASS FORMATION MECHANISM FROM POLYMERIC GELS

The transformation of a polymeric gel to a glass below the glass transition temperature T$_g$ can be described as a topotactic transformation. To be able to do so, the random network of a glass must already pre-exist in the gel, which is satisfied in silica polymeric gels where the random angular configuration of Si-O-Si polymers is similar to that in a glass. As mentioned previously, the structure of a polymeric gel differs from that of a glass because of the existence of a fine scale and high volume porosity. To transform a polymeric gel to a glass, it is simply necessary to densify the material before it can crystallize [53].

Polymeric silica gels which most easily make glasses below T_g are obtained with an acid catalyst. Their porosity is at an atomic scale and their polymeric molecular conformation can be very close to that of the glass network. They transform to silica glass between 600°C and 800°C [54,55]. The hydrolysis water ratio r_w is important as a low ratio favors the formation of linear polymers which are well suited to the synthesis of fibers and can be transformed to glass without any special precaution to avoid crystallization [14], contrary to the gels which are extensively crosslinked [56]. Pressure can also be used to help making a glass by direct conversion of a gel below T_g [28]. The parent gel structure, as well as its thermal history, determine the feasibility of a glass from a gel, below T_g. It would not be possible to make a glass below the glass transition temperature T_g from a crystalline phase. Hence a colloidal gel, even with very small crystallites could not be transformed to a glass without melting it first. For multicomponent silicate, the quality of a gel in terms of homogeneity determines the glass quality after transformation, as this has been shown in the systems TiO_2-SiO_2 [57,58] and SiO_2-B_2O_3-Al_2O_3-Na_2O-BaO [59].

A glass forms after a chemical evolution similar to that summarized in section 7.2. Exactly as in the transformation of a gel to a crystalline phase, the departure of residual organics and of chemical and adsorbed water, occurs first. Micropores are formed but they are largely balanced by the relaxation of polymer chains which makes a gel shrink, induces additional condensation reactions and creates new Si-O-Si bonds [14].

Infrared spectroscopy analysis of OH groups show that the major part of the residual water is removed below 250°C. However, some residual water still remains above this temperature and disappears during successive steps at increasing temperatures: 250°C, 800°C and even 1000°C. At each level where the residual water content decreases significantly, it is necessary to slow down the heating rate sufficiently so as to avoid microcracking or foaming, especially to synthesize a monolith.

In the case of silica xerogels, another aspect of the structural evolution observed during glass formation is that tricyclic siloxane appear in increasing proportion onto the gel surface, during heating up to 650°C [60]. This is due to the introduction of the surface silanols oxygen atoms in the gel network. However, these cyclic species are submitted to strong strains because the corresponding Si-O-Si bond angle is low. Hence, they tend to disappear at a higher temperature to reach a level comparable with that of fused silica near 900°C. Overall, the material structure as well as its porous texture change continuously and progressively even when the temperature is well below T_g and the viscosity in excess of 10^{13} poises.

GLASS FORMATION FROM GELS ABOVE T_g

Glasses can be obtained from gels, including colloidal gels, by sintering or hot pressing powders between T_g and the melting temperature T_m or by fusion above T_m [61].

Sintering between T_g and T_m, proceeds by viscous flow in the same time as residual dehydration [24,62,63]. Since the transformation temperature is higher than in the transformation of polymeric gels to glasses below T_g, it is important to choose a heat treatment schedule so as to avoid crystallization [64,65]. This depends on the system because the glass transition temperature T_g depends on the composition. For instance $T_g \approx 800°C$ in the system $SiO_2\text{-}B_2O_3$ and $\approx 1000°C$ in $SiO_2\text{-}P_2O_5$. In some systems such as $SiO_2\text{-}B_2O_3\text{-}P_2O_5$, crystallization of a BPO_4 phase is difficult to avoid [66] and the help of pressure is useful to convert such gels to glasses between T_g and T_m. The conversion to glasses also depends on the chemical processing. For instance, in the systems SiO_2, $La_2O_3\text{-}SiO_2$ and $B_2O_3\text{-}SiO_2$, gels made directly from alkoxides in organic solvents are known to crystallize less easily than gels made by aging hydrosols [63]. In the case of silica for instance, this is due to the fact that colloidal gels are made with a base catalyst and the size of the colloidal particles as well as their apparent density is variable [54,55]. Hence, the relative importance of viscous flow by comparison with dehydration is variable. In addition, residual OH catalyze the conversion of glasses to glass-ceramics. Drying also is important; monolithic gels dried by the supercritical method begin to shrink at a lower temperature than xerogels, because they have a different porosity.

If necessary, gels can also be melted to synthesize glasses, as in conventional glass processing [67]. Often, fusion produces a higher degree of chemical homogeneity and requires a shorter melting time with gels than with conventional powders [68-70].

GLASS COMPOSITIONS STUDIED BY SOL-GEL

A wide variety of glasses were synthesized by sol-gel. Roy [69] as well as MacCarthy et al. [71] examined the silicates and alumino-silicates containing Li, Na, K, Rb, Mg, Ca, Sr, Ba, Pb, Ga, Fe, Ln, Ti, Zr and Th, while Mukherjee et al. [67] examined binary and ternary systems formed with La, Al, Si and Zr. A first listing of binary and ternary silicate glass systems was gathered by Zarzycki et al. [72]. Other compositions include tungsten and titanium oxides [73], or boron as in the ternary system $SiO_2\text{-}B_2O_3\text{-}PbO$ [74]. Very complex compositions have also been studied. Some of them were made by mixing salts and alkoxides as in the system $SiO_2\text{-}B_2O_3\text{-}Al_2O_3\text{-}BaO\text{-}Na_2O$ [75,76]. A non-limitative list of the glass systems investigated by sol-gel is reported in table 7.4-1, it includes some glass-ceramics [46,77,78].

DIFFERENCES BETWEEN CONVENTIONAL GLASSES AND SOL-GEL GLASSES

Based on measurements of the oxygen deficiency in silica, it has been suggested that glasses made from gels below T_g are slightly different from conventional glasses, possibly because the slower transformation kinetics at these temperatures do not make it possible to reach exactly the atomic configuration as in glasses made above T_g [1]. However, data on the refractive index, density and thermal expansion coefficient, have shown no significant differences between glasses made by these two synthesis techniques, for a given composition [14]. All these data on the physical properties of glasses actually reflect an average state for an average composition; they do not depend on a local states.

Table 7.4-1 - Systems for which the gel to glass transition was studied.

Systems	References	Systems	References
Binary silicates		*Ternary silicates*	
SiO_2-B_2O_3	[63,66,77,79-82]	SiO_2-Al_2O_3-Na_2O	[104]
SiO_2-Al_2O_3	[64,83,84]	SiO_2-Al_2O_3-Li_2O	[77]
SiO_2-TiO_2	[58,64,85-88]	SiO_2-Al_2O_3-La_2O_3	[67]
SiO_2-ZrO_2	[84,89]	SiO_2-Al_2O_3-K_2O	[105]
SiO_2-SrO	[88,90]	SiO_2-Al_2O_3-Li_2O	[77]
SiO_2-SrO_2	[89]	SiO_2-Al_2O_3-MgO	[78]
SiO_2-CaO	[91]	SiO_2-Al_2O_3-ZrO_2	[106]
SiO_2-Na_2O	[65,77,92-96]	SiO_2-La_2O_3-ZrO_2	[67]
SiO_2-P_2O_5	[66,97,98]	SiO_2-B_2O_3-Na_2O	[68,77,107]
SiO_2-Fe_2O_3	[99,100]	SiO_2-B_2O_3-P_2O_5	[66]
SiO_2-La_2O_3	[63,67,101,102]	SiO_2-B_2O_3-PbO	[74]
SiO_2-oxynitride	[103]	SiO_2-ZrO_2-Na_2O	[84]
		SiO_2-TiO_2-La_2O_3	[67]
Higher order silicate systems		References	
SiO_2-Al_2O_3-BaO-B_2O_3		[108]	
SiO_2-Al_2O_3-B_2O_3-K_2O-Na_2O		[46]	
oxynitride glasses in the system			
SiO_2-Al_2O_3-BaO-B_2O_3-Na_2O		[75,107, 109-111]	
Systems other than silicates			
B_2O_3-P_2O_5-Na_2O		[112]	
BeO-CeO_2-ZrO_2		[113]	
TiO_2-W_2O_3		[73]	

On the other hand, kinetic properties such as the glass transition temperature and the viscosity at the softening temperature depend on local atomic configurations and they are significantly modified in sol-gel glasses; T_g is for instance slightly lower in glasses derived from gels, than in conventional glasses. Such properties are consistent with the existence of a residual water proportion of 0.01 to 0.1 % by mass. This residual water also explains a decrease by $\approx$ 100°C of the upper miscibility temperature, in the SiO_2-Na_2O system and the formation of a coarser heterogeneous texture [86,94]. The rates of nucleation and growth of crystalline phases are higher in gel-derived glasses, again because of residual OH and also because of a more uniform cation distribution in the cases where several cations are mixed [67,101].

Sol-gel processes are particularly interesting for compositions difficult to melt, such as in the systems TiO_2-SiO_2, Al_2O_3-SiO_2 and ZrO_2-SiO_2 [84], or where phase separation easily occurs during quenching, as in CaO-SiO_2 [91] or SrO-SiO_2 [88,90]. Sol-gel processes make it possible to bypass the difficulty to achieve a high cooling rate, as is necessary when homogeneous glasses are made by melting inside a miscibility gap. New compositions displaying new properties could be made by sol-gel, such as ZrO_2-SiO_2 with up to 48 % (by mass) ZrO_2, or Na_2O-ZrO_2-SiO_2 glasses with up to 33% ZrO_2 which show an improved alkaline resistance [89]. New oxynitride glasses could also be made; they show a softening temperature, a density, a refractive index, an elastic modulus and a hardness superior to those of simple oxide glasses and their thermal expansion is lower [103]. Nitrides made by conventional melting are prone to decompose and the oxide to be reduced. The amonolysis of sol-gel oxides makes it possible to circumvent this problem.

However, sol-gel processes have not made it possible to realize new single oxide component glasses, by comparison with conventional melting. This would tend to indicate that the phases transformation kinetics achieved with the sol-gel processes do not differ much from those feasible by melt quenching.

7.5 - CRYSTALLIZATION BY NUCLEATION AND GROWTH

ATOMIC DIFFUSION, THERMODYNAMICS AND NUCLEATION AND GROWTH

When gels are heat treated at increasing temperatures, the crystallization path sooner or later converges towards the most stable thermodynamic state. The stable diagram is reached at a temperature which depends on the type of ceramic and sol-gel process selected. It may be a high temperature, as in the synthesis of mullite by Mazdiyasni and Brown [114], where the X-rays diffraction lines of mullite appeared progressively between 1200°C and 1700°C. It can be a much lower temperature as in the zirconia synthesized by Mazdiyasni [30], where the stable monoclinic phase starts to form at 400°C, or as in the WO_3 made by Yamaguchi et al. [2] where the stable orthorhombic phase forms near 500°C. Once the stable thermodynamic state diagram has been recovered, the phase transitions observed during further temperature cycling are the same as in conventional processing, that is to say for WO_3:

$$
\begin{array}{ccccc}
 & 300°C & & 710\text{-}740°C & \\
\text{monoclinic} & \Leftarrow & \text{orthorhombic} & \Leftarrow \Rightarrow & \text{tetragonal} \\
 & & 645\text{-}675°C & &
\end{array} \qquad (7.5\text{-}1)
$$

In the cases where the stable crystalline phase first appears at a relatively low temperature, the formation mechanism is not always well established; it can very well be a topotactic transformation, as in the stabilized zirconia synthesized by Mazdiyasni [30] mentioned previously, where the metastable tetragonal phase transforms to the stable monoclinic phase at 400°C, while the grains remain smaller than 5 nm. This may also the case in a synthesis of complex titanates known as PLZT, which crystallized at 340°C while significant grain growth only begun at 500°C [115].

The mechanism of nucleation and growth of a new crystalline phase is usually considered to operate at a higher temperatures than those reported just before, for instance above 1000°C in the formation of α-Al_2O_3 from the transition θ-Al_2O_3 [116]. The crystalline phase α-alumina differs from the transition phases because it requires a change in the oxygen packing, from face centered cubic to hexagonal close-packing. When the oxygen anions are enough mobile for such a reorganization to become possible, α-alumina grains grow quickly at the expense of a previous population of small θ-alumina crystallites. The grain growth kinetics is so fast that each new α-alumina grain entraps a population of mesopores which were present in the previous gel network and which form a star-like network inside each α-alumina grain; a phenomenon which is consistent with a possible paracrystalline transition alumina structure already mentioned [9].

VARIATION OF THE SPECIFIC SURFACE AREA DURING NUCLEATION AND GROWTH

The pore size of sol-gel ceramics is relatively stable up to a critical temperature where it begins to drastically increase. This critical temperature largely depends on the material's nature; according to figure 7.5-1 it is respectively $\approx$ 500°C, 700°C and 1000°C for TiO_2, ZrO_2 and Al_2O_3.

Simultaneously, the specific surface area decreases moderately up to this critical temperature where it becomes much lower. A few specific surface area values at intermediate temperatures are given in Table 7.5-1. The value for silica aerogels is still high at 500°C, but it falls drastically at higher temperature where Al_2O_3 resists much better than ZrO_2.

On the other hand, the formation of a new phase by nucleation and growth from a sol-gel material is frequently preceded or occurs simultaneously with a drastic decrease of the specific surface area. For instance in a study in the [Ba,Ti,O] system, the specific surface area decreased from 350 m²/g to 11 m²/g when the phase $BaTi_5O_{11}$ crystallizes [48]. Simultaneously the crystallite size increased from 15 nm to 0.15 mm. During this evolution, the shrinkage is usually important: it reaches about 40% when α-alumina forms at 1200°C [5]. This indicates that the atomic mobility which makes the new nucleation and growth of a new phase possible, can enhance the coalescence of pores in a first stage, until a fast grain growth process can start to operate.

The close relationship between crystallization, the decrease of the specific surface area and the diffusion kinetics of atoms is well illustrated in the case of mullite. If conventional, hence relatively big particles of each oxide or hydroxide component are mixed with each other, crystallization of mullite does not occur before 1200°C as

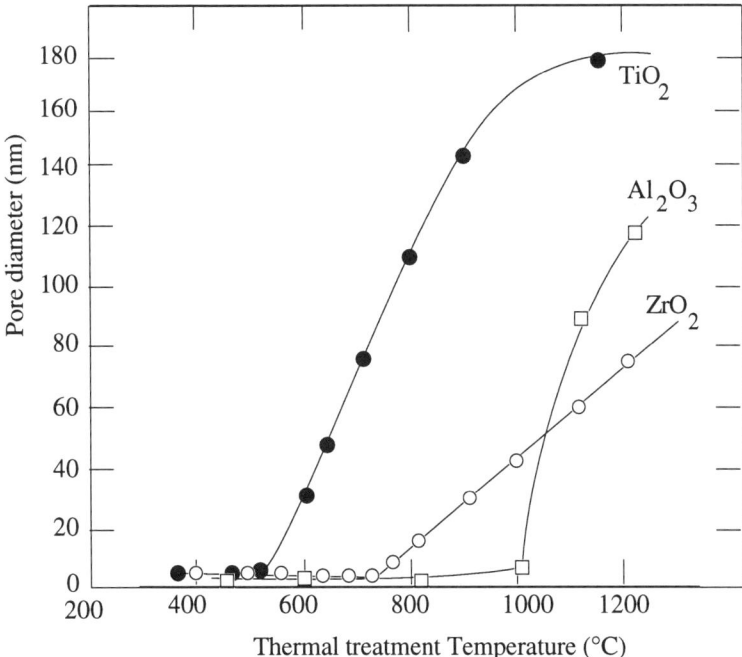

Figure 7.5-1 - Evolution of the pore size as a function of the temperature in TiO$_2$, ZrO$_2$ and Al$_2$O$_3$ xerogels. Adapted from After Larbot et al. [117].

Table 7.5-1 - Specific surface area of a few aerogels at intermediate temperatures. Adapted from Teichner et al. [118].

Aerogel	Temperature in air (°C)	specific surface area (m^2/g)
SiO$_2$	20	860
	300	1004
	500	800
Al$_2$O$_3$	20	464
	400	405
	600	239
ZrO$_2$	20	317
	300	254
	600	38

shown by Ghate et al. [119], because the distance on which Al and Si atoms must diffuse is large. On the other hand, the same material made by sol-gel by Barrett et al. crystallized at 850°C [120]. In another study on mullite made from mixed alkoxides, Mazdiyasni and Brown showed that the material remained amorphous up to 435°C and that the specific surface area, initially 550 m^2/g, decreased by a factor of 2 after one hour heat treatment at 600°C while crystallization kept progressing [114].

CRYSTALLIZATION OF GLASSES

Glasses crystallize by nucleation and growth, although there is some difference between the crystallization kinetics of glasses made by sol-gel glasses and those made by conventional processing [61,67,94,101]. As for the phase separation of a glass in two glassy phases with different compositions, the rates of nucleation and growth are higher for the crystallization of glasses made from gels than for conventional glasses. This was observed for instance in glasses of the systems SiO_2-La_2O_3 [101], Li_2O-$2SiO_2$ and Na_2O-$2CaO$-$3SiO_2$ [121]. Such a behavior can be partly attributed to the residual water which lowers the energy barrier to nucleation and growth. The homogeneity in the cation distribution is also important as polymeric gels crystallize at a higher temperature than colloidal gels, because they have less heterogeneity centers which can act as nucleation centers [122].

The formation of glass-ceramics from gels has been studied by many researchers, in particular Dislich [46], McCarthy et al. [123], Decottignies et al. [62,63], Gottardi et al. [28] and Sakka [14]. In the binary silicate systems, such as in Al_2O_3-SiO_2, TiO_2-SiO_2 and ZrO_2-SiO_2, the crystallization temperature decreases as the SiO_2 content decreases [21,89,124].

Moreover, the crystalline form depends on the composition [14]. For instance, a rutile structure is obtained for less than 10% TiO_2 and a metastable anatase type structure for a higher TiO_2 content, up to 90%. A gel with 3 % ZrO_2 by mass remains amorphous up to 1000°C and the $ZrSiO_4$ phase which appears for higher ZrO_2 compositions does not crystallize. Similarly, a silicate gel with 10% Al_2O_3 is still amorphous at 1400°C, while for an Al_2O_3 content ≈ 70% mullite crystallizes, and for content higher than 80% a crystalline phase related to the transition aluminas forms. In the SrO-SiO_2 system, $SrSiO_3$ or $SrSiO_4$ can crystallize, depending on the composition [125,126].

The silicate characteristics which is to remain amorphous up to a high temperature, is true for the non-oxide compounds of silicon. Amorphous Si_3N_4 made from $Si(NH)_2$ [43,44], crystallizes only at 1200°C to α-Si_3N_4 needles, at 1300°C to platelike particles and finally above 1450°C to β-Si_3N_4. Also, the latter phase only crystallizes at 1500°C in gels made by ammonolysis of $SiCl_4$.

Similarly, SiC fibers can be synthesized from polycarbosilane [14,127,128]. Their fabrication requires first to partially polymerize dodecamethyl cyclohexasilane in an autoclave at 400°C. The polymers with a molar mass higher than 1500 can be extracted by dissolution in benzene. This solution is then gelled and converted to SiC fibers by heat treatment at 1000°C under vacuum. A β-SiC fine grains phase crystallizes in the temperature range 1200°-1500°C.

7.6 - CONVERSION OF OXIDES TO NON-OXIDES

As mentioned in chapter 2, one method to synthesize ceramics which are not oxides by sol-gel consists in incorporating a source of the desire anion other than oxygen in an oxide sol and to favor the formation of the desired phase by chemical reaction at high temperature.

CARBIDES

The above principle is easy to apply to synthesize carbides by incorporation of a carbon source in an oxide sol. The carbon source can be for instance sucrose which yields $\approx$ 30% of its mass as carbon, petroleum pitch in solution in toluene or a phenolic resin in solution in acetone which yield $\approx$ 50% carbon. During heat treatment above 500°C in an inert atmosphere such as argon, the carbon source is pyrolyzed and a mixed silica-carbon material where each component is in a finely divided state, is produced [76]. The final carbide is formed by carbothermal reduction at high temperature, T $\approx$ 1600°C. The carbides which have been synthesized in this way comprise SiC powders made from colloidal silica gel where the colloidal particles were coated with some carbon derived from an organic precursor in solution [129]. Titanium carbide can be made in the same manner from a titanium oxide gel, according to a simple reaction which can be written [130]:

$$TiO_2 + 3C \Rightarrow TiC + 3\ CO \qquad (7.6-1)$$

It might seem preferable to start from organosilsesquioxanes gels containing -RSiO- chains to synthesize SiC. However these compounds do not form any SiC during pyrolysis under nitrogen at 900°C, in spite of an important mass-loss between 300 and 400 °C [131]. A mixture of carbon and silica can be obtained by calcination in an oxidizing atmosphere, then SiC is formed as previously according to the following reaction which does not occur below 1400°C.

$$SiO_2 + 3C \Rightarrow SiC + 2\ CO \qquad (7.6-2)$$

The amount of carbon incorporated as carbide does not increase with the alkyl chain length R. Rather, it increases when R is replaced by unsaturated aromatic groups such as the phenyls. The extent of crosslinking has an important influence since it conditions the number of Si-O-Si bonds which must be cut according to the reaction (7.6-2) [23,132].

OXYNITRIDES AND NITRIDES

Oxynitride glasses can be made by direct thermal treatment of porous oxide gels in a flow of ammonia gas. Aluminosilicate gels have been treated in this way, and only a small quantity of nitrogen was incorporated [110]. In silica gel fibers the nitrogen uptake did not exceed 0.4 % by mass when the fibers were made from TEOS [102] but it increased to 6% when replacing TEOS by the methylated derivative $CH_3Si(OC_2H_5)_3$. With the latter precursor, nitrogen started to incorporate in the material at 500°C, and the proportion increased with time and temperature up to 900°C where it reached its maximum value. At higher temperatures, the amount incorporated decreased again. Infra-red spectroscopy and X-rays fluorescence analysis indicated that nitrogen was incorporated as amine groups $Si-NH_2$ at low temperature and it partially replaced oxygen at 1000°C to build nitride bridges of the type :

$$\equiv Si-N-Si \equiv$$
$$\overset{|}{\underset{|||}{Si}}$$

(7.6-3)

This rather modest nitride formation comes from a concurrent crystallization reaction of silica to cristoballite [133] which begins at 1250°C, where cristobalite transforms partly to Si_2ON_2. On the other hand, at 1450°C, only cristobalite is formed and it does not transform back to an oxynitride compound. The presence of CO mixed in the ammonia somewhat lowers the onset temperature where nitrogen incorporates in the silica network and forms seeds of Si_2ON_2 .

A more efficient nitridation method consists in treating under nitrogen, a gel which contains a carbon precursor. In this way, Si_3N_4 can be synthesized by heat treating a SiO_2 gel which contains some elementary carbon under nitrogen at 1400°C [134]. Carbon reduces first SiO_2 to SiO, which reacts with N_2 to give Si_3N_4 according to the carbothermal reduction reaction:

$$3\ SiO_2 + 3C + 2\ N_2 \Rightarrow Si_3N_4 + 3\ CO_2$$

(7.6-4)

At lower temperature, in a range from 550°C to 1250°C, a mixture of Si_3N_4, SiC, C, SiO_2 and Si is obtained. This material is so amorphous that sometimes it is devoid of any broad X-ray diffraction line [135]. The same technique also makes it possible to produce titanium nitride according to the reaction [130]:

$$TiO_2 + 2C + 0.5\ N_2 \Rightarrow TiN + 2\ CO$$

(7.6-5)

BORIDES
Carbothermal reduction reactions can be used to produce borides. For instance, a convenient way to synthesize TiB_2 consists in transforming a $TiO_2.B_2O_3$ mixed oxide gel which contains the appropriate carbon content to completely combine with the oxygen atoms of the oxide. By heat treatment between 800 and 1500°C, the following carbothermal reduction occurs [130] :

$$TiO_2.B_2O_3 + 5C \Rightarrow TiB_2 + 5\ CO$$

(7.6-6)

Instead of carbothermal reduction, aluminothermy can also be used. This technique consists in incorporating a fine grain aluminum powder in a sol, before gelation. During heat-treatment at a temperature between 800 and 1600°C, the following reduction reaction occurs:

$$TiO_2.B_2O_3 + \frac{10}{3} Al \Rightarrow TiB_2 + \frac{5}{3} Al_2O_3 \qquad (7.6\text{-}7)$$

The resulting material is a TiB_2/Al_2O_3 composite where the segregation of an Al_3Ti compound can be avoided if a homogeneous dispersion of Ti and B has been achieved at the atomic level .

7.7 - REFERENCES

1 - Komarneni S., Roy R., Breval E., J. Am. Ceram. Soc., 68 (1985) C41-C42.

2 - Yamaguchi O., Tomihisa D., Kawabata H., Shimizu K., J. Am. Ceram. Soc 70 (1987) C94-C96.

3 - Klein L.C., Garvey G.J., J. of Non-crystalline solids 48 (1982) 97-104.

4 - Mazdiyasni K.S., Am. Ceram. Soc. Bull. 63 (1984) 591-594.

5 - Lanutti J.J., Clark D.E., Mat. Res. Soc. Symp. Proceed. 32 (1984) 369-381.

6 - Jabra R., Phalippou J., Zarzycki J., J. Non-crystalline Solids 42 (1980) 489-498.

7 - Mukherjee S.P., Debsikdar J.C., Am. Ceram. Soc. Bull. 62 (1983) 413, communication 33-G-83.

8 - Pierre A.C., Uhlmann D.R., Mat. Res. Soc. Symp. Proc. 32 (1984) 119-124.

9 - Yang X., Pierre A.C., Uhlmann D.R., J. of Non- Crystalline Solids 100 (1988) 371-377.

10 - Wilson S.J., Stacey M.H., J. of Colloid and Interface Science 82 (1981) 507-517.

11 - Günter J.R., J. Solid State Chem. 5 (1972) 354-359.

12 - Yamane M., Inoue S., Yasumori A., J. Non-Cryst. Solids 63 (1984) 13-21.

13 - Brinker C.J., Keefer K.D., Schaeffer D.W., Assink R.A., Kay B.D., Ashley C.S., J. Non-Cryst. Solids, 63 (1984) 45-59.

14 - Sakka S, "Gel method for making glass", in "Treatise on Materials Science and Technology", Edited by M. Tomazawa and R.H. Doremus., Vol 22 (1982) 129-167.

15 - Yoldas B.E., J. of Mat. sciences 21 (1986) 1087-1092.

16 - Barringer E.A., Bowen H.K., Com. American Ceram Soc. (1982) C199-201.

17 - Karkamar B., Ganguli D., Trans. Indian Ceram. Soc., 44 (1985) 10-14.

18 - Higuchi K., Naka S., Hirano S. S., Advanced Ceramic Materials 1 (1986) 104-107.

19 - Kindu D., Ganguli D., J. of Materials sciences Letters 5 (1986) 293-295.

20 - Bertoluzza A., Fagnano C., Morelli M.A., Ciamiclen G., Gottardi V., Guglielmi M., J. Non-Cryst. Solids 48 (1982) 117-128.

21 - Kamiya K., Sakka S., Mizutani M., Yogyo Kyokaishi 86 (1978) 552-559.

22 - Vance E.R., Ahmad F.J., Mat. Res. Soc. Symp. Proc., 15 (1983) 105-112.

23 - Seyferth D., Wiseman G.H., "Silazane precursors to Silicon Nitride", in "Ultrastructure Processing of ceramics, Glasses and Composites", Ed. Hench L.L., Ulrich D.R., Wiley, New-York (1984) 265-271.

24 - Wills R.R., Markle R.A., Mukherjee S.P., Am. ceram. Soc. Bull., 62 (1983) 904-911.

25 - Von Glemser O., Naumann P., Z. Anorg. Allg. Chem. 298 (1959) 134-151.

26 - Weber W.P., "Photolysis of Polysilanes", in "Ultrastructure Processing of ceramics, Glasses and Composites", Ed. Hench L.L., D.R.Ulrich D.R., Wiley, New-York (1984) 292-306.

27 - Hasegawa Y., Limura M., Yajima S., J. Mater. Sci. 15 (1980) 720-728.

28 - Dislich H., J. of Non-Crystalline Solids 57 (1983) 371-388.

29 - Liu X., Wang Y., J. of Non-Crystalline Solids 80 (1986) 564-570.

30 - Mazdiyasni K.S., Ceramics International 8 (1982) 42-56.

31 - Gitzen W.H., "Alumina as a ceramic material", the American Society, Columbus, Ohio, 1970.

32 - Wilson S.J., Mc Connell J.D.C., J. of Solid State Chemistry 34 (1980) 315-322.

33- Livage J., "Nature and Thermal evolution of Amorphous Zirconium Oxide Hydrate", Proc. of the 6th International Symposium on the Reactivity of Solids, J.W. Mitchell Ed., Wiley-Interscience (1969) 271-280

34 - Vicarini M.A., Nicolaon G.A., Teichner S.J.; Bulletin Soc. Chim. Fr. (1970) 1651-1664.

35 - Blesa M.A., Maroto A.J.G., Passagio S.I., Figliolia N.E., Rigotti G., J. of Mater. Sciences 20 (1985) 4601-4609.

36 - Garvie R.C., J. Phys. Chem. 69 (1965) 1238-1243.

37 - Whitney E.D., Trans. Faraday Soc., 61 (1965) 1991-2000.

38 - Livage J., Doi K., Mazieres C., J. Am. Ceram. Soc. 51 (1968) 349-353.

39 - Tani E., Yoshimura M., Somiya S., J. Am. Ceram. Soc. 66 (1983) 11-14.

40 - Debsikdar J.C., J. of Non-Crystalline Solids 87 (1986) 343-349.

41 - Yoldas B.E., J. of Mat. Sciences 21 (1986) 1080-1086.

42 - Davis B.H., J. Am. Ceram. Soc. 67 (1984) C168.

43 - Billy M., Ann. Chimie (Fr.) 4 (1959) 795-851.

44 - Mazdiyasni K.S., Cooke C.M., J. Am. Ceram. Soc. 56 (1973) 628-633.

45 - Zelinski B.J.J., Uhlmann D.R., J. Phys. Chem. Solids 45 (1984) 1069-1090.

46 - Dislich H., Angew. Chem. Internat. Edit. Engl. 10 (1971) 363-370.

47 - Naka S., Suwa Y., Abstr. Ann. Meet. Jpn. Ceram. Soc. (1972), p. 81. reference quoted by Sakka S., in "Gel method for making glass", in "Treatise on Materials Science and Technology", Edited by M. Tomazawa and R.H. Doremus., Vol 22 (1982) 129-167.

48 - Ritter J.J., Roth R.S., Blendell J.E., J. Am. Ceram. Soc. 69 (1986) 155-162.

49 - Blum J.B., Gurkovich S.R., J. of Mater. Sciences 20 (1985) 4479-4483.

50 - Rau R.C., J. Amer. Ceram. Soc. 47 (1964) 179-184.

51 - Rao C.N.R., Gopalakrishnan J., Vidyasagar K., Ganguli A.K., J. Mater. res. 1 (1986) 280-294.

52 - Dubois B., Doctorate Thesis, University of Bordeaux I, France (1984).

53 - Zarzycki J., J. Non-crystalline Solids 48 (1982) 105-116.

54 - Nogami M., Moriya Y., Yogyo-Kyokai-Shi 87 (1979) 37-42.

55 - Nogami M., Moriya Y., J. Non-Cryst. Solids 37 (1980) 191-201.

56 - Brinker C.J., Keefer D.W., Schaefer D.W., Ashley C.S., J. of Non-Crystalline Solids 48 (1982) 47-64.

57 - Yoldas B.E., J. Mater. Sci. 12 (1977) 1203-1208.

58 - Yoldas B.E., J. Non-crystalline Solids 38-39 (1980) 81-86.

59 - Brinker C.J., Mukherjee S.P., Conf. Glass Through Chemical Processing, March 19-21 (1980) Rutgers University, USA. Reference given by Dislich H., in "Glassy and Crystalline systems from gels: chemical basis and technical application", J. of Non-Crystalline Solids 57 (1983) 371-388.

60 - Brinker C.J., Bunker B.C., Tallant D.R., Ward K.J., J. de Chimie Physique, 83 (1986) 851-858.

61 - Mukherjee S.P., J. Non- Cryst. solids, 42 (1980) 477-488.

62 - Decottignies M., Mukherjee S.P., Phalippou J., Zarzycki J., C.R. Acad. Sc. Paris 285 C (1977) 289-292.

63 - Decottignies M., Phalippou J., Zarzycki J., J. Mater. Sci. 13 (1978) 2605-2618.

64 - Roy D.M., Roy R., Am. Mineralogist 40 (1955) 147-178.

65 - Hench L.L., Prassas M., Phalippou J., J. Non-Cryst. Solids 53 (1982) 183-193.
66 - Woignier T., Phalippou J., Zarzycki J., J . of Non-crystalline solids 63 (1984) 117-130.
67 - Mukherjee S.P., Zarzycki J., Traverse J.P., J. Mater. Sci. 11 (1976) 341-355.
68 - Mukherjee S.P., in "Materials Processing in the Reduced gravity Environment of Space", Ed. Rindone G. E., Mat. Res. Soc. Symp. Proc. 9 (1982) 321-331.
69 - Roy R., J. Amer. Ceram. Soc. 52 (1969) 344-345.
70 - Luth W.C., Ingamells C.O., Am. Mineralogist 50 (1965) 255-258.
71 - MacCarthy G.J., Roy R., Mackay J.M., J. Amer. Ceram Soc. 54 (1971) 639-640.
72 - Zarzycki J., Prassas M., Phalippou J., J. of Mat Sciences 17 (1982) 3371-3379.
73 - U.K. Atomic Authority , Brit. Pat. #1,367,736, Sept. 24,1974
74 - Thomas I.M., US Patent 3 799 754 , 26 Mar 1974.
75 - Brinker C.J., Mukherjee S.P., Thin Solid Films 77 (1981) 141-148.
76 - Segal D.L., J. Non-Crystalline Solids 63 (1984) 183-191.
77 - Phalippou J., Prassas M., Zarzycki J., J of Non-Cryst. Solids, 48 (1982) 17-30.
78 - Höland W., Plumat E.R., Duvigneaud P.H., J. Non-Cryst. Solids 48 (1982) 205-217.
79 - Yoldas B.E., J. Mater. Sci. 14 (1979) 1843-1849.
80 - Jabra B., Phalippou J., Zarzycki J., Rev. Chim. Miner., 16 (1979) 245-246.
81 - Nogami M., Moriya Y., J. Non-crystalline Solids 48 (1982) 359-366.
82 - Rabinovich E.M., Johnson Jr.D.W., MacChesney J.B., Vogel E.M., J. Non-Cryst. Solids 47 (1982) 435-439.
83 - Nogami M., Moriya Y., Yogyo Kyokai-Shi 85 (1977) 4448-454.
84 - Schroeder H., Gliemeroth G., Naturwissenschaften 57 (1970) 533-541.
85 - Kamiya K., Sakka S., Tashiro N., Yogyo Kyokaishi 84 (1976) 614-618.
86 - Sakka S., Kamiya K., J. Non-crystalline Solids 42 (1980) 403-422.
87 - Kamiya K., Sakka S., Yamanaka I., Proc. X Int. Cong. Glass 13 (1974) 44-49.
88 - Jabra R., Phalippou J., Prassas M., Zarzycki J., J. Chim. Phys. 78 (1981) 777-780.
89 - Kamiya K., Sakka S., Tatechimi Y., J. Mater. Sci. 15 (1980) 1765-1771.
90 - Yamane M., Kojima T., J. Non-Cryst. Solids 44 (1981) 181-191.
91 - Hayashi T., Saito H., J. Mater. Sci. 15 (1980) 1971-1977.
92 - Puyane R., James P.F., Rawson H., J. Non-crystalline Solids 41 (1980) 105-115.
93 - Prassas M., Phalippou J., Hench L.L., Zarzycki J., J. Non-crystalline Solids 48 (1982) 79-95.
94 - Weinberg M.C., Neilson G.F., J. Mater. Sci. 13 (1978) 1206-1216.
95 - Neilson G.F., Weinberg C., in "Materials Processing in the reduced gravity environment of space", Ed. Rindone G., Elsevier, New- York, Mat. Res. Soc. Symp. Proc. 9 (1982) 333-341.
96 - Swiecki Z., Miksiewicz C., Szklo Ceramica 31 (1980) 95-98.
97 - Stol R.J., Van Helden A.K., De Bruyn P.L., J. Colloid Interface Sci. 57 (1976) 115-131.
98 - Thomas I.M., U.S.Patent 3 767434 , 23 oct 1973.
99 - Guglielmi M., Principi G., J. Non-Cryst. solids 48 (1982) 161-175.
100 - Yoshio T., Kawaguchi C., Kanamura F., Takahashi K., J. Non-Cryst. Solids 43 (1981) 129-140.
101 - Mukherjee S.P., Zarzycki J., J. Am. Ceram. Soc. 62 (1979) 1-4.
102 - Mukherjee S.P., Zarzycki J., Badie J.M., Traverse J.P., J. Non-Cryst. Solids 20 (1976) 455-458.
103 - Kamiya K. Ohya M., Yoko T., J. of Non-Cryst. Solids 83 (1986) 208-222.
104 - Carturan G., Gottardi V., Graziani M., J. Non-Cryst Solids. 29 (1978) 41-48.
105 - Klein L.C., Garvey G.L., J. Non-crystalline Solids 38-39 (1980) 45-50.
106 - Makishima A., Oohushi H., Wakakuwa M., Kotani D., Shimohira T., J. Non-crystalline solids 42 (1980) 545-552.

107 - Nogami M., Hayakawa J., Moriya Y., J. Mater. Sci. 17 (1982) 2845-2849.
108 - Brinker C.J., Scherer G.W., in "Ultrastructures Processing of ceramics, Glasses and Composites". Gainesville, Florida. Ed. Hench L.L., Ulrich D.R., Wiley, New-York (1984) 43-59.
109 - Brinker C.J., Mukherjee S.P., J. Mater. Sci. 16 (1981) 1980-1988.
110 - Brinker C.J., Haaland D.M., J. Am. Ceram. Soc. 66 (1983) 758-765.
111 - Brinker C.J., Haaland D.M., Loehman R.E., J. Non-Cryst. Solids 56 (1983) 179-184.
112 - Gottardi V. , Scarini G., Carturan G., Marchetti A., Frosni V., in "Thermal Analysis", Ed. WIederman H., Birkauser Verlag, Basel (1980) 493-497.
113 - Meriani S., Longo V., Festa D.R., Guglielmi M., Trans. Br. Ceram. Soc. 80 (1981) 87-90.
114 - Mazdiyasni K.S., Brown L.M., J. Am. Ceram. Soc. 55 (1972) 548-552.
115 - Brown L.M., Mazdiyasni K.S., J. of Am. Ceram. Soc. 55 (1972) 541-544.
116 - Rooksby H.P., J. of Appl. Chem. (London), 8(PL-1) (1958) 44-49.
117 - Larbot A., Fabre J.P., Guizard C., Cot L., J. Am. Ceram. Soc., 72 (1989) 257
118 - Teichner S.J., Nicolaon G.A., Vicarini M.A., Gardes G.E.E, Adv. Colloid Interface Sci., 5 (1976) 245 .
119 - Ghate B.B., Hasselman R.H., Spriggs R.M., Am. Ceram. Soc. Bull. 52 (1973) 670-672.
120 - Barrett W.T., Sanchez M.G., Smith J.G., Advances in Catalysis 9 (1957) 551-557.
121 - Gonzales-Oliver C.J.R., Johnson P.S., James P.F., J. Mater. Sci., 14 (1979) 1159-1169.
122 - Phalippou M., Zarzycki J., Lalanne J.F., Ann. Chim. Fr. 3 (1978) 99-105.
123 - MacCarthy G.J., Roy R., Mackay J.M., J. Amer. Ceram Soc. 54 (1971) 637-638.
124 - Sakka S., Kamiya K., "Amorphous materials produced from metal alcoholates", in "Process. Kinet. Prop. Electron. Mag. Ceram. Proc. US- Jpn. Semin. Basic Sci. Ceram.", W.Komatsu, R.M. Fulrath, Y.Oishi, M. Koizumi, and S. Somiya, Eds. 1975, 135-143.
125 - Yamaguchi O., Matsumoto K., Shimizu K., Bull. Chem. Soc. Jpn. 52 (1979) 237-238.
126 - Yamaguchi O., Ito Y., Shimizu K., Bull. Chem. Soc. Jpn. 53 (1980) 275-276.
127 - Yajima S., Hayashi J., Omori M., Chem. Lett. (1975) 931-934.
128 - Yajima S., Okamura K., Hayashi J., Omori M., J. Amer. Ceram. Soc. 59 (1976) 324-327.
129 - Wei G.C., Kennedy C.R., Harris L.A., Amer. Ceram. Soc. Bull. 63 (1984) 1054-1061.
130 - Sane A.Y., "Refractory metal borides, carbides and Nitrides, and composites containing them", European Patent 0115745- Eltech Systems Corporation, 15. 08 (1984).
131 - White D.A., Oleff S.M., Boyer R.D., Budinger P.A., Fox J.R., Advanced Ceramic Materials 2 (1987) 45-52.
132 - Anderson C.H., Warren R., Composites, 15 (1984) 16-24.
133 - Petrovski P., Jelacic C., Silicates Industriels 46 (1981) 85-90.
134 - Zhang S.-C.,Cannon W.R., J. Am. Ceram. Soc. 67 (1984) 691-695.
135 - Winter G., Verbeek W., Mansmann M., "Production of shaped Articles of Silicon carbide and Silicon Nitride", Ger. Offen. 2,243,527, May 16, 1974 (U.S. Patent 3,892,583).

SINTERING
SOL-GEL CERAMICS

8.1 - INTRODUCTION

Simultaneously with the transformation of a gel to ceramics, it was mentioned in chapter 7 that the specific surface area of sol-gel materials decreases during thermal treatments. The progressive elimination of the porosity is an evolution named sintering. In terms of thermodynamics, sintering originates from the specific surface area S_a a porous material, which introduces a positive contribution G_s due to the surface of pores, to the Gibbs free energy of a material :

$$G_s = \gamma S_a \qquad (8.1\text{-}1)$$

Since states with a lower Gibbs free energy are more stable at a given temperature and pressure, the specific surface area should tend to decrease, an evolution which can proceed according to two types of pore evolution:
- (1) by changing the shape of pores but not their volume.
- (2) by eliminating the pores.
The first type of evolution does not produce any material densification, contrary to the second one which makes a material shrink without losing any mass.
Moreover, sintering is not the only transformation which involve Gibbs free energy of surface origin. Grain boundaries constitute another type of surfaces which contribute to another Gibbs free energy term G_{gb}. In order to reduce this free energy contribution, the specific grain boundary area S_{gb}, must decrease. Hence another evolution named grain growth, enters in competition with sintering.
Sol-gel ceramics just after drying and even after heat treatments at intermediate temperatures often have a very high specific surface area and an extremely small grain size. Hence, both sintering and grain growth tend to be vigorous. Theories have been developed to understand the sintering of ceramics and its competition with grain growth, but they usually apply to conventional ceramics. The applicability of these classical mechanisms to sol-gel ceramics can provide some valuable understanding, but it must be applied with caution because of their very fine scale initial texture.

8.2 - POSSIBLE TEXTURE EVOLUTION

THERMODYNAMICS
Inside a solid particle, near an external surface such as a pore surface, mechanical stresses are introduced by the surface tension each time the surface is curved. If the external pressure in the pores outside the solid is p_0 and the two principal curvature radii of the surface are r_1 and r_2, the local pressure inside the solid, just under the surface is

$$p = p_0 + \gamma \left(\frac{1}{r_1} + \frac{1}{r_2}\right) \tag{8.2-1}$$

When the local surface is convex , that is to say when the solid is on the same side as the center of curvature, a radius r must be taken with a positive value. On the other hand, when the local surface is concave, that is to say when the solid is on the opposite side to the center of curvature, the radius r must be taken with a negative sign. For sake of simplicity, it is possible to consider the case where $p_0 = 0$ (treatment under vacuum). In such conditions, when two spherical particles of radius r are linked by a neck with principal radii x (convex radius) and ρ (concave radius) at the neck tip and such that in absolute value $\rho << x$, (figure 8.2-1), the mechanical stresses states inside the solid near the external surface are:

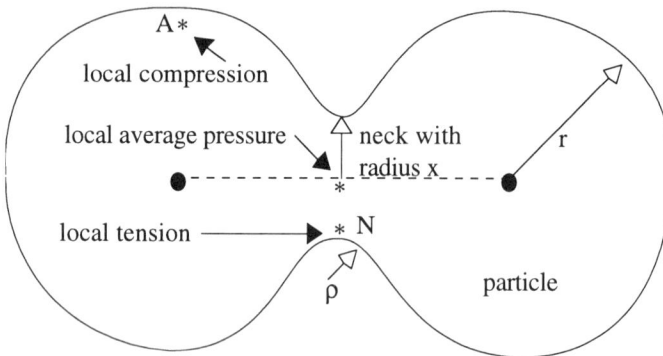

Figure 8.2-1 - Initial sintering stage of two spherical particles. Adapted from Kingery, Bowen, Uhlmann [1].

- a compression far from the neck, at point A where:

$$p_A = \sigma_c = \frac{2\,\gamma}{r} > 0 \tag{8.2-2}$$

- a tension near the neck at point N where:

$$p_N = \sigma_t = \gamma\left(\frac{1}{x} - \frac{1}{\rho}\right) \approx -\gamma\frac{1}{\rho} < 0 \tag{8.2-3}$$

This mechanical stress difference between points A and N is equivalent to a difference in the local Gibbs free energy of mechanical nature and it tends to attenuate by transfer of matter. Finally, the surface energy happens to be at the origin of atomic transport.

TEXTURAL TRANSFORMATION KINETICS

The transport of atoms to sinter a material can proceed along several different types of paths. Some of these paths actually result in densification, in particular when the atoms originate from sources, that is to say local parts of the material located along dislocations and grain boundaries, or when atoms are transported by a cooperative phenomenon known as viscous flow. Other transport paths only change the pore shape, in particular when the matter comes from the pore surface.

Herring's scaling laws-Principle

The atomic transport paths are usually compared with each other by reference to typical kinetic laws named Herring's scaling laws [2], which consist in comparing the evolution of two samples where all initial linear dimensions differ by a same multiplication factor L. That is to say all lengths in a sample scale proportionally to L, any surface proportionally to L^2 and any volume to L^3. For such samples, it is possible to compare how the time t to reach a given relative transformation, for example a given percentage of the theoretical density, increases with L; t is a function proportional to L^h where h is a fractional exponent which depends on the transport mechanism considered.

A porous material is usually described by a simple geometrical model. For the initial stage of sintering, the most frequent model is composed of spheres linked by necks. For the intermediate sintering stage, a geometrical shape often selected is the tetrakaidecahedron illustrated in figure 8.2-2. The pores have a cylindrical shape and they are located along the edges of a tetrakaidecahedron, so that they form an interconnected network [3]. With this model, the final stage of sintering can be described by placing isolated spherical pores at the corners of the tetrakaidecahedra.

Example of scaling law

When densification occurs, theoretical derivations show that the absolute value of the relative decrease $\dfrac{\Delta V}{V_0}$ of the sample volume, where V_0 is the initial sample volume, is a function of the particle size or grain size r_g and time t, which can be written

$$\frac{\Delta V}{V_0} = f\left(\frac{D}{T}\right) r_g^{-\alpha} t^\beta \qquad (8.2\text{-}4)$$

where exponents α and β are positive exponents and $f\left(\dfrac{D}{T}\right)$ is a function of a diffusion coefficient D and the temperature T.

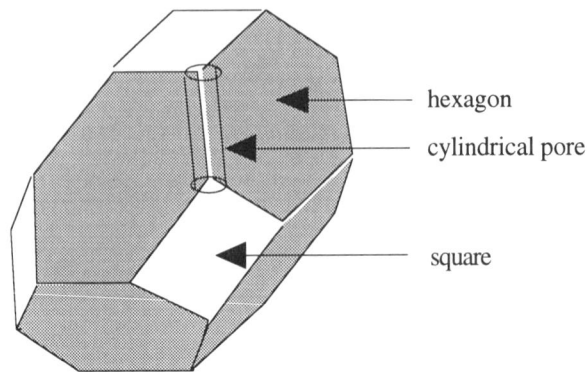

hexagon

cylindrical pore

square

Figure 8.2-2 - Tetrakaidecahedron model often used for the intermediate and final sintering stages. Adapted from Coble [3]

In terms of Herring's scaling law $r_g^{-\alpha}$ scales as $L^{-\alpha}$ so that the time t increases with L as

$$t \approx g\left(\frac{T}{D};\frac{\Delta V}{V_0}\right) L^{\alpha/\beta} \qquad (8.2\text{-}5)$$

where $g\left(\frac{T}{D};\frac{\Delta V}{V_0}\right)$ is a function of the ratio $\frac{D}{T}$ and the relative densification state selected. For instance, densification by atomic diffusion along grain boundary leads to a law where the exponents are $\alpha=\frac{6}{5}$ and $\beta=\frac{2}{5}$. The detailed form of the law depends on the geometrical model chosen. However, the exponents are independent of this model, and they remain the same for the initial, intermediate and final sintering stages, which constitutes the strength of the Herring's scaling law approach. For the mechanism chosen as an example, the time for a given relative sintering increases with L as

$$t \approx g\left(\frac{T}{D};\frac{\Delta V}{V_0}\right) L^3 \qquad (8.2\text{-}6)$$

Comparison of a few mechanisms
In the case of grain growth, one looks at the time t for a given relative grain growth $\frac{\Delta r_g}{r_{g0}}$ instead of a given relative volume shrinkage $\frac{\Delta V}{V_0}$. Overall, the scaling laws for a few atomic transport mechanisms, both for grain growth and for densification, can be compared, and a few results are gathered in table 8.2-1.

Influence of the texture scale
Table 8.2-1 shows that all processes, sintering as well as grain growth, accelerate for a finer texture, that is to say a smaller particle or grain size r_g. As an example,

sintering in a material where atom transport occurs by lattice diffusion with a diffusion coefficient $D = 10^{-15}$ m^2 sec.$^{-1}$, a similar relative densification is reached respectively in 1.1 hour and in 4 sec when the grain size decreases from 1 μm to 0.1 μm. This remark is especially important for sol-gel material which have a much finer texture than conventional materials, that is to say in which all textural changes are considerably enhanced .

Table 8.2-1 - Comparison of the times for a given textural change during sintering and grain growth.

Textural transformation	Mechanism	time t as a function of the scaling factor L, for a given textural change
Sintering	- Lattice diffusion (coefficient D) from grain boundary towards the neck	$\dfrac{L^3}{D}$
	- Grain boundary diffusion (coefficient D_{gb}) from grain boundary towards neck	$\dfrac{L^4}{D_{gb}}$
	- Viscous flow (viscosity η)	$L\,\eta$
Grain growth	- surface diffusion (coefficient D_s) from convex surface (sphere) towards neck	$\dfrac{L^4}{D_s}$
	- Evaporation condensation, mean gas free path limited by:	
	- gas pressure (Diffusion coefficient D_g)	$\dfrac{L^3}{D_g}$
	- pore size	L^2

Also, as L increases with time for a given material, this means that all transformation processes are faster in their initial stage than in their final stage: each kinetic process slows down with time.

COMPETITION BETWEEN GRAIN GROWTH AND SINTERING
The different scaling laws such as shown in table 8.2-1 make it possible to compare the relative importance of each textural modification mechanism, as a function of the scale L.

As a first example, if densification is controlled by lattice atomic diffusion and grain growth by atomic diffusion at the surface of pores, the densification rate is proportional to $\dfrac{1}{L^3}$, and the grain growth rate to $\dfrac{1}{L^4}$ (figure 8.2-3a). In this case grain growth dominates at small scales L, that is to say at small grain sizes r_g.

As a second example, If densification is controlled by atomic diffusion at grain boundaries and grain growth by atomic diffusion at the surface of pores, both rates are proportional to $\dfrac{1}{L^4}$. In practice, one of these texture evolution dominates at all grain sizes: it may be coarsening or densification depending on the values of the diffusion coefficients (figure 8.2-3b). Changing the grain size has no effect to improve densification.

As a last example, if densification is controlled by atomic diffusion at grain boundaries, and grain growth by atomic diffusion in the vapor phase inside pores, the densification rate is proportional to $\dfrac{1}{L^4}$ and the grain growth rate to $\dfrac{1}{L^2}$. In this case densification is favored by small grain sizes (figure 8.2-3c).

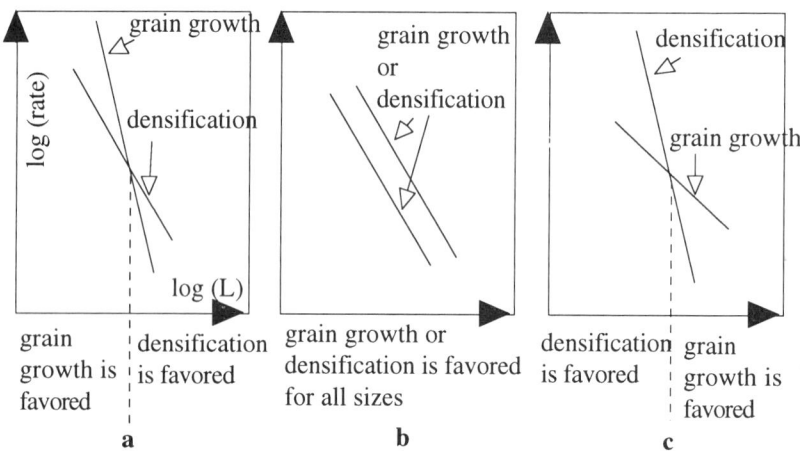

Figure 8.2-3 - Comparison of the densification and growth rates for various mechanisms; (a) densification controlled by atomic lattice diffusion and grain growth by atomic diffusion at the surface of pores; (b) densification controlled by atomic diffusion at grain boundaries and grain growth as in (a); (c) densification as in (b) and grain growth controlled by atomic diffusion in the vapor phase.

Hence, no unique trend exists with respect to the influence of the initial powder size on densification. In some materials, densification is enhanced by selected very small colloidal particles. In other materials, the initial particles must not be too small. In some materials at last, the initial powder size does not matter. This general situation also applies to sol-gel materials in which surface atomic diffusion becomes relatively more important than in conventional materials, especially in the early stages of sintering when necks between particles begin to develop, because of the high initial specific surface area.

This complexity is increased even more by the fact that it is often difficult to discriminate between the various densification mechanisms according to the previous scaling exponents, due to the inaccuracy in their experimental determination. Scaling exponents have values too close from each other in regards with the data dispersion. However, the scaling laws approach is interesting because it makes possible to estimate the effect of some processing variables. It provides an explanation why some compounds such as ZrO_2 are easier to densify when made by sol-gel, than other ones such as alumina. In the latter oxide, densification may become more difficult when the initial particles have a size below a certain value. In addition to the size and time effects, the scaling law models point out to the importance of the temperature and nature of the species which diffuse, through the magnitude of their diffusion coefficient.

GEL NETWORK TRANSFORMATION DURING SINTERING

Conventional sintering models usually focus on the texture evolution of a couple, or at most a very small number of initial particles. On the other hand, in gels, it would be interesting to know how the porous solid network architecture transforms, at least during the early stages of sintering when the specific surface area remains high. For this purpose, it is necessary to study the texture evolution of packing comprising a large number of spherical particles, for instance. However, studies in this direction are much more scarce. A few aspects are described in the next paragraphs.

Effect of unsymmetrical particles

Most traditional sintering models address the sintering of well defined geometrical particles shape, such as spheres, cubes, tetrakaidecahedra. Petzow and Exner have examined the behavior of two unsymmetrical particles [4] and they have shown that with respect to their neck (figure 8.2-4), the particles rotate so as to close their sharpest angle while the wider one opens.

Evolution of strings of particles

On the other hand, studies on three particle models have shown that the angle between the lines joining the spheres centers first begins to increase, then it decreases (figure 8.2-5). At last, computer simulations by Ross et al. [5] have shown that loose packing comprising a large number of particles tend to align strings of particles, so as to form "star-like" networks of particles (figure 8.2-6).

Effect of particle segregation by size

If the particles are not free to move, any lack of symmetry introduces compressive, tensile or shear stresses in the necks where free rotation is prohibited. For instance, when segregation of particles with a different size occurs, successive layers of large and smaller particles can be deposited (figure 8.2-7). Since the sintering driving force is larger for small particles which tend to shrink more than larger ones, tensile stresses develop in the small particles' layer and compressive ones in the large particles' layer. This can bend a sample when a coating is deposited onto a thin substrate.

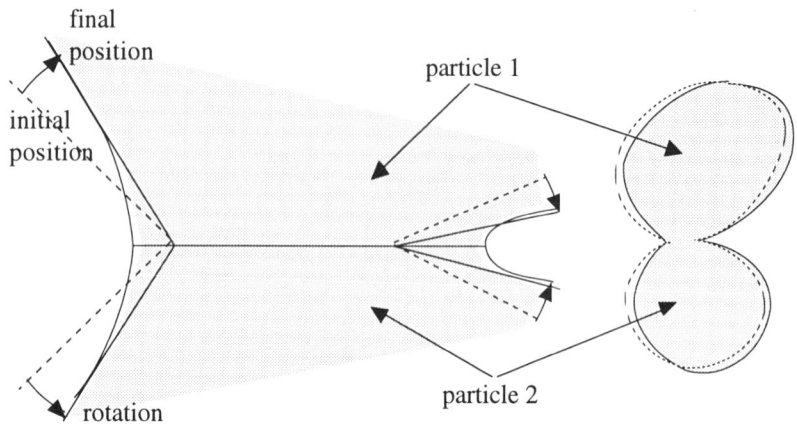

Figure 8.2-4 - Evolution of an unsymmetrical neck during the sintering of two particles. Adapted from Petzow and Exner [4].

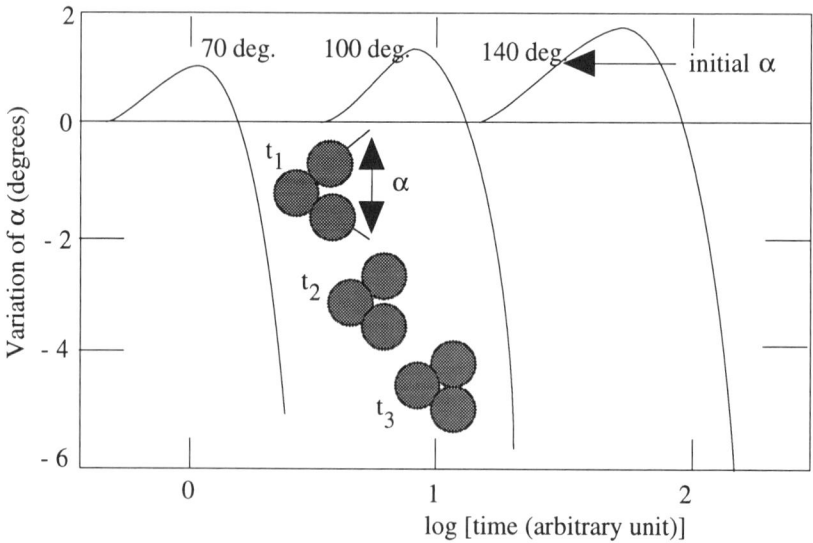

Figure 8.2-5 - Evolution of the angle α between the lines joining the centers of three copper spheres : (t_1) initially; (t_2) after 4h sintering at 920°C; (t_3) after 64h sintering at 920°C; for initial angles of 70deg., 100deg. and 140deg. Adapted from Petzow and Exner [4].

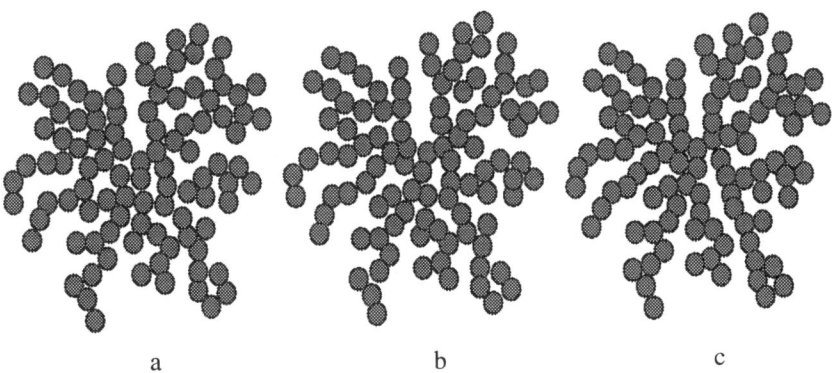

a   b   c

Figure 8.2-6 - Sintering behavior of a two dimensional random loose packing of equal-size spheres after interparticle distance shrinkage of: (a) 0%; (b) 4%; and (c) 8%. Adapted from Ross et al. [5].

On the other hand, if the deposited coating is constrained by an underlying substrate surface and if the bonding between the coating and the substrate is good, differential shrinkage can be eliminated at the substrate-coating interface and densification can proceed smoothly [6]. Residual tensile stresses can remain in the necks between the smaller particles; their magnitude is inversely proportional to the size of these particles.

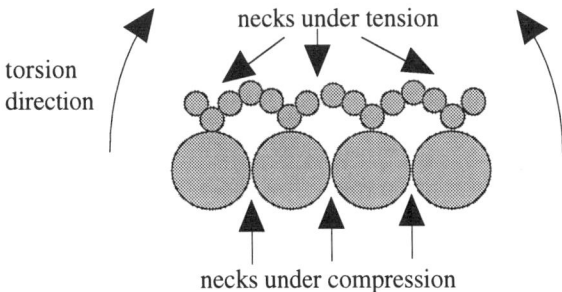

Figure 8.2-7 - Stresses with segregation of fine and large particles. After Nii [7].

Similarly, mechanical stresses develop in packing with unequal pore size; their magnitude is inversely proportional to the pore size [8,9]. To attenuate this problem, it is possible to eliminate the smallest pores in the powder by performing a partial densification of the fine particles, before sintering the coating [6]: in alumina coatings deposited on sapphire substrate, bigger necks are achieved by heat treating first the samples in a temperature range where grain growth dominates over densification (1000°C). Bigger grains grow, which makes it possible to avoid extensive cracking at higher temperatures. It is also possible to look for a composition where sintering proceeds by the intermediate of a liquid phase, which favors the formation of stronger necks and makes it possible to avoid cracking, such as in $ZnO-Bi_2O_3$ compounds.

8.3 - ATOMIC TRANSPORT MECHANISMS OPERATING DURING SINTERING

ATOMIC DIFFUSION

Case of compounds
When sintering proceeds by atomic diffusion in the case of a ceramic which comprises several types of ions, each ion has to diffuse in proportion to its stoichiometry in the ceramic. Consequently, sintering is controlled by the slowest ion transported along its fastest path. In a compound comprising one type of anions "a" and one type of cation "c", this effect is well taken into account by using an effective diffusion coefficient D_{eff} given by:

$$D_{eff} = \frac{D_a D_c}{D_a + D_c} \qquad (8.3\text{-}1)$$

where D_a is the diffusion coefficient of anions along their fastest path, and D_c that of the cations.

If one species, for instance the anions, has the smallest diffusion coefficient, it is possible to accelerate its diffusion with the help of cations additives which have a lower valence and a size close to that of the host cation. To balance the electric charges in the compound, anion vacancies are then often created, which accelerates anions diffusion. In oxide, this concerns the oxygen anions which diffuse significantly at high temperature only. The formation of oxygen vacancies increases their mobility and can help sintering. The proportion of additives must remain below but can be close to the solid solution limit.

Atomic diffusion in sol-gel materials
It is possible to question whether transport mechanisms such as atomic diffusion can be extrapolated from the conventional ceramics, to the colloidal sizes.
Atomic transport by diffusion in the vapor phase rests on an evaporation-condensation process due to differences in the equilibrium vapor pressure near a solid surface, depending on the local surface curvature. According to Kelvin's equation, the equilibrium vapor pressure is lower near a neck between two particles where the solid surface is essentially concave, than far from a neck, at atomic scales. The resulting

vapor pressure gradient induces a vapor phase diffusion which keeps feeding the neck. It modifies the pore shape but does not induce densification which in a gel often occurs between 900 and 1200 °C. At these temperatures, the equilibrium vapor pressure of oxygen is very low for an oxide. On the other hand, the residual hydration water usually absent in conventional ceramics is much more volatile at relatively low temperatures. Hence, this mechanism cannot be completely rejected for the sol-gel processes, as water is known to be a general sintering aid.

Surface atomic diffusion does not induce densification neither. However, it is likely to operate at a lower temperature than usual, because of the extremely high specific area of sol-gel and also because of the special nature of the gel surface which easily adsorbs water molecules or OH groups.

Overall, the atomic transport mechanisms which make the pore shape become spherical without producing any densification are likely to be common in sol-gel materials. If this is the case, these mechanisms can be strongly modified by tailoring the nature of a sol-gel solid surface, for instance by adsorbing hydrophobic molecules. Hence the liquid sol-gel chemistry can have a marked influence on the sintering behavior.

As seen previously, lattice atomic diffusion can actually densify a material, depending on the source from where the atoms originate. In a crystalline material such as an oxide, this mechanism begins to significantly operate when the atomic mobility of oxygen atoms is high enough that is to say at high temperatures. It requires the presence of atomic point defects such as vacancies, interstitialcies, or dislocations. However gels are far from being very well crystallized; they have at best very small grains with a high defects density. Consequently, depending on the compound, lattice diffusion can often operate at a much lower temperature than in conventional ceramics. Actually, a distinction between the different atomic diffusion paths, such as on a surface, inside a lattice, along dislocations or along grain boundaries may be difficult in sol-gel materials, in so far as it becomes difficult to distinguish between dislocations, grain boundaries and surfaces. It would seem more appropriate to consider a global diffusion process through a very disordered gel or transition phase structure.

Sintering and crystallization in sol-gel ceramics

In so far as densification can be considered to occur in sol-gel ceramics by global atomic diffusion, the corresponding diffusing species are also likely to enhance the crystallization by nucleation and growth of the most thermodynamically stable crystalline phase, as seen in chapter 7. Densification of sol-gel ceramics often occurs concurrently with their crystallization [10]. Even in compounds where sintering occurs at a relatively low temperature, the crystallization of a stable phase usually occurs at a slightly lower but very close temperature.

Sol-gel zirconia is a typical material where densification is easy; it often begins at 800°C [11]. The crystalline monoclinic zirconia structure which is stable at this temperature also forms at low temperature; often at 500°C. Moreover, densification is faster at higher temperatures but it reaches a maximum rate in the temperature range 1300-1400°C and it slows down again when transformation to the tetragonal phase occurs. It has also been shown, in Y_2O_3 stabilized ZrO_2, that atomic diffusion along grain boundaries dominate the initial stage sintering. In agreement with this view, Herring's scaling law correctly predict that using smaller size particles is beneficial to sintering [12, 13].

Water vapor also enhances the sintering process. For instance, ThO_2 gel spheres could be densified up to 99% of their theoretical density by sintering them in air containing up to 99.7% humidity (figure 8.3-1). Simultaneously, the grain size increased drastically [14].

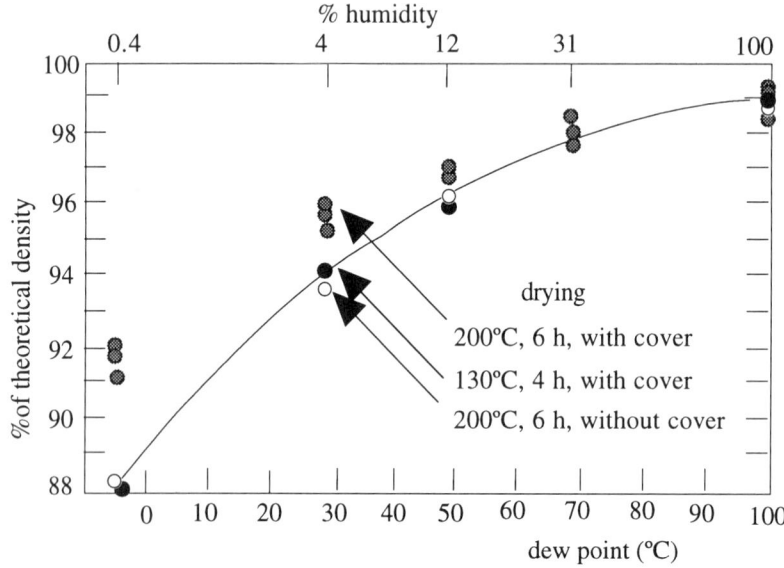

Figure 8.3-1 - Density of ThO_2 spheres with 0.5 mm diameter sintered in air-steam atmospheres. Adapted from Yamagishi and Takahashi [13].

A similar phenomenon was observed in UO_2 pellets [15]. The mechanisms proposed rest on the formation of non-stoichiometric compounds UO_{2+x} and ThO_{2-x} which is consistent with a dark-grey coloration of the materials [16].

Sol-gel ceramics can themselves be quite effective sintering additives. Sol-gel ZrO_2 has been used with non-oxide powders such as carbides and borides, and made it possible to realize dense particulate composites [17]. With a nitride such as Si_3N_4 made by reaction of $SiCl_4$ with liquid ammonia, according to a sol-gel type process, alkoxide sol-gel additives such as $Y(OH)_3$ and $Mg(OH)_2$ have made it possible to improve the sintering behavior [18], as the density was increased from 2.85 g/cm³ to 3.2 g/cm³.

As ZrO_2, sol-gel TiO_2 crystallize and sinter at a low temperature [19]. Powders with an average grain size of 0.08 mm made from titanium isopropoxide, could be densified at 800°C to 99% of their theoretical density [20]. However, this behavior is not specific to sol-gel processing as TiO_2 coating were also prepared by thermal decomposition of titanium alkoxides on a hot substrate at 600°C and sintered directly at this temperature to give a hard thin transparent TiO_2 layer [21].

In contrast with the previous materials where densification is enhanced by the sol-gel state, the poor sintering performance of sol-gel Al_2O_3 was already mentioned. Actually, densification begins in the same range of temperature where the stable phase, α-alumina, crystallizes, as for other materials. However, the pores are not eliminated, they end up being trapped inside large grains, which again points out to a complex relationship between densification grain growth and pores.

To sinter compounds which comprise several cations, such as mullite $3Al_2O_3.2SiO_2$, the inter-diffusion of cations is necessary. Hence, as for the simple oxides, crystallization often occurs simultaneously with sintering. Sol-gel processing makes it possible to directly obtain powders which easily sinter, while in conventional ceramic processing, lengthy grinding times are necessary to make small size particles so as to lower the temperature where diffusion can efficiently operate [22].

Examples concern mullite which can be sintered at 1200°C instead of 1400°C by conventional processing [23,24], spinel which can be densified between 600 and 800°C [25] and lead zirconium titanates, or PLZT, which can be densified in oxygen at 1175°C [26,27]. For the latter compound, sintering between 860°C and 1150°C proceeds with the help of a liquid phase composed of PbO, TiO_2 and ZrO_2 in which the solid particles packing could reorganize. At higher temperature and up to 1300°C, dissolution recrystallization occurred and made the growth of large grains possible. Finally, at still higher temperatures, evaporation-condensation was the main densification mechanism and it required a low PbO vapor pressure.

Other important technological titanates and zirconates which densify well when made by the sol-gel process include $BaTiO_3$, prepared by mixed alkoxide techniques by Mazdiyasni et al. [28] and sintered at 1300°C to dense translucent bodies, as well as $SrTiO_3$, $SrZrO_3$ [29] and 6 %(by mole)Y_2O_3-ZrO_2 powder which can be sintered to dense transparent ceramics at a temperature of the same order [30].

All the hydrolysis-condensation conditions of sol-gel synthesis have an effect on the sintering rate [11], in the same way as they are responsible for a large variability in the nature of the first phases which crystallize.

VISCOUS FLOW SINTERING

Instead of moving individually as in diffusion, atoms can move by a cooperative displacement known as viscous flow. This kinetic process is well known to operate in glasses made by quenching of melted materials and it also operates largely in the densification of sol-gel glasses. It makes it possible to densify SiO_2 polymeric gels to glasses at 1000°C, instead of melting them at 2000°C. The main difference between gels and glasses is that during the initial stages of densification, at least, the viscosity of gels is much lower than that of bulk glasses and therefore densification is faster [31]. This property is due to the presence of residual OH. After some time, the viscosity increases due to dehydration and also to a greater extent of polymerization and to structural polymeric relaxation. This explains that for thin coatings, the densification is relatively slower than for bulk gels, because the contraction of the gel is impeded and modified by the substrate.

The viscous flow mechanism concerns mainly the silicate systems for which the end products are glasses. They have been studied by Sakka and Kamiya [33], Yamane et al. [32], Brinker and Mukherjee [33]. However, viscous flow models have also been applied to non-silicates, such as monolithic gels obtained from a double alkoxide of Pb and Ti by Blum and Gurkovich [34].

Viscous flow densification models take implicitly into account that gels such as silica gels are not Newtonian fluids but visco-elastic materials, which can keep their monolithic shape. When submitted to a mechanical stress, they first show an immediate elastic response, then a delayed elastic response and finally a viscous deformation. However, densification is almost entirely caused by viscous flow and the densification rate depends on the morphology, the composition, and the chemical reactivity.

General description of viscous flow sintering
The Gibbs free energy decreasing rate due to the surface energy γ, in a porous material, can be derived from its specific surface area S_a decreasing rate by:

$$\frac{dG}{dt} = \gamma \frac{dS_a}{dt} \qquad (8.3\text{-}2)$$

For a material cylinder of volume V and radius r, this rate of energy dissipation can also be related to the strain rates $\frac{d\varepsilon_i}{dt}$ and stress σ_i where i= r, θ or z in cylindrical coordinates, by:

$$\frac{dG}{dt} = V \left(\sigma_r \frac{d\varepsilon_r}{dt} + \sigma_\theta \frac{d\varepsilon_\theta}{dt} + \sigma_z \frac{d\varepsilon_z}{dt} \right) \qquad (8.3\text{-}3)$$

In the hypothesis that a gel behaves as a glass in terms of atomic transport by viscous flow, and for more simplicity if it can be considered to behave as a Newtonian liquid with a high viscosity η flowing in direction z, it is then possible to show that:

$$\frac{d\varepsilon_z}{dt} = - \frac{\gamma}{3\eta r} \qquad (8.3\text{-}4)$$

which indicates that the importance of viscous flow increases for fine gel networks.

Sintering models
The exact laws to describe the viscous sintering of a gel, depend on the geometrical model used to describe the network. The most simple of these models which was applied to gels is by Frenkel [35]. It simply describes the merging of spheres of radius r by viscous flow. The viscous flow lines are illustrated in figure 8.3-2. Any linear dimension L(t) in the network decreases linearly with time as:

$$\frac{L(t)}{L(0)} = 1 - \frac{3\gamma t}{8\eta r} \qquad (8.3\text{-}5)$$

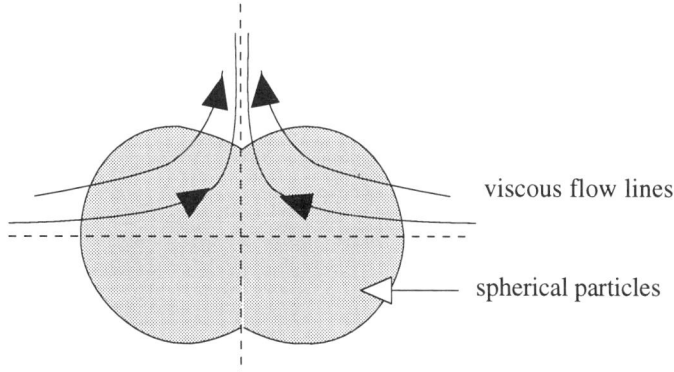

viscous flow lines

spherical particles

Figure 8.3-2 - Viscous flow lines in the densification model of spheres of Frenkel. Adapted from Scherer [36].

Frenkel's was modified by Mackenzie and Shuttleworth to address the final stage of densification. For this purpose, interconnected spherical particles were replaced by spherical and closed pores [37].
A more elaborate model was developed by Scherer. It describes a gel network by cylindrical solid rods placed along the edges of cubes, so that the pores form a very open network during the initial stage of sintering [38] (figure 8.3-3).

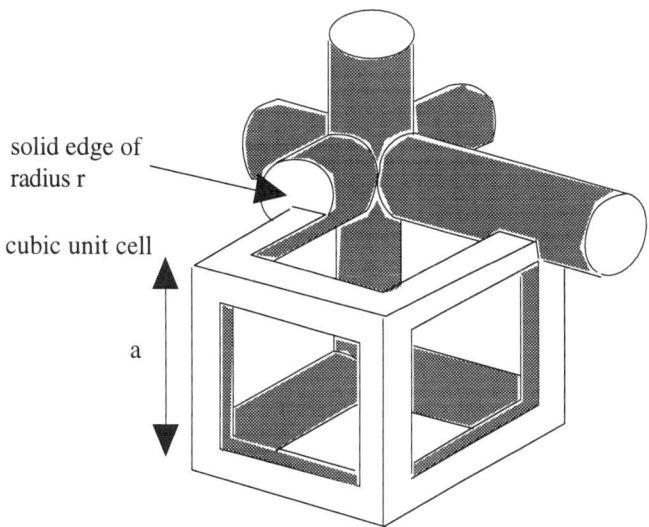

solid edge of
radius r

cubic unit cell

a

Figure 8.3-3 - Gel network for Scherer's densification model by viscous flow. Adapted from Scherer [36].

This model was generalized to include a distribution of pore sizes [39] and it was applied to colloidal silica gels [36]. The results can be represented, with reduced variables, by functions of the type:

$$\frac{\rho}{\rho_{th}} = f\left(\frac{\gamma}{\eta \, r_{p0}} \left(\frac{\rho_{th}}{\rho_o}\right)^{1/3} (t-t_o) \right) \qquad (8.3\text{-}6)$$

in which f designates a function of the complex variable between parenthesis termed the reduced time, ρ_0 is the initial apparent density, ρ is the apparent density at time t, ρ_{th} is the theoretical solid density and r_{p0} is the initial pore size. For n identical pores per unit volume, this model gives:

$$\frac{\rho}{\rho_{th}} = f\left(\frac{\gamma n^{1/3} t}{\eta}\right) \qquad (8.3\text{-}7)$$

This leads to a variation of the reduced density $\frac{\rho}{\rho_{th}}$ as a function of the reduced time, of sigmoidal shape, by opposition with Frenkel's model (figure 8.3-4).

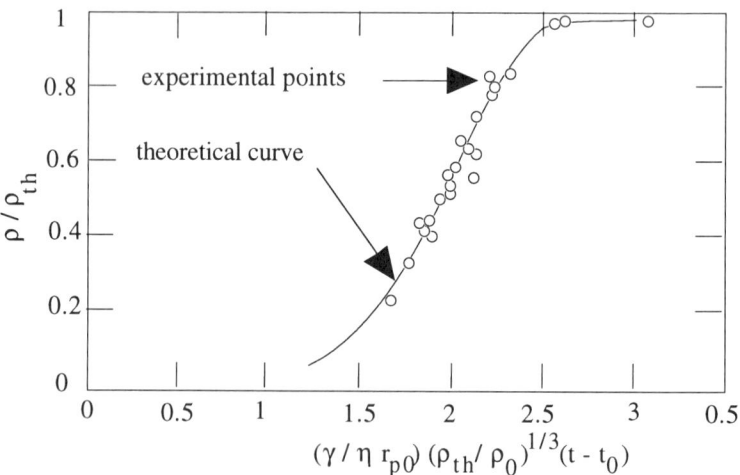

Figure 8.3-4 - Relative density as a function of the reduced time for the cubic model in figure 8.3-3. Adapted from Scherer [36].

Densification of gels depending on their structure;

Colloidal silica gels begin to densify at temperature above 900°C, higher than in polymeric gels, because of the coarser network structure [40]. For instance, colloidal silica gel made by Nogami and Moriya, which consisted of spheres with a diameter $\approx$ 20 nm and a very low density $\approx$ 0.9g/cm^3, followed Frenkel's viscous flow model above 1050°C with the same activation energy as that of fused silica (40.7 kJ/mole) [41]. Compacts of sol-gel silica powders sinter to translucent glass at 1150°C [42]. The type of network built by the particles is important. Densely agglomerated particles densify more easily and at lower temperatures than gels with particles interconnected linearly. Large pores with a size comparable to the size of agglomerates slow down the densification process exactly as for the other densification mechanisms.

On the other hand, silica aerogels dried in hypercritical conditions densify below 1100°C, in spite of their extremely open structure. In this case, the pores have a radius which is extremely small so that they can start to sinter at 525°C, according to a viscous flow mechanism [43].

The chemical composition has an important influence on the densification behavior, In particular, if more than 3000 ppm of OH remain present after drying, a gel starts to foam at 1200°C. This can be avoided by treating gels in Cl_2, which eliminates the OH but somewhat rises the densification temperature. Residual organics and entrapped CO_2 must also be eliminated at intermediate temperatures. As mentionned previously, the viscosity of gels keeps increasing with time during densification because their chemical composition keeps changing, in particluar they lose residual OH.

8.4 - GRAIN GROWTH

BASIC MECHANISM
As mentioned previously, grain growth is also associated with a decrease in interfacial energy, due to the grain boundaries. For a curved grain boundary such as illustrated in figure 8.4-1, Laplace's equation indicates that a greater mechanical pressure exists at point A, that is to say on the side of the grain boundary located towards the curvature center, than at point B on the opposite side.

$$p_A - p_B = \gamma_{gb} \left(\frac{1}{r_1} + \frac{1}{r_2} \right)$$
(8.4-1)

where γ_{gb} is the grain boundary interfacial energy and r_1 and r_2 the two main curvature radii of this grain boundary.

The relative compression at A and tension at B can simultaneously be released by moving some atoms from the grain on the A side, to the grain on the B side. This just requires a moderate atomic displacement or more precisely a realignment of these atoms. Hence, grain B grows and grain A shrinks: a grain boundary moves towards its center of curvature. Another practical consequence is that a planar interface should not move.

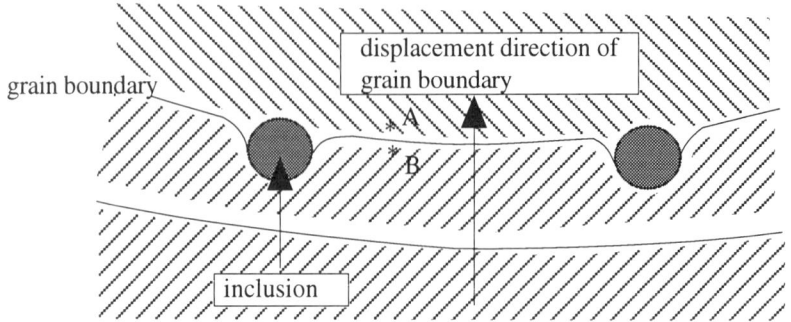

Figure 8.4-1 - Motion of grain boundaries and their pinning by inclusions. Adapted from Kingery, Bowen, Uhlmann [1].

GRAIN GROWTH MODELS

The most simple model for grain growth is based on atomic diffusion across the grain boundary, which corresponds to atoms jumping from grain A to grain B in figure 8.4-1. This results in a parabolic growth law :

$$r^2 - r_o^2 = 2 \frac{D_{gb}^* \gamma V_m \; t}{RT \; w} \qquad (8.4\text{-}2)$$

where D_{gb}^* is the diffusion coefficient for atomic jumps across a grain boundary of thickness w.

A more accurate model by Hillert [44] distinguishes the grains which grow, if their size is larger than a critical size r_{cr} and the smaller grains which shrink. The critical grain size increases with time according to the equation:

$$\frac{dr_{cr}^2}{dt} = \frac{D_{gb}^* \gamma V_m}{2 \; RT \; w} \qquad (8.4\text{-}3)$$

and the grain size evolution is given by:

$$\frac{dr}{dt} = M_b^o \frac{V_m \; \gamma}{w} \left(\frac{1}{r_{cr}} - \frac{1}{r} \right) \qquad (8.4\text{-}4)$$

In the previous equation M_b^o is the grain boundary mobility which is related to its displacement velocity V_b when submitted to a force F_b , by:

$$M_b^o = \frac{V_b}{F_b} \qquad (8.4\text{-}5)$$

GRAIN BOUNDARIES PINNING BY IMPURITIES

Impurities, such as atoms or particles, induce lattice stresses when they are introduced in a host lattice. The corresponding mechanical energy adds up to the free energy of the system. It can be released by moving the impurities to the grain boundaries where the atoms are less tightly packed. Consequently, some energy is required to separate the grain boundaries from the impurities, when these boundaries move during grain growth. Practically, the displacement of grain boundaries can be slowed down by foreign inclusions (figure 8.4-1), an effect which increases as the foreign inclusion size decreases [45]. The impurities exert a drag force F_b^d and the total force necessary to move a grain boundary is:

$$F_b^{total} = F_b^0 + F_b^d = \frac{V_b}{M_b^0} + F_b^d \qquad (8.4\text{-}6)$$

If the impurities can diffuse fast enough to keep up with the grain boundary, the conditions are such that the drag force exerted by the impurities is important. However, if the grain boundary displacement velocity V_b is larger than the time τ for an additive to diffuse a unit distance, this additive starts to separate from the grain boundary and its drag force decreases (figure 8.4-2).

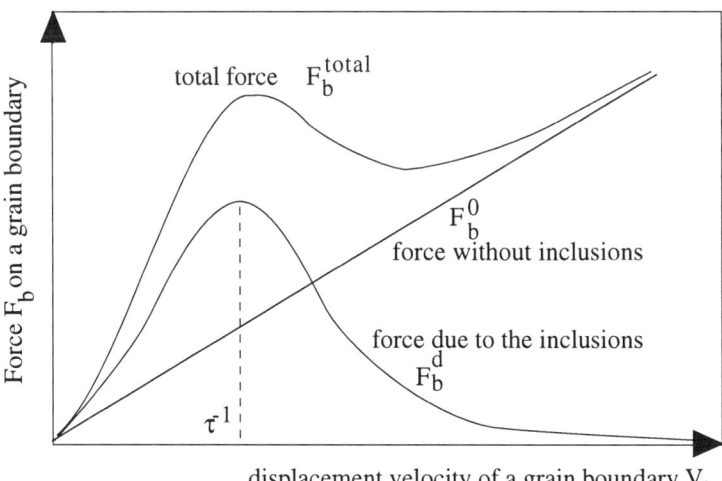

Figure 8.4-2 - Force to move a grain boundary with segregated impurities, as a function of the grain boundary velocity.

Pores themselves can play a role similar to the additives with respect to grain boundary migration. This is the reason why grain growth often occurs during the last stages of sintering when less pores are present to slow down the migration of grain boundaries. Pores must be eliminated during densification, but the way they interfere with the sintering process is very complex. Some aspects of this interaction are examined in the next section.

8.5 - INTERACTION OF PORES WITH THE SINTERING PROCESS

Pores are actors in the sintering process. According to Laplace's equation, a spherical pore of radius r_p introduces a tensile mechanical stresses in the atoms near point A in the pore surface given by :

$$\sigma_t = p_A = \frac{-2\gamma_s}{r_p} \qquad (8.5\text{-}1)$$

To release these mechanical stresses, pores tend to segregate at the grain boundaries.

Pores can also move, disappear, merge with each other or remain stable. These properties of pores are addressed first, and their consequence for the grain growth and densification evolutions are examined in a further section. The effect of particular pore distributions such as due to the formation of agglomerates or packing of particles with unequal sizes, are reviewed last.

POSSIBLE PORE TRANSFORMATIONS

Kinetic stability of a pore
The shape of a pore shape is an important characteristic. It is determined by the number N of grains which surround the pore and the dihedral angle ϕ derived from the grain boundary surface tension γ_{gb} and pore surface tension γ_s, by (figure 8.5-1):

$$\gamma_{gb} = 2 \gamma_s \cos \frac{\phi}{2} \qquad (8.5\text{-}2)$$

When ϕ becomes equal to the dihedral angle θ of the regularly inscribed polyhedron, for a pore located at the junction between several grains, this pore happens to have planar surfaces and densification stops [46]. As an example, the polyhedron is a tetrahedron with a dihedral angle is $\theta = 70.5$ deg for a pore at the junction between 4 grains. When the angle Φ is larger than this value, the pore grows, for a smaller Φ angle the pore shrinks. Planar surfaces are naturally present in boehmite platelet particles such as made in the sol-gel process of alumina, which can explain the difficulty to sinter this material.

Mobility of a pore
A pore can move by the transport of matter which can diffuse in the surface, in the lattice, or in the gas phase inside the pore (figure 8.5-2). For each transport

mechanism, it is possible to derive an expression for the pore mobility, which is defined as the ratio of the pore velocity to the force which makes it move :

$$M_p = \frac{V_p}{F_p} \qquad (8.5\text{-}3)$$

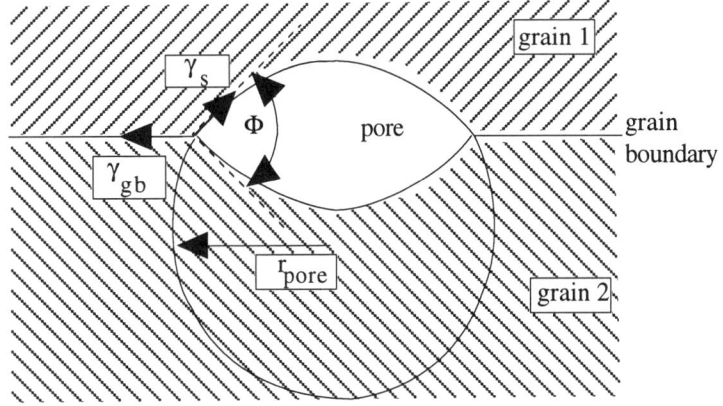

Figure 8.5-1 - Dihedral configuration and curvature of a pore.

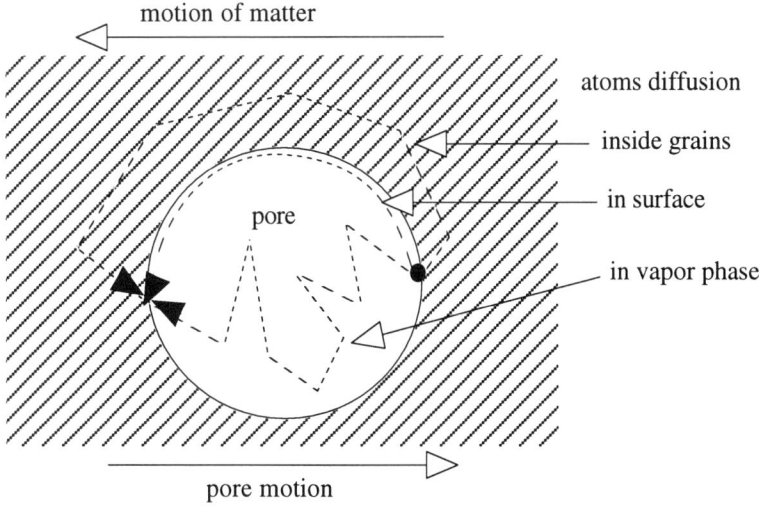

Figure 8.5-2 - Possible transport mechanisms for the migration of a pore.

Different expressions for the mobility of a pore can be derived, depending on the transport mechanism which operates [47].

For surface diffusion, in a thin surface layer of thickness δ_s, with the diffusion coefficient D_s in a material with molar volume V_m :

$$M_p^{SD} = \frac{D_s \delta_s V_m}{RT \, \pi \, r_p^4} \qquad (8.5\text{-}4)$$

where R is the universal gas constant. For lattice diffusion with the diffusion coefficient D_L :

$$M_p^{L} = \frac{D_L V_m}{RT \, \pi \, r_p^3} \qquad (8.5\text{-}5)$$

For diffusion in the vapor phase when the mean free path of the gas molecules is controlled by the gas pressure p_v and the diffusion coefficient is D_g :

$$M_p^{g} = \frac{D_g p_v V_m}{(RT)^2 \, 2\pi \, r_p^3} \qquad (8.5\text{-}6)$$

Pore coarsening

Figure 8.5-3a, illustrates a grain boundary where two spherical pores with a different radius are located; the radii are r_{pS} for the small pore and r_{pL} for the large one. The tensile mechanical stress in the matter at the surface of the pores is larger in absolute magnitude (more negative) near the small pore at point S than near the large pore at point L, because

$$\frac{-2\gamma_s}{r_{pS}} < \frac{-2\gamma_s}{r_{pL}} \qquad (8.5\text{-}7)$$

To release the tension gradient between points L and S, atoms will diffuse from point L towards point S. Consequently, the small pore will shrink while the bigger pore will grow. This mechanism is known as Ostwald ripening when it concerns the growth of solid particles at the expense of other smaller solid particles. Presently it occurs with the pores, which tend to grow, a phenomenon also termed pore coarsening.

The phenomenon of coarsening also concerns grains, as the larger grains tend t grow at the expense of the smaller ones. This is illustrated in figure 8.5-3b where the small center grain slowly disappears and induces a coarsening of the pores located at the border between three grains. These pores will merge with each other when the small center grain will disappear.

Overall, in spite of the sintering process, pores tend to grow and therefore to become less mobile.

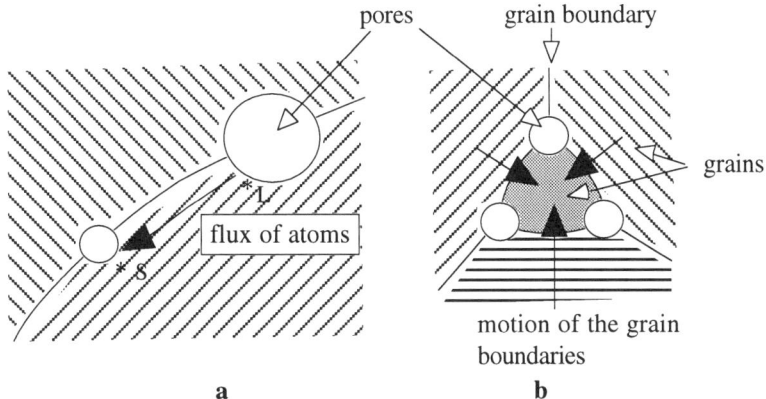

Figure 8.5-3 - Coarsening of pores by: (a) Ostwald ripening; (b) grain-growth.

ACTION OF PORES ON THE GRAIN BOUNDARY MOBILITY

The displacement velocity V_b of a grain-boundary which drags a population of N pores per unit surface area , can be written :

$$V_b = M_b^0 (F_b - NF_p) \qquad (8.5-8)$$

where F_b is mechanical force exerted on the grain boundary due to its curvature, and F_p is the drag force necessary to move a pore (Figure 8.5-4).

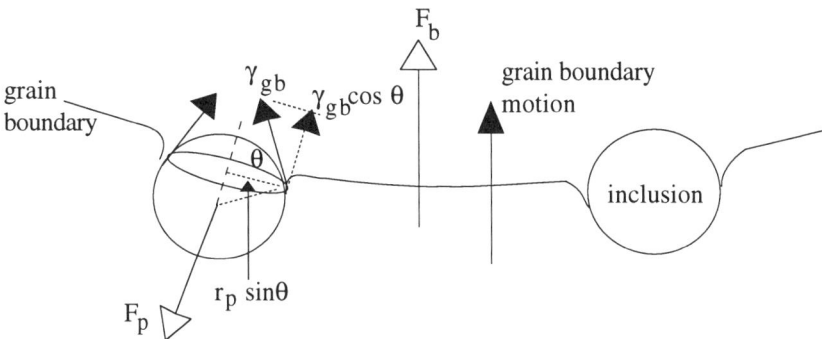

Figure 8.5-4 - Dragging force exerted by pores on a grain boundary

When pores located at a grain boundary and the grain boundary move together:

$$V_b = V_p = M_p \, F_p \tag{8.5-9}$$

By equating the two expressions in (8.5-8) and (8.5-9) for V_b, it is possible to extract an expression for F_p as a function of F_b and to report it in equation (8.5-6). This leads to :

$$V_b = V_p = \frac{M_b^0 \, F_b}{1 + N \dfrac{M_b^0}{M_p}} = M_b^p \, F_b \tag{8.5-10}$$

where M_b^p is the effective mobility of the grain boundary with its pores. Two extreme situations can then be considered. On one hand, when many pores are present, each with a mobility much lower than that of the grain boundary:

$$M_b^0 \gg M_p \tag{8.5-11}$$

then the effective mobility of the grain boundary is

$$M_b^p \approx \frac{M_p}{N} \tag{8.5-12}$$

That is to say, the pores control the grain growth, or coarsening process.
On the other hand if there are few pores which have a mobility much higher than that of the grain boundary:

$$M_b^0 \ll M_p \tag{8.5-13}$$

then the effective mobility of the grain boundary is practically identical to its mobility without pores

$$M_b^p \approx M_b^0 \tag{8.5-14}$$

That is to say, the migration of atoms across the grain boundary controls the coarsening which is not affected by the pores.
An approximate borderline between these two extreme situations is given by

$$M_b^0 \, N = M_p \tag{8.5-15}$$

In a diagram where the grain size r_g is plotted as a function of the pore size r_p, this corresponds to a line of equal mobility of the pores and grain boundaries, which separates a domain where the pores control grain growth by slowing down the motion of grain boundaries, and a domain where the pores do not control grain growth but remain at the grain boundaries (figure 8.5-5). The position of this line depends on the most efficient displacement mechanism of the pores.

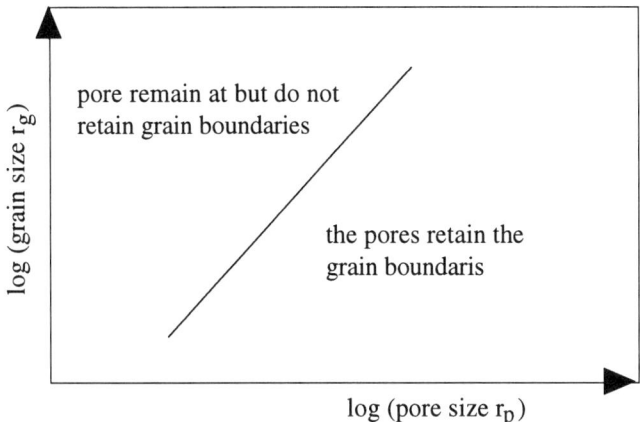

Figure 8.5-5 - Distinction of the domains where pore retain or do not retain the grain boundary migration.

ABNORMAL GRAIN GROWTH [45]

As densification proceeds, pores slowly disappear so that the drag force they can exert on the motion of grain boundaries gradually decreases. Hence, at a sufficiently advanced stage of densification, pores are eventually no longer able to limit grain growth.

Then three cases must actually be considered. A first possibility is that the residual pores remain small enough and are enough mobile to keep up with the grain boundaries; they simply do not control grain growth anymore and this was addressed in the previous section. However a second possibility is that these residual pores have begun to grow enough and their mobility has decreased too much to keep up with the grain boundaries, so that they separate from them. A third possibility is that pores which are separated from the grain boundaries could catch up again these grain boundaries.

Pore separation from grain boundary

After separation, no drag force is exerted on the grain boundaries so that some grains start to grow very fast, a regime known as abnormal grain growth or recrystallization (Figure 8.5-6).

In this case the grains grow linearly with time according to a law of the type :

$$r_g - r_{g0} = kt \qquad (8.5\text{-}16)$$

where k is a constant and r_{g0} the average grain size when abnormal grain growth begins.

Such a phenomenon occurs when the grain boundary velocity with its pores reaches the maximum pore velocity. At the moment of separation

$$V_b > V_p \qquad (8.5\text{-}17)$$

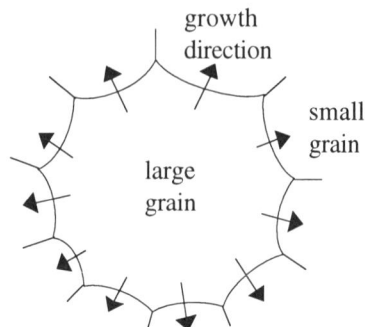

Figure 8.5-6 - Abnormal grain growth

Replacing V_b and V_p by their respective expressions in equation (8.5-10) and (8.5-9)

$$\frac{M_b^0 \, F_b}{1 + N \, \dfrac{M_b^0}{M_p}} > M_p \, F_p \tag{8.5-18}$$

or $\qquad F_b > \dfrac{M_p F_p}{M_b^0} + N \, F_p \tag{8.5-19}$

In practice, pore separation can occurs in conditions when equation (8.5-11) in the previous section prevails, that is to say when $M_b^0 \gg M_p$. In this case, equation (8.5-19) simplifies to

$$F_b > NF_p \tag{8.5-20}$$

For instance if a grain boundary moves by atomic diffusion across the grain boundary, with the diffusion coefficient D_{gb}^* the grain boundary mobility is

$$M_b^0 = \frac{D_{gb}^*}{RT} \tag{8.5-21}$$

If the distance between pores d_{pp} is proportional to the grain size

$$d_{pp} \approx \frac{r_g}{a} \tag{8.5-22}$$

Then the pore density $\qquad N \approx \dfrac{a^2}{r_g^2} \tag{8.5-23}$

The local curvature of the grain boundary between two pinning pores can be estimated to be $\approx \frac{r_g}{a^2}$ so that the driving force F_b to move the boundary according to Laplace's equation is

$$F_b \approx \frac{2\gamma_{gb}\, a^2}{r_g} \tag{8.5-24}$$

The dragging force F_p applied to a pore by a grain boundary is (Figure 8.5-4)

$$F_p = 2\,\pi r_p \gamma_{gb} \sin\theta \cos\theta \tag{8.5-25}$$

Its maximum value is

$$F_{p,max} = \pi r_p \gamma_{gb} \tag{8.5-26}$$

If F_p, F_b and N are reported in the inequality (8.5-20), this gives

$$r_g > \frac{\pi r_p}{2} \tag{8.5-27}$$

Pores catching up grain boundary
In this phenomenon, the grain boundaries have already separated from the pores but they themselves begin to slow down sufficiently, so that the pores could again join them. This last event occurs especially when impurities also contribute to slow down the motion of grain boundaries. In this case the mobility of the free grain boundary would be lower than that of a grain boundary with pores:

$$\frac{M_p}{N} > M_b^0 \tag{8.5-28}$$

In this case equation (8.5-19) becomes

$$\text{or} \quad F_b < \frac{M_p\, F_p}{M_b^0} \tag{8.5-29}$$

It has been shown that this corresponds to the following relationship where k is a constant

$$r_g = k.r_p^3 \tag{8.5-30}$$

Sintering maps
The two phenomena of pore separation and pore catching up the grain boundaries are summarized in the sintering map of figure 8.5-7. On a log scale, they are represented by straight lines. These two border lines between pore separation from grain boundaries and pore catching up grain boundaries, intersect on the borderline from figure 8.5-5, which separates two domains where pores control or do not control grain growth respectively, but remain at the grain boundaries.
The domains where pore separates from the grain boundaries must be avoided during processing, because pores which are trapped inside a grain can only be eliminated by

lattice diffusion, which is very slow. On the other hand, pores located at grain boundaries are more easily eliminated by atomic diffusion along the grain boundaries, which is a faster process. For practical purposes, the density reached in a ceramic when the grain boundaries separate from the pores can be considered as the final density.

Prevention of abnormal grain growth
Two ways of preventing abnormal grain growth are also sketched in figure 8.5-7: rapid densification and addition of impurities.
Rapid densification can be applied when the activation energy for densification ΔH_d is higher than that for grain growth ΔH_g, hence the diffusion coefficient of the mechanism responsible for densification increases faster than the one responsible for coarsening when the temperature increases. This technique simply consists of rapidly heating the material to a high temperature. It is easy to apply to small samples.
As this was previously mentioned, it is also possible to exert a drag force on the grain boundaries with impurities, either atoms termed dopants, or foreign particles. In fact, without dopants, abnormal grain growth would occur sooner or later.

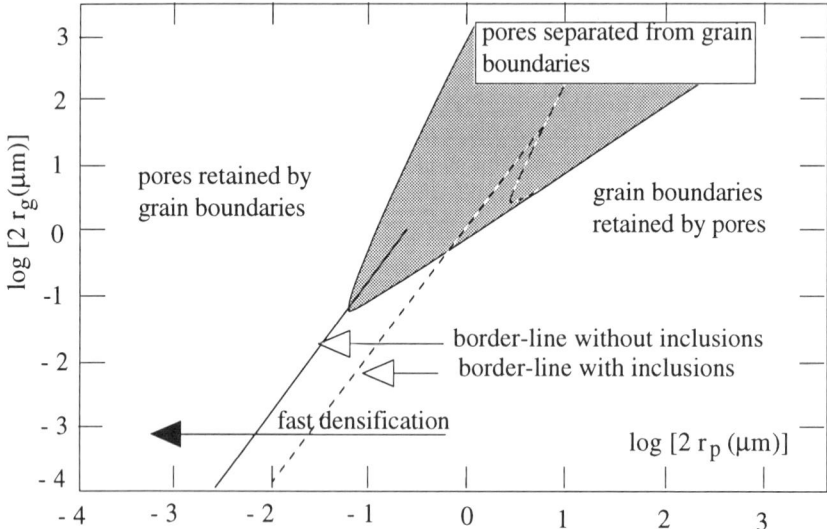

Figure 8.5-7 - Dependence of pore-boundary interaction on microstructural parameters in a system when the pores move by surface diffusion. Adapted from Brook [45].

Dopants increase the minimum grain size r_g^* above which pores separate from grain boundaries. If pores move by surface diffusion and grain boundaries by diffusion across the boundary for instance, r_g^* is given by:

$$r_g^* = \sqrt{\frac{16 \; \pi^3 \; D_S \; w_s \; V_m}{D_{gb}^* \; r_p^2}} \qquad (8.5\text{-}31)$$

Dopants have an effect on all diffusion coefficients. Acting on the surface diffusion coefficient D_s increases r_g^* in (8.5-31), which is profitable to avoid abnormal grain growth. However it also accelerates normal growth so that even dense ceramics end up having big grains, which is detrimental to many properties such as the mechanical properties. The best goal consists in increasing the lattice diffusion coefficient D_L and the diffusion coefficient along grain boundaries D_{gb} to favor rapid densification and to lower the diffusion coefficient across grain boundaries D_{gb}^* so as to reduce the domain where the pores separate from the grain boundaries.

To prevent abnormal growth, it is also important to select an initial powder with the most narrow particle size distribution possible, because pore separation from the grain boundaries first occurs with the largest grains. Yan et al. [48] have shown that the percent of theoretical density which can be attained with monosized particles is 99.3%, instead of 90.6% for a material where the maximum grain size $r_{g,max}$ is more than twice the average grain size $r_{g,av}$. The minimum dopant amount necessary to avoid abnormal growth is also much lower, about 14 times, with the former monosized distribution than with the latter one.

Abnormal growth in sol-gel ceramics
In sol-gel process, abnormal grain growth can occur at a low temperature, depending on the ceramics. As an example, in sol-gel ZrO_2 which sinters well, abnormal grain growth occurs at 1200°C, at a much lower temperature than by conventional processes [27]. This has the effect of limiting the final density when sintering is carried out at these temperatures. As with conventional ceramics, the absence of impurities at grain boundaries often explains this behavior [22].

The pores can have an effect similar to that of the impurities to slow down the grain boundary migration. However, for this mechanism to operate, they must be extremely small given the small crystallite size.

Some sol-gel ceramics have real densification problems, such as sol-gel α-alumina which nucleates above 1100°C from the transition aluminas [49]. In this case, the initial pores derive directly from the topotactic transformation of boehmite gel and they are too large in comparison with the grain size, to stop the grain boundary migration. Moreover they have a planar texture. Consequently they immediately remain trapped inside large α-alumina grains (figure 8.5-8) and are difficult to eliminate further on [50]. A solution to the sintering problem of these materials consists of adding foreign impurities to help pinning the grain boundaries. This is the technique used by Kumagai and Messing which could densify α-alumina from boehmite sols at 1200°C, with the help of α-alumina seeds [51]. Moreover, these seeds could act as α-alumina nucleation centers. Similarly, the good sintering behavior reported in Al_2O_3-ZrO_2 composites can be due to the fact that the ZrO_2 particles have a size which fits at 4-grains Al_2O_3 junctions [52]. Such seeding can be made by mixing stable sols of each oxide.

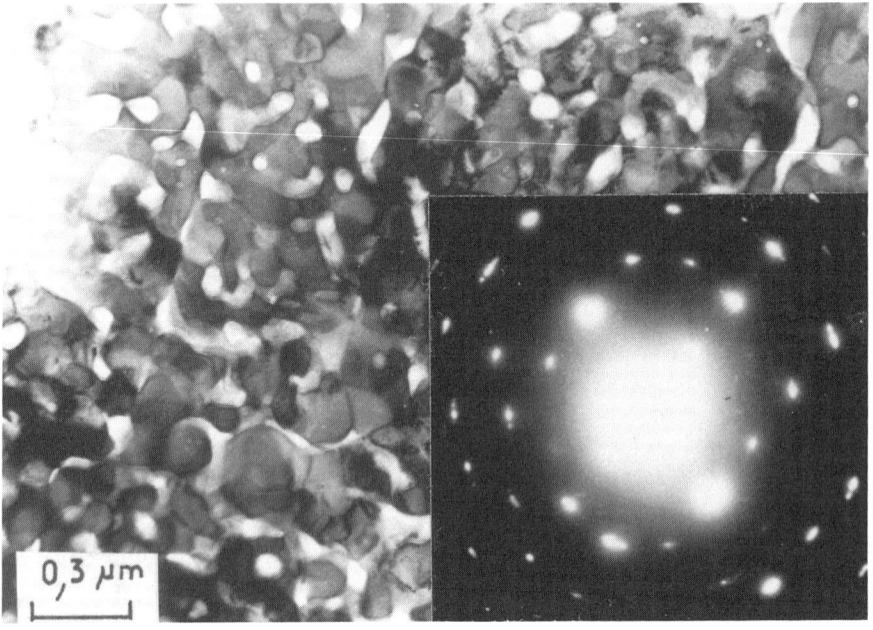

Figure 8.5-8 - Abnormal grain growth of α-alumina grains from boehmite gels which trapped a radial network of pores. From Pierre et al. [49].

To avoid or limit grain growth in sol-gel ceramics, it is also possible to chose a thermal treatment schedule adapted to the phase transformation mechanisms. For instance, it is possible to consider making heat treatments at temperatures lower than 750°C [53] for materials which densify at higher temperatures when made by conventional processing. This can offer the best compromise to maximize the relative densification rate with respect to the grain growth rate.

PORES DUE TO INITIAL POWDER PACKING
As shown by equation (8.5-7), small pores are submitted to a higher densification driving force than large pores. Hence, the pore size distribution is very important. This explains that densification often proceeds by two or more successive steps, each step corresponding to a different pore population. In this section, special pore configurations due to powder packing artifacts are examined in more details.

Case of agglomerates

In conventional ceramic powder processing, it is well known that the achievement of a good densification requires to eliminate the agglomerates and to start from a high packing density before sintering, that is to say a high green density. In a green part, at least two different pore populations are usually present: a first one inside agglomerates and another one between agglomerates.

In gel powder packing the situation is even more complex, as we must add the porosity internal to the gel. Agglomerates can form however, and their size distribution depends on the preparation method [54]. They can impart severe fluctuations on the sintering driving force. It was shown for instance that ThO_2 gels comprising un-agglomerated single particles sinter well at relatively low temperature. On the contrary, particle agglomerates retain their porosity during gelation and do not sinter to dense oxides before high temperatures. A density of 98% of theoretical was reached at 1100°C with un-agglomerated particles, instead of 1600°C with conventional agglomerated powders [55]. Un-agglomerated packing of titanate powder doped with rare earths also sinter below 1200°C [56].

Particles can be de-agglomerated by washing them in various solvents such as alcohols. As an example the size of Y_2O_3 stabilized ZrO_2 agglomerates could be reduced by half by washing them in alcohol [13]. Such washing is also known to be efficient even after calcination of the gels [57,58]. On the other hand, the advantages of pressing a green packing are moderate. In the case of Y_2O_3 stabilized ZrO_2 particles, it was found that pressing had no beneficial effect above 70 MPa [12].

In sol-gel materials, grain growth occur in the same time as agglomeration, so that it becomes difficult to discriminate experimentally between the so called primary particles, which consist of small grains or crystallites, and the secondary particles which are agglomerates of the primary particles [53]. As reported in chapter 7, the specific surface area of sol-gel ceramics begins to slightly decrease in the temperature range where transition phases are formed. It then accelerates when the stable phases begin to crystallize. Actually, the specific surface area decreases in the same time as grains grow. For instance, in a study on $HfO_2.Y_2O_3$, the initial crystallite size was extremely small (≈ 1 nm). It increased to ≈ 30 nm after thermal treatment between 550 and 600°C [59]. In the rare earth oxides Gd_2O_3, Dy_2O_3, Er_2O_3 and Y_2O_3, the drastic decrease of the specific surface area upon calcination at 800°C coincided with a grain growth from a few nm to a several tens of nm [60]. A similar result was found in $BaTiO_3$ [61]. Even for a non-oxide material such as ZnS powder made by thiolysis of Et_2Zn in a H_2S saturated toluene solution, the grain size increases from 0.02 to 0.1 μm during heat treatment at 800°C under vacuum [62]. It seems, therefore, that grain growth first affects the primary particles, until the secondary particles become each one single grain.

Monodispersed powder packing

Sol-gel processing is the first powder synthesis technique which made it possible to easily prepare particles with a monodispersed size distribution, and to study the effects of such monodispersion on green powder packing and sintering.

The experimental results showed that perfectly monosized spheres pack in perfectly ordered "colloidal" domains, which also comprise defects similar to the dislocations and vacancies observed in usual atomic crystals (Figure 8.5-9). Each colloidal domain, or grain, densifies very well. However, a macroscopic sample generally comprises a large number of such colloidal grains limited by grain boundaries and Liniger and Raj showed that these grain boundaries opened during the sintering process to form flaws which are difficult to eliminate further on [63]. Monodispersed particles make it difficult to achieve a monodisperse pore size distribution which is the important characteristic for a good sintering behavior. Liniger and Raj also showed that these packing artifacts could be eliminated by replacing monodispersed powders by appropriate polydispersed powders as explained after.

Polydispersed powder packing
It would seem at first that packing small particles in the interstitial sites located between larger particles would decrease the size of the pores to eliminate, and hence would accelerate densification. However, the situation is more complex than expected.

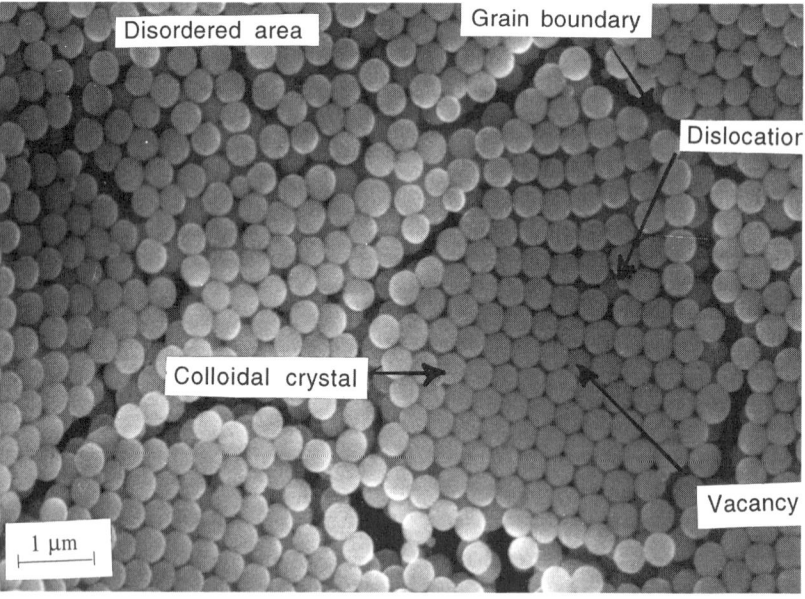

Figure 8.5-9 - Ordering of monosized silica spheres made by the Stöber process [64] and defects created.

If two types of powders with a different particle size are mixed with each other and one powder has much smaller particles than the other one, the smaller particles fit well in between the larger particles. However they densify much faster than the larger particles and they introduce sharp flaws which finally hinder the complete sintering of a green part. A large magnitude difference in the size of particles here again is responsible for the formation of an uneven pore size distribution (figure 8.5-10).

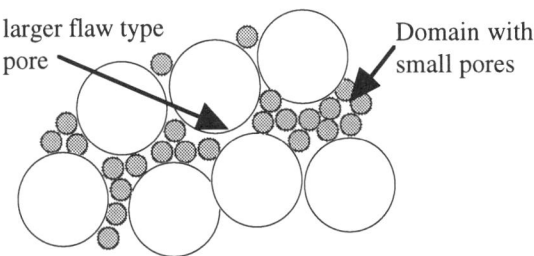

Figure 8.5-10 - Packing obtained with bidispersed spheres having a large size difference.

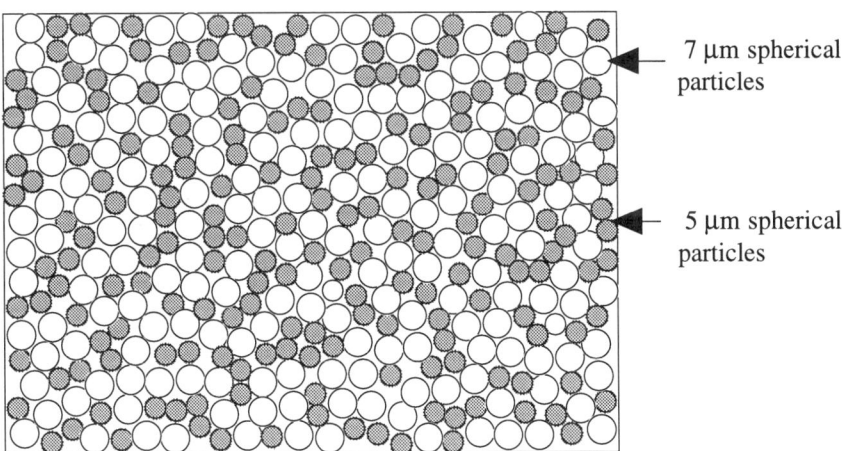

Figure 8.5-11 - Packing obtained when mixing 50% spheres of 7 μm diameter, with 50% spheres of 5 μm diameter. Adapted from Liniger and Raj [63].

On the other hand, Liniger and Raj [63] studied the influence of the size ratio and the number ratio of bimodal sphere distributions (two sizes of spheres), on their ordering and densification. They found that the best theoretical packing should be realized for a radius ratio of 0.26 [65]. However, when they attempted to make such packing from spherical particles dispersed in a liquid, the mobility of the small spheres was too high in comparison with the mobility of the large ones so that the two types of spheres segregated by diffusion. Experimentally, they found that the best sintering packing was achieved for a radius ratio of 5/7. Moreover, when varying the number ratio of each type of spherical particles from 0% to 100%, they found that the monocrystalline packing defects observed for each extreme number ratio, had disappeared for a 50%-50% mixture of the two types of spheres. With the latter number proportion, a complete random liquid-like packing was formed without any grain boundaries and dislocations (figure 8.5-11). The pore size distribution of such packing was monodispersed and it densified very well.

8.6 - HOT-PRESSING

Hot pressing is a technique which is especially useful for materials in which the coarsening rate is high. Generally, the densification rate increases with the applied pressure P_a as P_a^n , where the magnitude of the exponent n ranges from 1 for diffusion, to more than 3 for viscous deformation. When the grain size r_g increases, the densification rate decreases as r_g^{-m} where m = 2 for atomic transport by lattice diffusion and m=3 for grain-boundary diffusion.

For hot pressing to be efficient, it is necessary to apply a pressure higher than the driving force necessary to make the pores shrink. That is to say :

$$P_a \gg \frac{2\gamma}{r_{por}} \qquad (8.6-1)$$

For industrial applications, the most efficient hot pressing technique is hot isostatic pressing which is expensive but can be scaled up and makes it possible to eliminate most processing flaws.

In sol-gel materials, hot-pressing was first applied to silicate systems because they densify by viscous flow. It was used to make alumino-boro-silicate glass disks from gel powders between 650 and 700°C [66]. The viscous flow sintering model of Mackenzie and Shuttleworth was modified to take into account the applied pressure P_a, . Its result is that the relative density $D(t) = \frac{\rho}{\rho_s}$ follows a densification law expressed by:

$$\ln(1-D(t)) = \ln(1-D_0) + \frac{3P_a t}{4\eta} \qquad (8.6-2)$$

This model was further modified by Decottignies et al. [67] to take into account the evolution of viscosity η with time, that is to say the heating rate. Actually, the viscosity changes by an important extent during heating, as this was mentioned before and as this is illustrated in figure 8.6-1. Vasilos [68] showed that the densification of glass powders by hot pressing between 1100 and 1200°C obeyed Murray's law [69]:

$$\frac{d(1-D(t))}{dt} = \frac{3\ P_a}{4\eta}\ \frac{1}{\dfrac{dT}{dt}} \tag{8.6-3}$$

where $\dfrac{dT}{dt}$ is the heating rate. This result has been confirmed by other hot pressing investigations on SiO_2, La_2O_3-SiO_2 and B_2O_3-SiO_2 [67,70,71]. Silica gels derived from alkoxides densify more easily than gels derived from hydrosols and fused glass (figure 8.6-2). The sintering temperature of multicomponent glasses made from gel is significantly lower than those made by conventional processing, because the components are already mixed on a very fine scale. For instance, Dislich could make transparent pyrex plates by hot-pressing under 2800 atm between 650 and 700°C, instead of 1600°C by conventional techniques [66]. The behavior of mullite sol-gel powders is somewhat more difficult. Vacuum hot pressing is necessary to produce samples with a density of 99 to 99.5% of the theoretical density and the final materials are not transparent [24].

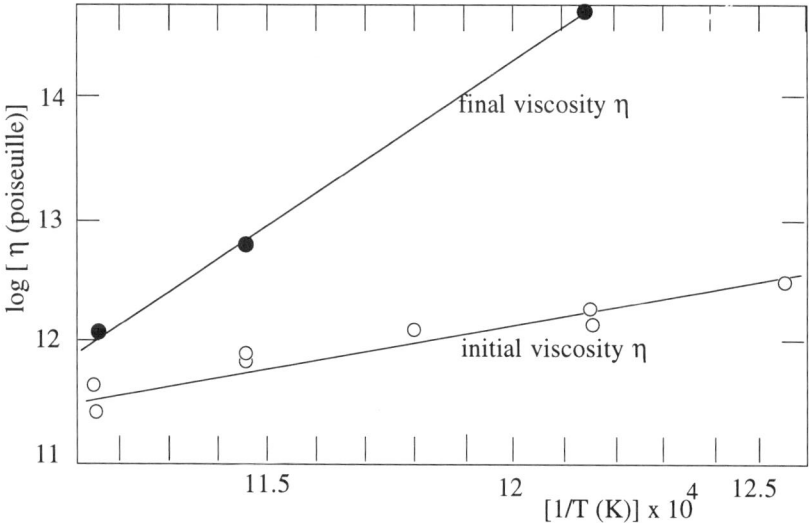

Figure 8.6-1 - Evolution of the viscosity as a function of the heat treatment time in a silica gel. Adapted from Decottignies et al. [67].

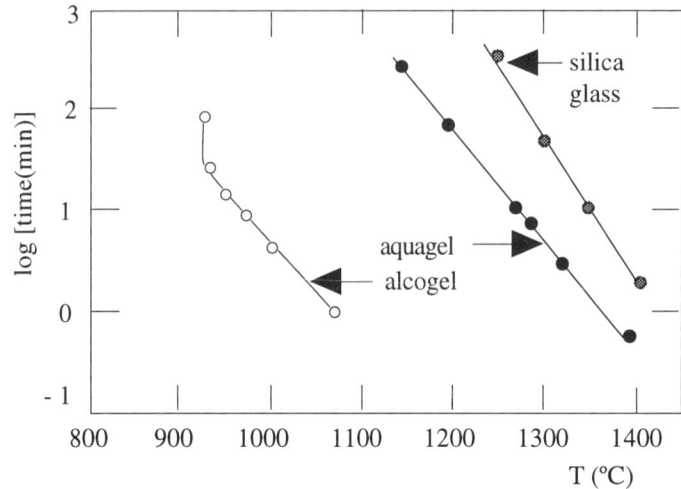

Figure 8.6-2 - Pressing time to make dense sample under a pressure of 42.6 MPa, as a function of temperature, for three types of silica powders. Adapted from Decottignies et al. [67].

In non-silicate systems, the application of high pressures at a given temperature improves densification, to some extent. Haertling and Land [72] have studied the case of the lanthanum modified lead zirconate ceramics, or PLZT. Hot pressing was applied at a temperature increasing from 900°C at the beginning of densification, and up to 1200°C at the end. The final material had a good optical transparency on dimensions of the order of 1.3 cm. in height and 3 cm in diameter.

Hot pressing is necessary with the non-oxide ceramics such as the carbides (e.g., SiC), even when the powder has been made by sol-gel from a carbon precursor in a SiO_2 gel. Some improvement could be achieved with the sol-gel powder since 99.5% of the theoretical density was reached after hot-pressing at 2000°C with 0.6% BN as the dopant [55], instead of 96% for conventional powder [73].

Overall, many studies remain to be done to understand the hot pressing behavior of non-silicate gels.

8.7 - REFERENCES

1 - Kingery W.D., Bowen H.K., Uhlmann D.R., "Introduction to Ceramics', Wiley, New-York (1975).
2 - Herring C., J. of Applied Physics 21 (1950) 301-303.
3 - Coble R.L., J. of Applied Physics, 32 (1961) 787-799.
4 - Petzow G., Exner H.E., Z. Mettalkde 67 (1976) 611-618.

5 - Ross J.W., Miller W.A., Weatherly G.C., Acta Met. 30 (1982) 203-212.

6 - Garino T.J., Bowen H.K., J. Am. Ceram. Soc., 70 (1987) C315-C317.

7 - Nii K.H., Z. Metallkde 61 (1970) 935-941.

8 - Scherer G.W., Garino T.J., J. Amer. Ceram. Soc., 68 (1985) 216-220.

9 - Bordia R.K., Raj R., J. Am. Ceram. Soc. 68 (1985) 287-292.

10 - Uhlmann D.R., Klein L., Onorato P.I.K, Hopper R.W., "The formation of lunar breccias: Sintering and crystallization kinetics", Proc. 6th Lunar Sci. Conf. , Ed. Merril R.B., Pergamon Press, New York, Vol. 1. (1975) 693-705.

11 - Yoldas B.E., J. of Mat. Sciences 21 (1986) 1080-1086.

12 - Rhodes W.H., J. Amer. Ceram. Soc. 64 (1981) 19-22.

13 - Yamagishi S., Takahashi K., J. of Nuclear Materials 144 (1987) 244-251.

14 - Daniels A.U., "Densification and sintering of thoria and thoria gels", University microfilms No 66-4752 (1966) 208 pages.

15 - Stuart W.I., Adams R.B., J. Nucl. Mater. 58 (1975) 201-204.

16 - Banister M.J., J. Am. Ceram. Soc. 58 (1975) 10-14.

17 - Davidge P.W., Woodhead J. L., "Ceramic materials by sol-gel route", U.S. Patent 4,429,051- United Kingdom Atomic Energy Authority - 31 jan. 1984.

18 - Shaw T.M., Pethica B.A., J. Am. Ceram. Soc. 69 (1986) 88-93.

19 - Komarneni S., Roy R., Breval E., J. Am. Ceram. Soc., 68 (1985) C41-C42.

20 - Barringer E.A., Bowen H.K., Comm. Amer. Soc. (1982) c199-c201.

21 - Mazdiyasni K.S., Lynch C.T., Smith II J.S., J. Am. Ceram. Soc. 48 (1965) 372-375.

22 - Colomban P., L'industrie Céramique 792-3 (1985) 186-196.

23 - Ghate B.B., Hasselman R.H., Spriggs R.M., Am. Ceram. Soc. Bull. 52 (1973) 670-672.

24 - Mazdiyasni K.S., Brown L.M., J. Am. Ceram. Soc. 55 (1972) 548-552.

25 - Livage J., Lemerle J., Ann. rev. Mater. Sci. 12 (1982) 103-122.

26 - Brown L.M., Mazdiyasni K.S., J. Am. ceram. Soc., 55 (1972) 541-544.

27 - Colomban P., L'industrie Céramique 697 (1976) 531-535.

28 - Mazdiyasni K.S., Dolloff R.T., Smith II J.S., J. Amer. Ceram. Soc., 52 (1969) 523-526.

29 - Smith II J.S., Dolloff R.T., Mazdiyasni K.S., J. Amer. Ceram. Soc. 53 (1970) 91-95

30 - Mazdiyasni K.S., Lynch C.T., Smith II J.S., J. Amer. Ceram. Soc. 50 (1967) 532-537.

31 - Brinker C.J., Scherer G.W., "Relationships between the sol-to-gel and the gel-to-glass conversion", Proc. Conf. Ceram. Process. Gainesville, Florida. Ultrastructures Processing of ceramics, Glasses and Composites. Gainesville, Florida. Ed. Hench L.L., Ulrich D.R., Wiley, New-York (1984) 43-59.

32 - Yamane M., Aso S., Okano S., Sakaino T., J. Mater. Sci. 14 (1979) 607- 611.

33 - Brinker C.J., Mukherjee S.P., Conf. Glass Through Chemical Processing, March 19-21 (1980) Rutgers University, USA. Reference given by Dislich H., in "Glassy and Crystalline systems from gels: chemical basis and technical application", J. of Non-Crystalline Solids 57 (1983) 371-388.

34 - Blum J.B., Gurkovich S.R., J. of Mater. Sciences 20 (1985) 4479-4483.

35 - Frenkel J., J. Exptl Theoreti. Phys. (Russian) 9 (1939) 1238-1244, also J. Phys. USSR (English) 2 (1940) 49-54.

36 - Scherer G.W., "Glass from Colloids", Proc. XIII Intern. Cong. Glass, Hamburg, Germany, Glastech. Ber. 56 (1983) 834-838.

37 - Mackenzie J.K., Shuttelworth R., Proc. Phys. Soc. 62B (1949) 833-852.

38 - Scherer G.W., J. Am. Ceram. Soc. 60 (1977a) 236-239.

39 - Scherer G.W., J. Am. Ceram. soc. 60 (1977b) 243-245.

40 - Brinker C.J., Keefer D.W., Schaefer D.W., Ashley C.S., J. of Non-Crystalline Solids 48 (1982) 47-64.

41 - Nogami M., Moriya Y.,   J. Non-Cryst. Solids 37 (1980) 191-201.
42 - Sakka S, "Gel method for making glass", in "Treatise on Materials Science and Technology", Edited by M. Tomazawa and R.H. Doremus., Vol 22 (1982) 129-167.
43 - Fricke J., Journal of non-Crystalline solids 100 (1988) 169-173.
44 - Hillert M., Acta Met. 13 (1965) 227-238.
45 - Brook R.J., J. Amer. Ceram. Soc. 52 (1969) 56-57
46 - Kingery W.D., François B., "Sintering of crystalline oxides. I. Interactions between grain boundaries and pores", in "Sintering and Related Phenomena", Proc. Int. Conf. 2nd., Gordon and Breach, Ed. Kuczynski G. C., (1967) 471-498.
47 - Shewmon P., Trans. AIME, 230 (1964) 1134.
48 - Yan M.F., Cannon R.M.J., Bowen H.K., Chowdry U., Materials Sc. and Eng. 60 (1983) 275-281.
49 - Yang X., Pierre A.C., Uhlmann D.R., J. of Non-Crystalline Solids 100 (1988) 371-377.
50 - Dynys F.W., Ljunberg M., Halloran J.H., Mat. Res. Soc., 32 (1984) 321-326.
51 - Kumagai M., Messing G.L., J.Am. Ceram. Soc. 68 (1985) 500-505.
52 - Lange F.F., Hirlinger M.M., J. Amer. ceram. Soc., 67 (1984) 164-168.
53 - Mazdiyasni K.S., Ceramics International 8 (1982) 42-56.
54 - Rhodes W.H., Haag R.M., "High purity fine particulate stabilized zirconia (Zyttrite ")", Report No. AFML-TR-70-209 prepared by Avco Systems Division for the U.S. Air Force Materials Lab., Wright-Patterson AFB, Ohio, (1970).
55 - Segal D.L., J. Non-Crystalline Solids 63 (1984) 183-191.
56 - Mazdiyasni K.S., Am. Ceram. Soc. Bull. 63 (1984) 591-594.
57 - Hoch M., Nair K.M., Ceramurgica International 2 (1976) 88-97.
58 - Haberko K., Ceramurgica international 5 (1979) 148-154.
59 - Mazdiyasni K.S., Brown L.M., Amer. ceram. soc. 53 (1970) 585-589.
60 - Mazdiyasni K.S., Brown L.M., J. Amer. Ceram. Soc. 54 (1971) 479-483.
61 - Brown L.M., Mazdiyasni K.S., Anal. Chem. 41 (1969) 1243-1250.
62 - Johnson C.E., Hickey D.K., Harris D.C., Mat. Res. Soc. Symp., 73 (1986) 785-789.
63 - Liniger E., Raj R., J. Am. Ceram. Soc. 70 (1987) 843-849.
64 - Stöber W., Fink A., Bohn E., J. Colloid Interface Sci., 26 (1968) 62-69.
65 - Frost H.J., Raj R., J. Amer. Ceram. Soc. 65 (1982) C19-C21.
66 - Dislich H., Angew. Chem. Internat. Edit. Engl. 10 (1971) 363-370
67 - Decottignies M., Phalippou J., Zarzycki J., J. Mater. Sci. 13 (1978) 2605-2618.
68 - Vasilos T., J. Am. Ceram. Soc., 43 (1960) 517-519.
69 - Murray P., Rodgers E.P., Williams A.E., Trans. Br. Ceram. Soc. 53 (1954) 474-510.
70 - Mukherjee S.P., Zarzycki J., J. Am. Ceram. Soc. 62 (1979) 1-4.
71 - Jabra B., Phalippou J., Zarzycki J., Rev. Chim. Miner., 16 (1979) 245-246.
72 - Haertling G.H., Land C.E., Ferroelectrics 3 (1972) 269-280.
73 - Wei G.C., Kennedy C.R., Harris L.A., Amer. Ceram. Soc. Bull. 63 (1984) 1054-1061.

APPLICATIONS
OF SOL-GEL PROCESSING

9.1 - INTRODUCTION

Sol-gel processes are interesting from a scientific point of view as well as for new or potential industrial applications. In some rare examples, sols or gels can be used as end products. Most often however, they must be transformed to an appropriate ceramic phase with interesting properties and they must also be densified. In detail, a sol-gel fabrication process is made according to a chronology or a post-gelation treatment which depends on the end product. Those products retained in the present chapter are sols and gels as final materials, coatings, fibers, monoliths, hybrid organic-inorganic materials and high specific surface area products such as filtering membranes and catalysts. On the other hand, powders and glasses usually are intermediate products for the applications just mentioned before. They were largely addressed in chapter 3 and 7 and they are not reconsidered in the present chapter .

9.2 - APPLICATIONS IN THE SOL OR IN THE GEL STATE

SOLS
The current applications of ceramics as finished products in the sol state concern essentially the Ferrofluids[R], commercialized by the Ferrofluid corporation, and mentioned in chapter 3. These products were developed around the year 1960 by NASA to control the flow of rocket fluids in zero gravity environment [1,2]. These sols comprise magnetic colloidal particles such as Fe_3O_4 which can move when an appropriate magnetic field is applied to the sol. As a sol is kinetically stable, the fluid in which the magnetic colloidal particles are dispersed move altogether with these particles, hence they can be used as transmission fluids in non gravity.

GELS

Wet gels

A typical application in a wet gel state concerns V_2O_5 , as coating because of the special semiconducting properties of this oxide [3,4]. This gel can be deposited as a paint in thin layers on large areas, with a thickness of 50 to 1 000 nm. The semiconducting properties of the V_2O_5 gels are due to the fact that vanadium can have two different valence states; +4 and +5. Hence, the electrons can be transferred between the corresponding energetic levels by optical [5] or by thermal [6] activation. The mechanism of electrical conduction is typical of small polarons. These gels have been patented for antistatic coatings on the dorsals of photographic films [7]. They are appreciated for their lower sensitivity to humidity.

They can also be applied in electrical switching devices [8]. When a voltage higher than a critical voltage V_t is applied to the device, the electrical resistance of a gel layer is low and the switch is in the "ON" state. On the other hand, the electrical resistance is high and the switch is in the "OFF" position if $V<V_t$. The switch also returns to the "OFF" position when the electrical current falls below a threshold I_h. For such applications, a gel layer thickness is typically of the order of 1 μm, the threshold current I_h is above 50 mA and the value of V_t ranges from 10 to 20 volts depending on the $[V^{4+}]$ concentration. The relative proportion of tetravalent vanadium, $\frac{[V^{4+}]}{[V]}$, is comprised between 0.01 and 0.04. The lowest relative $[V^{4+}]$ proportion corresponds to the highest V_t values. For a 50 Hz ac current, the gel characteristics remain stable for at least several days. At last it is possible to intercalate organic molecules between the V_2O_5 atomic layers and to modify the gel properties [4,8-11].

Lamellar $WO_3.nH_2O$ (n=1 or 2) gels present similar properties. In "ON" state the gel is transparent and in the other one the gel is blue [12]. Hence these gels can also be used in optical displays. Multi-layer coatings could be made from such gels, they keep behaving well after many cycles and their memory is durable [13].

In a different domain, wet SiO_2 gels have found an application as media for the growth of crystalline particles [14], as this was mentioned in chapter 3, and in chromatography.

Dry xerogels and aerogels

Aerogels often have a very high specific surface area [15], up to 800 m^2/g. This property makes them outstanding thermal and acoustic insulation materials. For instance, they can be applied in acoustic delay lines and in piezoactive antireflective acoustic coatings of thickness $\frac{\lambda}{4}$. They are also used in Cerenkovo counters in high energy physics [16], because they make it possible to cover a continuous range of refractive index from 1.015 to 1.06. Such refractive indices were achieved with gases and liquids before and aerogels are much more convenient. Tiles of appropriate dimensions, such as 20cm x 20cm x 3cm, can easily be made by hypercritical drying.

SiO_2 aerogel monoliths can be very transparent, hence they are being investigated for insulation windows but their cost is prohibitive when made by supercritical drying. The cost can be drastically decreased when similar materials are made by drying with surfactants, which offers a real potential for such applications.

9.3 - COATINGS AND THIN FILMS

This is presently the domain where sol-gel processing is the most used, because drying stresses can be easily overcome in thin coatings.

FUNCTIONS OF SOL-GEL COATINGS

Sol-gel coatings can have many functions. The most frequent ones are optical functions, because oxide are transparent to visible light wavelengths. They can transmit, absorb, or reflect, radiations with a given wavelength. They can also be used to protect a substrate against corrosion, abrasion, or scratch, be chemically and thermally stable, or even stable against some radiation.

The chemical applications include the possibility to improve the chemical durability of surfaces, for instance glass surfaces, by coating them with β-Al_2O_3 or α-Al_2O_3 [17,18]. Sol-gel CeO_2 coatings can also be made to protect gas cooled reactors [19].

Some sol-gel glass coatings are used to protect against hydrolysis some special glasses sensible to water used in laser filters. A multicomponent oxide glass made by sol-gel and corresponding to reaction (9.3-1) was the first to be commercialized, for the latter application [20]:

$$mSi(OMe)_4 + nAl(OBu^s)_3 + aP_2O_5 + pMg(OMe)_2$$
$$\Rightarrow \qquad Si_mAl_nP_aMg_pO_{(4m+3n+5a+2p)/2} \qquad (9.3-1)$$
$$(H_2O , T)$$

They can constitute protection barriers in both directions, for instance, by stopping alkali migration towards the surface of glass sheet made by the float process. Protection barriers against the alkalis have been made from the butoxide of Sb, B, Ti, Al, and deposited inside electric bulbs. SiO_2-GeO_2 coatings make it possible to protect non-oxide ceramics against oxidation at a high temperature with a thermal expansion adjusted to that of the substrate. The oxidation resistance of stainless steels has been improved by superposition of silica and ceria layers in which alumina colloidal particles were added [21]. The coating thickness was from 0.1 to 2 μm.

Sol-gel coatings can be deposited on substrates other than ceramics, such as plastics and metals and, as another non-negligible advantage, they are economical at least on simple shapes such as plates or tubing. A summary of possible functions of sol-gel coatings, depending on the substrate nature is reported in table 9.3-1. Plastic sheets with an appropriate surface treatment give the best results in terms of thickness, homogeneity and bending strength.

Table 9.3-1 - Effects of thin film coatings made from alkoxides. After Sakka [22].

Substrate	Function of coatings
Glass	chemical durability
	alkali resistance
	mechanical strength
	reflectivity control
	coloring
	electrical conduction
Metal	corrosion resistance
	oxidation resistance
	insulation
Plastic	surface protection
	reflectivity control

OPTICAL COATINGS

Sol-gel coatings have been investigated by many researchers such as Schroeder [17,23] and Dislich [24,25], more particularly for optical and electrical applications. Several review articles have addressed the subject [26,27] and a partial list is reproduced in table 9.3-2. Another list of single oxide films with their optical absorption characteristics is given in table 9.3-3 and it includes the chemical precursor from which they were made.

In the applications of sol-gel to make antireflective films, the film thickness can be correctly adjusted by the technique of dip-coating presented further on, while the sol composition makes it possible to monitor the refractive index. For instance, TiO_2 coatings are valuable as solar shields on glass [26,28]. It is possible to adjust the visible light absorbency without changing the antireflective qualities by embedding palladium in the TiO_2. The TiO_2-SiO_2 system was well studied [29], such materials have been used by the "Deutsche Spezialglas AG" company to make coatings, since 1964. More recently, the Sandia and Westinghouse companies have applied them to silicon solar cells, while the Battelle Institute has applied them on laser lenses to control nuclear fusion [30]. The coatings made by sol-gel are even more resistant to bombardment than the previous coatings and their resistance to alkalis can be increased by incorporating some ZrO_2, itself made from alkoxide.

The first wide-band antireflective coatings from multicomponent oxides were made in the SiO_2-Na_2O-B_2O_3 system where phase separation occurred. The most soluble phase was leached out with the help of an acid. The final microporous material is outstanding to concentrate high power laser beams in nuclear fusion experiments and can withstand a laser bombardment of 21 J/m^2 instead of 5 J/m^2 for the same material made by conventional techniques. Similar results were achieved with the composition SiO_2-Al_2O_3-CaO-MgO-Na_2O, by the Schott company [20]. Brinker and Pettit also quoted the composition BaO-Al_2O_3-B_2O_3-SiO_2 [33]. Antireflective sol-gel laser coatings have also been developed in France, by the CEA society. They consist of porous silica monolayer coatings with a thickness of $\frac{\lambda}{4}$ or two-layer coatings comprising a methylsilicone layer and a porous silica layer [34].

Table 9.3-2 - Partial list of sol-gel coatings made from alkoxides.

Composition	reference
TiO_2	[17]
SiO_2	[17]
SiO_2 + 20% (Cu, Co, Ni, Ce)	[22]
$TiO_2.SiO_2$	[23]
$TiO_2.SiO_2$ + 20% (Cu, Co, Ni, Fe)	[22]
$TiO_2.4SiO_2$	[31]
88%SiO_2,5%B_2O_3,2.6%Al_2O_3,3.7%Na_2O,0.7%K_2O	[30]
60%SiO_2,25.2%Al_2O_3,0.02%MgO,9%P_2O_5,2.6%B_2O_3,0.1%As_2S_3	[30]
62%SiO_2,30%PbO,8%Na_2O	[30]
80%SiO_2,10%Al_2O_3,10%Na_2O	[30]
$ZrO_2.4SiO_2$	[31]
$Al_2O_3.4SiO_2$	[31]
0.5%Al_2O_3,93.5%SiO_2	[32]

Tin doped Indium oxide coatings (or ITO) can also be made by sol-gel [25]. They are very transparent to visible solar radiation but they reflect the long waves infra-red radiation. Consequently, they can be used in heat mirrors. They let the visible solar light enter through a glass window but they keep the heat inside. Their thickness is $\approx$ 1μm at most and the sol-gel processes make it possible to prepare them more simply and more rapidly than other techniques.

Other valuable optical coatings made by sol-gel include PLZT films for optoelectronic applications and transparent electrically conductive films of Cd_2SnO_4, made from Cd acetate and Sn alkoxide. The sol-gel technique makes it possible to avoid the formation of phases such as SnO_2, CdO and $CdSnO_3$, which limit the conductive properties. The formation of these phases cannot be avoided in conventional techniques, including the technique of cathodic sputtering.

Table 9.3-3 - Optical characteristics of single metal oxide coatings.
After Schroeder [23].

Oxide	Favored precursor	refraction index n	Absorbing below (nm)	Structure
Al_2O_3	$Al(NO_3)_3.9H_2O$	1.62	$\approx$250	Amorphous
	$Al(OBu^s)_3$			Crystalline
CeO_2	$Ce(NO_3)_3.6H_2O$	2.11	400	Crystalline
HfO_2	$HfOCl_2.8H_2O$	2.04	$\approx$220	Crystalline
In_2O_3	$In(NO_3)_3$	1.95	420	Crystalline
La_2O_3	$La(NO_3)_3$	1.78	220	
Nd_2O_3	$Nd(NO_3)_3$	Inhom.		
PbO	$Pb(CH_3COO)_2$	Inhom.	$\approx$380	Amorphous
Sb_2O_4	$SbCl_5$	1.90	340	
SiO_2	$Si(OR)_4$	1.455	$\approx$205	Amorphous
SnO_2	$SnCl_4$	Inhom	350	Crystalline
Ta_2O_5	$TaCl_5$	2.1	310	
ThO_2	$ThCl_4$	1.93	$\approx$220	Crystalline
	$Th(NO_3)_4$			
TiO_2	$TiCl_4$	2.3	380	Crystalline
	$Ti(OR)_4$			
Y_2O_3	$Y(NO_3)_3$	1.82	$\approx$300	
ZrO_2	$ZrOCl_2$	1.72	340	Crystalline

FABRICATION TECHNIQUES

Several different techniques can be used to produce coatings from solutions, such as [23]:
- Dip-coating, which consists of dipping the part to coat in a precursor solution, then to pull it out. The object to be coated can also be fixed and the solution container is moved successively up and down.
- Spreading a film of the precursor solution on the part to coat which is spinned. This technique applies well to cylindrical surfaces and disks.
- Spraying a solution on the surface to coat. However, it is a little more difficult to maintain a uniform film thickness in the desired range by this technique.
The technique of dip coating is the most used: it is applied on 12 m^2 plates by the Schott Company [30]. Hydrolysis is performed in an atmosphere with controlled humidity once the plates have been pulled from the sol. Next, they are heat treated at a temperature between 400°C and 500°C. It is necessary to control the diffusion,

first of the incoming water used for hydrolysis, secondly of out coming condensation products, to achieve uniform, homogeneous and dense coatings.

The sol or solution in which the part to coat is dipped can consist of a polymeric solution derived from alkoxides, a solution of metal salts with volatile anions in an organic solvent, or a colloidal sol. For a good process control, it is desirable to chose a precursor solution with the following qualities: a high solubility of the precursors, a good fluidity, constant properties in time, a good wetting property on the substrate to be coated, a gelation behavior without heterogeneous precipitation, a transformation to an oxide film with a good adherence to the substrate.

The surface to be coated must be cleaned and a constant pulling rate must be maintained. A coating operation is realized at a relatively low temperature, <500°C, so that the substrate is usually not modified. Adherence of the coating to the substrate is made possible by the presence of OH radicals on the surface to be coated. For instance on glass, a sol-gel oxide MO_x can be linked to the substrate by a reaction of the type :

$$\text{glass}-\overset{|}{\underset{|}{Si}}-OH + RO-M \;\Rightarrow\; \text{glass}-\overset{|}{\underset{|}{Si}}-O-M \; + ROH$$

$$(9.3\text{-}2)$$

To achieve a coating without flaws and to avoid peeling from the substrate, the bonding between the hydrolyzed precursor and the substrate must be established before gelation occurs. For instance, it is possible to deposit on glass slides, SiO_2 and TiO_2-SiO_2 glass coatings which, after heat treatment between 200 and 500°C, give transparent and uniform films with an adhesive strength of 100 kg/cm^2.

The coating thickness can be controlled by monitoring the viscosity of the solution. Usually, it is comprised between 0.1 µm and 0.3 µm after one dipping operation (figure 9.3-1).

The parameters which make it possible to control the coating thickness ε are the sol viscosity η, its density ρ and the velocity v at which the surface to be coated is pulled out of the sol. These parameters are related to the coating thickness ε by the relation [35]:

$$\varepsilon = K \; \frac{\eta v}{\rho g} \qquad\qquad (9.3\text{-}3)$$

where g is the gravity and K is a constant which depends on the material.

When the viscosity of the solution is too high, the film is thick and contains many defects. However, this depends on the material. A thickness of 1 µm with a good quality can be obtained with β-alumina made from the alkoxides $NaOC_3H_7$ and $Al(OC_4H_9)_3$ [36]. To increase the film thickness, additives can be added to the sol or the dipping operation can be repeated several times, such as in a study on complex coating with composition 6.5% Al_2O_3, 93.5% SiO_2 by Shimbo et al. [32].

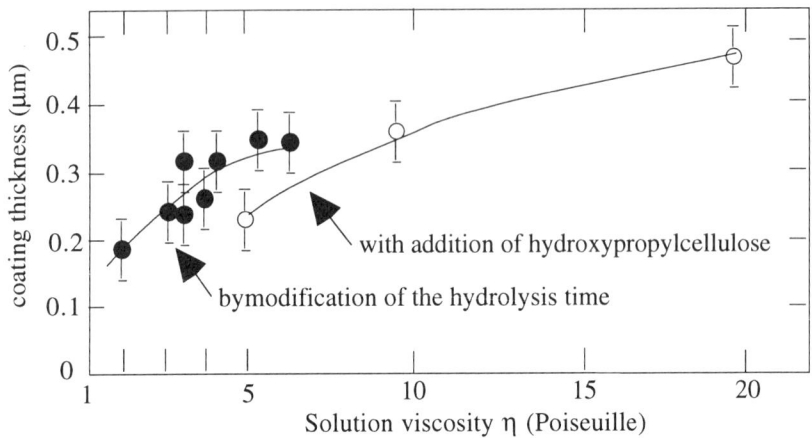

Figure 9.3-1 - Variation of the thickness of coating films with the viscosity of Si(OEt)$_4$ solutions. Adapted from Sakka [22].

FREE STANDING FILMS

To the important domain of coatings, one must add the synthesis of free-standing films, summarized by Sakka [22]. Such films could be used as insulators, electric capacitors, magnetic materials and as fillers for reinforced plastics. Their fabrication techniques include:
- Spreading an alkoxide film over water and scraping off the gel film formed.
- Gelation over a liquid unmiscible with a sol. Mercury can be used as the liquid but it introduces environmental problems. The Wood metal, which melts at 65°C, is a convenient substrate. Tetrabromoethane also is a liquid immiscible with most sols. It has a low reactivity, a high density and a high surface tension.
- Gelation over a non-wetting surfaces such as Teflon®,
- Pulling a sol out of a slit.
- Pulling off a film held by capillary tension to a wire, from a sol which has an appropriate viscosity.
The latter technique has been applied to the synthesis of films without support in SiO$_2$; SiO$_2$ + 8 to 70 mole % TiO$_2$ and SiO$_2$ + 2 to 70 mole% ZrO$_2$ [22]. The best adapted solutions comprised a molar hydrolysis water proportion r_w < than 4, which corresponds to the formation of polymeric gel structures. The radial shrinkage during the glass conversion is 50% to 60% and the film thickness ranges from 10 to 40 μm.

9.4 - FIBERS

MAIN COMPOSITIONS

Sol-gel processes are easily applicable to the formation of oxide fibers, composed of a single or of several components, polycrystalline or glassy. Their composition includes many silicates but also other compounds such as TiO_2 or Al_2O_3, a non-limitative list is reproduced in table 9.4-1.

Table 9.4-1 - Ceramic fibers made by sol-gel.
Adapted from Zelinski and Uhlmann [26].

System	References
Glass fibers:	
SiO_2	[37,38,39]
SiO_2-Al_2O_3	[40]
SiO_2-TiO_2	[20,40,41]
SiO_2-ZrO_2	[42,43]
SiO_2-ZrO_2-Na_2O	[42,43]
Polycrystalline fibers	
α-Al_2O_3	[44]
Al_2O_3-B_2O_3	[45]
Al_2O_3-B_2O_3-SiO_2	[45]
Al_2O_3-Cr_2O_3-MO_2 where M= Si, Ti, Zr or Sn	[46]
Al_2O_3-SiO_2 (or P_2O_5) B_2O_3 (or Cr_2O_3)	[47]
TiO_2	[20]
TiO_2-CaO	[20]
TiO_2-MgO	[20]
TiO_2-Al_2O_3	[20]
TiO_2-CaO-Al_2O_3	[20]
SnO_2	[20]
ZrO_2	[20]
VO_2	[20]
ThO_2	[20]

The mechanical properties of these fibers is not as good as those of fibers made by melt quenching. However some of these fibers have been applied in special applications such as the alumina Saffil[®] fibers made by the ICI company, used in the space shuttle tiles. The sol-gel techniques to make fibers become interesting in the case of glasses with a high melting temperature, or when melting is impossible

for some other reason, such as for fibers of composition ZrO_2-SiO_2 with a high ZrO_2 content. The latter fibers show a remarkable resistance to alkalis but they are hardly feasible by the technique of drawing from a melt. On the other hand they can be synthesized by hydrolysis and polycondensation of $Zr(OPr)_4$ and $Si(OEt)_4$ alkoxides [48].
Sol-gel fibers made are not limited to oxides, they include phosphates [20] and carbides or nitrides made by incorporation of a carbon or a nitrogen precursor such as an organic polymer in the gel, which is decomposed during heat treatment [49]. Such fibers have been made in the mixed Si_3N_4-SiC system; their reported fracture stress was ≈ 2.8 GPa and their elastic modulus ≈ 200 GPa [50].

FABRICATION TECHNIQUES
The main techniques to produce fibers by sol-gel are spinning, pulling out from a solution, extrusion and directional freezing. In all cases, the viscosity and the structure of the initial solution, sol or gel, are important. Sol gel silica fibers made by Sakka are shown in figure 9.4-1.

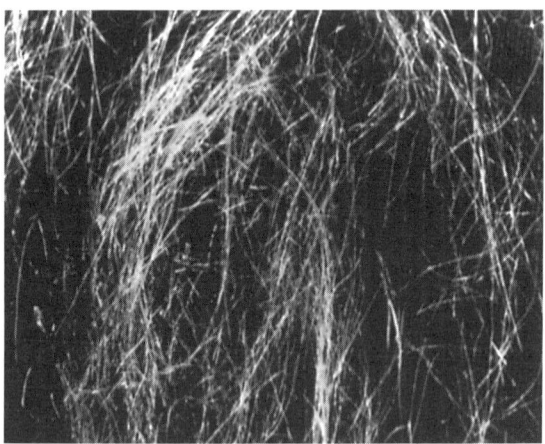

Figure 9.4-1 - Silica fibers made form gels. By Sakka [22] with permission.

For a good spinnability, solutions must contain linear polymeric species. Hence, it is necessary to hydrolyze the silica precursors with a water proportion lower than 2, such as in the system with mass composition 10 to 70 % TiO_2 - 90 to 30 % SiO_2 [51]. In this case, the appropriate sol viscosity is between 1 and 10 poiseuilles (SI units or $N.s.m^{-2}$) [26]. This target can often be achieved by hydrolysis in contact with water vapor dispersed in air, as this has been done for fibers of composition Al_2O_3- SiO_2, ZrO_2-SiO_2 and Na_2O-ZrO_2-SiO_2 [40,41,42]. Spinning must be performed before gelation, which is achieved by heating.

Fibers are frequently made by immersion of a glass rod in a polymeric solution and the glass rod is slowly pulled up. The solution must present the same linear structure as used for spinning and gelation is achieved in the same way [42,52]. Eventually, it is possible to add a gelling agent such as polyvinyl alcohol. The process is very sensitive to chemical conditions such as the pH [27] which can have a considerable effect on the viscosity. Depending on its value, a sol can be used to pull short staple fibers or of long fibers. Short fibers can be used in thermal insulation, an application where a good mechanical integrity is not required and for which sol-gel is relatively cheap because of the low processing temperature.

Conversion to glass does not create special problems because the fibers have a fine diameter < 1 μm. The main difficulty, in comparison with the technique of pulling from a melted oxide at very high temperature, is due to the slow diffusion kinetics of the hydrolysis water and condensation products. The problem arises mainly when the technique is applied industrially in a continuous process [20].

Also, even if the drying stresses do not break the fibers, because of their small diameter, it can be difficult to avoid the formation of flaws which limit the mechanical resistance of sol-gel fibers in thermo-structural applications.

The technique of extrusion makes it possible to use gelled sols or solutions, and it is mainly used for crystalline materials such as α-Al_2O_3 , mullite ($3Al_2O_3.2SiO_2$), aluminum borosilicate ($3Al_2O_3.B_2O_3.3SiO_2$) and Al_2O_3-Cr_2O_3-SiO_2 compounds [44,46,47]. Sol-gel powders can be used with this technique to make optical fibers preforms in SiO_2 and Al_2O_3 with a high transmission coefficient [53].

Fibers and plates with a specific surface area of 800 to 1200 m^2/g can been prepared by unidirectional freezing silicic acid solutions [38].

9.5 - MONOLITHS

Monoliths can be made either by direct transformation of gel monoliths, or by sintering sol-gel powder green parts.

TRANSFORMATION OF GEL MONOLITHS

This first method is more difficult to use than when making fibers, because the stresses which develop during drying and phase transformation currently break a sample [20]. To avoid such fracturation, the drying and glass transition or crystallization stages must be performed slowly and carefully controlled. However, new drying techniques such as supercritical drying, or drying with a micellar surfactant, have brought some significant progress.

One advantage of making sol-gel monoliths comes from the purity of the products which are obtained when they are made from alkoxides. Besides, monoliths with flaws can be used as preforms for further transformation steps, such as silica communication fibers which can be melted in a further stage from a preform. A silica sol-gel monolith preform was presented as early as 1982 at the annual conference of the "American Ceramic Society"; its dimensions were 12 cm in length and 3 cm in diameter. The communication fibers drawn from it had a low loss factor $\approx$ 6 dB/km [54]. For the same application, coatings from alkoxides of Ba, Ti, Sn, Al, Si, Mg, Pb, can also be deposited inside a SiO_2 tubing so that the optical fiber densified from the preform has a sol-gel made core. Ruptured lumps can eventually be melted at a temperature often lower than by conventional processes and with a better homogeneity and purity.

Many groups have attempted to make crack free monoliths [20]. Some monoliths were glassy such as with SiO_2, others were crystalline such as with Al_2O_3. The best success has been pieces with a thickness of the order of one cm made from SiO_2 and TiO_2 xerogel monoliths [22]. Similar small xerogel alumina monoliths have been made by Yoldas [55] and by Ozaki and Hideshima [56]. Often, the appellation monolith designates samples with a thickness of the order of a few mm, such as in the partial list given in table 9.5-1.

The best success from classical xerogels were generally obtained by hydrolysis in excess water followed by slow drying [55,58,66].

Table 9.5-1 - Examples of small glass monoliths prepared by hydrolysis of alkoxides. After Sakka [22].

Molar composition	Size (mm)	References
SiO_2	10 x 8 x 0.8	[39,51]
	25 x 15 x 1	[57]
	15 x 15 x 2	[57]
	15 (diam) x 5	[58]
	12 (diam) x 20	[59]
	23 (diam) x 2	[60]
	65 (diam)	[61]
$1\%B_2O_3,99\%SiO_2$	15 (diam)	[52,62]
$50\%Al_2O_3,50\%SiO_2$	10 x 5	[63]
$CaO.4SiO_2$ and $CaO.9SiO_2$	10 x 8 x 2	[64]
$5\%TiO_2,95\%SiO_2$	8 x 8 x 5	[51]
$7.9\%TiO_2.92.1\%SiO_2$ and $10\%TiO_2,90\%SiO_2$	20 x 20 x 4	[55]
doped SiO_2	$\approx$ 100 (diam)	[65]

Much better results have been achieved by mixing organic additives, termed drying control chemical additives (DCCA), to the sols. By this technique, Hench and

Nogues could make near net shape finished optical lenses [65]. As mentioned in a previous chapter, these organic additive decrease the surface tension of the liquid, and the drying stresses. They also make a more uniform pore texture such as this occurs with micelles, which limitates cracking due to buckling. In a related procedure, a colloidal ludox sol was mixed with potassium silicate and an organic additive; formamide [67]. After gelation, the monoliths obtained dried to a gel with large pores ($\approx$ 60 nm) and were transformed to glass by sintering at a high temperature.

By the supercritical drying method, rods about 1 cm in diameter and several cm in length or blocks $\approx$ 10 cm x 10 cm x 1 cm, are easy to realize in the silicate systems such as SiO_2, B_2O_3-SiO_2, SiO_2-P_2O_5 and SiO_2-B_2O_3-P_2O_5 [68]. This also is a faster technique than natural evaporation, but much more expensive.

MONOLITHS FROM SOL-GEL POWDER

Gels can be granulated and sintered or hot-pressed. In this way, borosilicate glasses have been synthesized by Dislich [30], and SiO_2, La_2O_3-SiO_2 and B_2O_3-SiO_2 glasses were made by Decottignies et al. [69].

This is a technique much easier to apply than the previous ones to manufacture crystalline materials industrially. The first industrial application dates back to the years 1960 for nuclear fuels and was summarized by Segal [27]. The fuels included UO_2 pellets containing 2 to 5% of ^{235}U and mixed $(U,Pu)O_2$ for the fast breeders.

These pellets were composed of identical spheres with a size > 1 mm. An important advantage of this technique, at the origin of its choice, was to avoid pollution by dispersion of radioactive matter. Later on, it was found that these spheres densified at a temperature lower and gave better products than the conventional powders. As a result the process was extended to materials of composition ThO_2, ThO_2-UO_2, UO_2-ZrO_2, ZrO_2, ThC_2 and ThC_2-UC_2 [70-75].

Application examples of this technique in the so-called field of technical ceramics is given in table 9.5-2. The titanates occupy an important place. Samples of lead titanate ($PbTiO_3$) with a diameter of 1.5 cm could also be directly obtained by transformation of monoliths [76]. The sol-gel product was more pure, more homogeneous and obtained at a lower temperature than with conventional processes. However its crystallinity was not perfect. The technique of sol-gel powder alleviated this inconvenience. For instance, it made it possible to synthesize samples of the complex PLZT titanate, completely dense and transparent [77]. These complex titanates of composition $Pb_{1-1.5}La_xZr_{1-y}Ti_yO_3$, are ferroelectric and birefringent.

All secondary phase segregation at the grain boundaries was eliminated, the grain size was submicronic, the porosity low, the homogeneity with a good stoichiometry was excellent; all these qualities explained the exceptional optical transparency of materials made by sintering sol-gel powders.

The processing of sol-gel powders has provided better quality dielectrics, which opened the way to the miniaturization of capacitors, transducers, very-high frequency wave guides and microwave containing equipment [96]. Materials for high-capacitance devices must have a high dielectric constant and a low loss factor in small sample. A titanate such as $BaTiO_3$ is well suited to this function. Concerning piezoelectric materials used in memory package applications, a low coercive force and a wide variation of the effective birefringence as a function of the remanent polarization are needed. In the domain of microwave applications, a high relative dielectric constant of at least 10 and if possible in the range of 100, a low loss tangent and a good constancy of the relative dielectric constant as a function of the temperature, must be combined with a high thermal conductivity necessary to minimize any local heating. For this purpose, $BaTiO_3$ with a dielectric constant of 5000 at room temperature, and a loss tangent coefficient of the order of 4 x 10^{-3} , is again well suited [96]. Other phases in the same $BaO-TiO_2$ system, which have a dielectric constant of 15000 at room temperature together with a high dispersion of the conductivity as a function of the frequency, are only feasible by sol-gel. The use of such materials has now been extended to radars and communications.

Table 9.5-2 - Monolithic ceramics made by sol-gel methods.
After Colomban [78].

composition	Type of monolith			references
	ceramic	transparent ceramic	glass	
P.L.Z.T		x		[77,79-82]
$BaTiO_3$	x			[83,84]
$SrTiO_3$	x			[83,84]
$SrZrO_3$	x			[83,85]
$MgAl_2O_4$	x			[30]
Mullite	x			[83]
β-eucryptite	x		x	[30,86,87]
$K_2ZrSi_2O_7$	x			[88]
NASICON	x		x	[87-93]
Na-$βAl_2O_3$	x			[78]
K-$βAl_2O_3$	x			[78]
ZrO_2:Y_2O_3		x		[94]
HfO_2	x	x		[83]
TiO_2	x			[95]
WO_3	x			[11]
Nb_2O_5	x			[12]

Also, some ferroelectric titanates which have a high ac and/or dc loss factor when they are in the semi-conducting state, are used as "positive temperature coefficient" (PTC) thermistors, for instance in electrooptic holograms mass memory, page composers and transparent optical data processing equipment.

To this list of applications, one can add permanent magnets with a high coercive force, which require ceramics in which each grain comprises one single magnetic domain. The optimum grain size is of the order of 900 nm in the case of ferrites [97]. In all these applications, the sol-gel processes constitute original techniques which permit a significant amelioration.

Table 9.5-2 mentions a compound termed NASICON. This material is an ionic conductor with sodium and silicon with a composition such as $Na_3Zr_2Si_2PO_{12}$ [98]. Its synthesis by conventional process leaves traces of zirconia at the grain boundaries which deteriorate its physical, electrical, and mechanical properties. Such inconveniences can be avoided with the sol-gel processes. A glassy state without grain boundaries can even be obtained between 600°C and 700°C. Its electrical properties are outstanding in spite of a non-negligible residual porosity. This glass crystallizes near 900°C.

Thermo-structural ceramics can also be synthesized by sol-gel process. As an example, yttria stabilized zirconia powders with a size of 20 to 70 μm made by sol-gel have been used as the feeding powder in plasma spraying [99]. Sol-gel made tetragonal ZrO_2, which is a metastable phase at low temperature, has been studied for use in the fabrication of refractory monoliths [66]. The fabrication of abrasives is an application where the product performance can be improved by sol-gel techniques: α-Al_2O_3 based abrasives with a better quality than those made by fusion above 2000°C can be made by sol-gel [100]. This improved result was achieved by homogeneous mixing of colloidal boehmite particles with zirconia, hafnia and spinel. It produced a texture with a fine grain size below 0.3 μm and a random grain orientation.

COMPOSITE MATERIALS

Ceramic composites can be divided in particulate composites, short fiber composites and long fiber ones. For each type of composite, sol-gel processes can be used to synthesize the dispersed phase (particles, short or long fibers), or the matrix.

The "transformation toughened" ceramic, which consist of tetragonal zirconia particles dispersed in a ceramic matrix, constitute an interesting class of particulate ceramic composites, used for instance as abrasives. Other abrasives were made with particles synthesized by carbothermal or aluminothermal reductions of sol-gel materials and they presented properties superior to those of abrasives made by conventional methods. Examples of such materials are $TiC/TiB_2/Al$ composites used in cutting and grinding tools and $ZrB_2/SiC/C$, $ZrB_2/Al/Al_2O_3$ and $TiB_2/Al_2O_3/Al$ composites which can resist to oxidation at high temperature [101].

To these particulate composites, one must add special applications for which sol-gel glasses are well suited. Among them, one can mention composites in which a glass matrix can encapsulate radioactive wastes to avoid radioactive pollution [102]. Hollow glass spheres with a diameter of 80 to 100 μm and a wall with a thickness of 1 μm can also be made by inflating gel particles with the gases originating from the gel. These hollow spheres can be used to store nuclear fusion products such as Deuterium and Tritium [103].

Gels in the wet state can also be used in the same manner as thermosetting or thermoplastic polymers to make composites with short or long fibers. After heat treating the gel to transform it to a ceramic at a high temperature, a ceramic-ceramic composite is realized. A very special application already exists for the synthesis of thermal protection tiles for the NASA space shuttle. A gel made from TEOS is used to bind silica or mullite fibers altogether. Then, this gel is converted to silica by thermal treatment. For structural applications, a ceramic-ceramic composite with a strength of 630 MPa was made by the company Babcock & Wilcox. It was synthesized by infiltrating a SiC Nicalon® fibers weaving with a mullite sol-gel, followed by sintering [104]. Other studies were undertaken on systems comprising an alumina matrix made by sol-gel and SiC or alumina fibers. The results were not up to the level of expectations [105,106] but more progress may be anticipated with further research.

9.6 - FILTRATION MEMBRANES

Sol-gel processes offer much flexibility to tailor the fabrication of membranes with particular porous properties. In the present section, ceramic membranes made by sol-gel with their applications and their techniques of fabrication are reviewed.

POROUS MEMBRANES
A membrane can be defined as a selective barrier between 2 fluid media to separate their constituents. The constituents which can be separated are solid particles from a liquid suspension or from a gas, liquid molecules form a liquid, gas molecules from a gas. The incoming fluid medium before filtration is termed the influent, the permeate or the retentate. The retentate designates the influent which is left after filtration. The out coming filtered medium is termed the effluent, the filtrate or the concentrate (Figure 9.6-1).

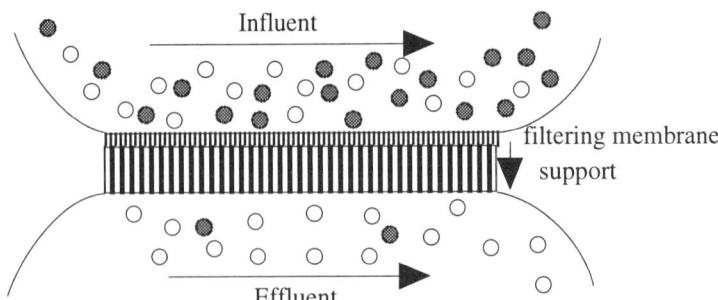

Figure 9.6-1 - Membrane as a selective barrier between two homogeneous medium. Adapted from Keizer et al. [107]

A membrane can be non-porous and it leads to separation by molecular diffusion, or porous and it leads to separation by size exclusion or differences in diffusion transport inside small channels. Membranes with pores below 1nm are applied in filtration by reverse osmosis. Membranes with pores in the size range from 1 to 100 nm correspond to ultrafiltration. Membranes with pores in the range from 100 nm to 10 μm are used in microfiltration. At last, membranes with pores bigger than 10 μm simply correspond to filtration. Very often, a membrane is characterized by its molecular weight cut off (MWCO) which is the molar mass of the species retained at 90% by the membrane. Historically [108], the role of membranes in osmosis was discovered in 1748 by Nollet. Later, the laws of diffusion were established from studies on membranes by Ficks and Van't Hoff established the laws for osmosis on membranes made by deposition of cupric ferrocyanide in unglazed porcelain.

Separation occurs under a pressure gradient, a chemical gradient or an electric potential gradient. In pervaporation, evaporation occurs through the membrane and filtration occurs because the equilibrium vapor pressure is different for the solute and for the main solvent. Electric potential gradients are applied to membranes in electrodialysis. In this process, ions migrate from one electrode to the other one and they are transferred to the electrodes by chemical reactions which produce species with different oxidation states. The liquid flow through a porous medium can be considered to be laminar and the filtration flux density J is related to the number N of pores with diameter d_p per unit cross-section area of the membrane, and to the thickness e of the membrane, by Poiseuille's equation [109]

$$J = N \frac{d_d^4}{e} \tag{9.6-1}$$

And as
$$N \approx d_p^{-2} \tag{9.6-2}$$

$$J \approx d_p^2 \tag{9.6-3}$$

Hence the rate of filtration decreases drastically as the pore size d_p decreases. This filtration rate becomes particularly low for ultrafiltration which explains why it can be useful to compensate this effect by applying a pressure difference ΔP between the 2 sides of the membranes. ΔP is termed the transmembrane pressure. However, the effect of ΔP depends on the equilibrium composition profile in the influent, near the interface with the membrane. Colloidal particles usually deposit on the influent side of a membrane occur, which can be measured by a resistance to filtration R_d due to the deposit. This resistance adds up to a resistance R_m due to the membrane itself. Overall, the filtration flux can be written

$$J = \frac{\Delta P}{R_d + R_m} \tag{9.6-4}$$

If the colloidal layer is compressible it can be shown that [109]

$$J = \frac{\Delta P^{1-s}}{e} \tag{9.6-5}$$

In microfiltration, s < 1 and the filtration flux increases with the transmembrane pressure ΔP. On the other hand in ultrafiltration s = 1 and the transmembrane pressure has no effect.

If the deposit thickness e keeps increasing because the colloidal particles in the influent cling to the membrane, the previous formula shows that the filtration flux progressively decreases and fouling occurs. Such a phenomenon is particularly frequent in through flow filtration where the influent is forced to pass through the membrane. In cross flow filtration, as this illustrated in Figure 1, the influent flows laterally on the membrane and an equilibrium concentration profile establishes. However, in cross-flow ultrafiltration, the equilibrium interface profile depends on the influent velocity and the filtration flux is known to increase with the influent velocity.

CERAMIC MEMBRANES

The use of membranes for microfiltration and ultrafiltration, in the world, is estimated to be $200,000m^2$ and the proportion of ceramics membranes is about 10% [110,111]. However, the field of ceramic membranes is recent. This field really begun to develop in the 1980's leading to the so called third generation membranes, to distinguish them from the previous organic membranes [112].

Ceramic membranes offer many advantages over organic membranes. They are stable at high temperature. They are incompressible, they can be mechanically loaded under large transmembrane pressures and they do not show significant creep. They also are chemically stable, especially in organic solvents. They are insensible to bacterial action. They do not age significantly so that their lifetime is long. Ceramic membranes can be submitted to repeated and rigorous cleaning operations such as steam sterilization and intense backflush to avoid or to limit deposit buildup. The electrochemical activity and electrocatalytic properties of ceramic surfaces can be used advantageously in filtering operations, to modify the interfacial zone in the influent side of a membrane. For instance, studies showed that the filtering flux of SiO_2 colloidal sols by Al_2O_3 membranes is lower near the isoelectric point of Al_2O_3 (pH $\approx$ 9) and higher near the isoelectric point of SiO_2 (pH $\approx$ 2.5) [113]. By selecting an appropriate pH, the filtering flux can be maintained to a high level because of attenuated fouling. The pore dimension and pore size distribution can be controlled during the membrane fabrication. Ceramic membranes can be brittle but this problem can conveniently be solved by special configurations and supporting systems. The main inconvenience of ceramic membranes is their high cost, both for installation capital and for eventual modifications in case of defects. Their fabrication technology can sometimes be complex, for instance to alleviate high temperature sealing problems.

Overall, ceramic membranes are superior for applications in gas separation, in food and in biotechnology, such as for dairy products, fruit juices, protein concentration, fermented alcoholic beverage, and in the filtration and treatment of water and waste water. Ceramic membranes can also work as porous chemical reactor with catalytic effects, such as in dehydrogenation reactions or in methane reforming [110]

The ceramics most extensively studied for applications in membranes are alumina, zirconia, titania, silica, various binary systems such as cordierite and mullite, silicon carbide and silicon nitride. Sol-gel processing is usually the preferred fabrication technique for oxides, in particular for very thin ceramic ultrafiltration membranes which make it possible to increase the filtration flux, can be made by sol-gel. A list of commercial ceramic membranes was provided by Hsieh [114]

The effectiveness of a membrane for a certain application depends on its detailed morphology and microtexture. Ceramic membranes usually are available in three possible configurations: flat plates, tubular and honeycomb. The tubular geometry offers the best mechanical support and it is the most widespread. The honeycomb configuration is complex in design, more expensive but it provides more filtration area.

Porous ceramic membranes can be prepared in various ways which depend on the nature of the ceramic, the desired pores texture, the membrane thickness and the type of application. The main techniques include :
- Leaching a phase separated glass [115].
- Anodic oxidation of thin metal foils in an acid electrolyte [116].
- Irradiation of a dense film protected by a negative polymer film mask as in large scale integrated computer chips [117].
- Pyrolysis of organic polymers films [118].
- Physical and chemical vapor deposition [114,119].
However the sol-gel technique is the most convenient one.

SOL-GEL CERAMIC MEMBRANES

Self supported sol-gel ceramic membranes
Unsupported membranes are presently laboratory curiosities; they are mostly for research purposes. The techniques to make them are the same as for free standing films in section 9.3. A few one-component and two-components self-supported membranes with their pore characteristics is provided in Table 9.6-1. The membrane thickness is typically of the order of 100 µm, with a pore size which ranges from of a few nm to a few times 10nm and the pore volume is in the range of 40 to over 60%.

Supported sol-gel ceramic membranes.
The principle of typical asymmetrical supported tubular membrane is illustrated in Figure 9.6-2. The filtering membrane itself is one the inner cylinder side; it has a thickness of 10 to 20 µm. The support is on the outer side and has a thickness of 1 to 2 mm. An intermediate layer may have a thickness of 10 to 50 µm. Such multilayered membranes are often made by the successive packing of ceramic particles with different size. Filtration occurs by percolation through the layer composed of the smallest particles. The layer with most coarse particles acts as a support. In tubular filters such as in Figure 9.6-2, the outer support layer is made by conventional powder sintering. The pores texture is uneven, but their size is large, in the range from 1 to 15 µm [121]. The intermediate layer is made by slip casting a fine conventional powder, its pore size in the range 100 to 1500 nm. The slip is cast inside the support cylinder and can be sucked into it. The capillary tension makes the slip penetrate the large pores of the support, so that its particles are retained and mostly deposited near the entrance of these pores [122], on the filtering side of the membrane.

Table 9.6-1 - Structural properties of unsupported membranes.
Adapted from Burggraaf et al. [120].

Material	heat treatment		Pore diameter (nm)	Pore volume (%)	BET surface area (m²/g)
	T (°C)	time (h)			
γ-AlOOH	200	34	2.5	41	315
	300	5	5.6	47	131
γ-Al₂O₃	500	34	3.2	50	240
	550	5	6.1	59	147
	800	34	4.8	50	154
θ-Al₂O₃	900	34	5.4	48	99
α-Al₂O₃	1000	34	78	41	15
TiO₂	300	3	3.8	30	119
	400	3	4.6	30	87
	450	3	3.8	22	80
	600	3	20	21	10
CeO₂	300	3	2	15	41
	400	3	2	5	11
	600	3		1	1
Al₂O₃-CeO₂	450	3	2.4	39	164
	600	3	2.6	46	133
Al₂O₃-TiO₂	450	3	2.5	38-48	220-260
Al₂O₃-ZrO₂	450	5	2.6	43	216
	750	5	2.6	44	179
	1000	5	≥ 20		

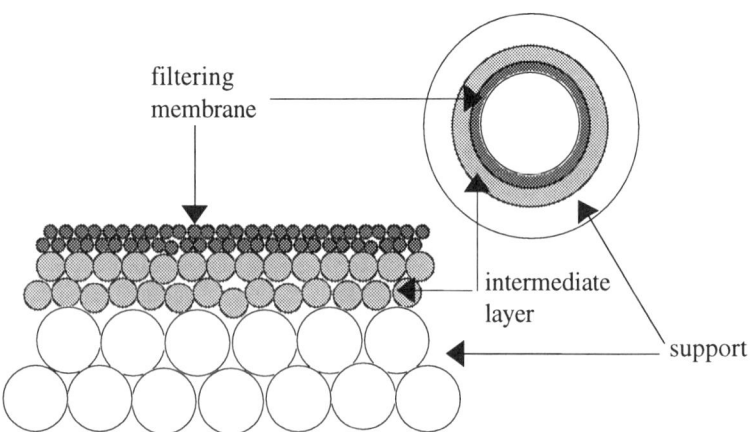

Figure 9.6-2 - Cross-section of an asymmetrical ceramic membrane. Adapted from Cot et al. [111].

The sol-gel process is used to deposit the filtering layer on top of the intermediate slip cast layer, by the same technique. To do so, the ceramic tube can be filled with a colloidal sol, and a pressure of at least 300 kPa can be applied to force the sol through the support [123]. That is to say, ultrafiltration is applied to the colloidal sol itself. The colloidal particles penetrate in the pore entrance and create the filtering membrane on the internal surface of the cylindrical support [124].

Additives in the slip help to improve the bonding between the support and the filtering deposited layer [123].

The sol-gel process provides a narrow size distribution, in a range which can go from 1 to 15nm and which makes the membrane very selective. The main sol-gel oxides with are deposited by sol-gel are alumina, titania, zirconia and the binary system titania ruthenia [111,125,126]. Moreover, the sol-gel coating is thin so that the effluent flow rate is high. As for all coatings (section 9.3) the deposited thickness e increases with the time t as [127]

$$e = K \sqrt{t} \qquad (9.6\text{-}6)$$

Where K is a constant which depends on the rheological characteristics of the sol such as its concentration, on characteristics of the substrate such as its pore size and on wetting properties of the sol on the support. Data for TiO_2 and ZrO_2 membranes are reported in Figure 9.6-3.

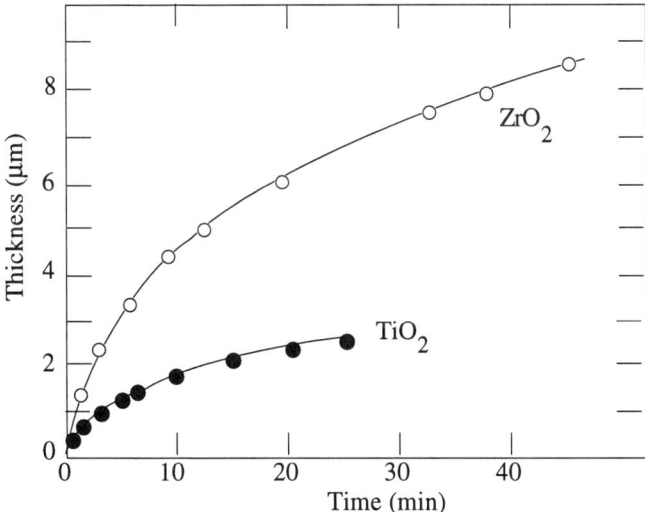

Figure 9.6-3 - Membrane thickness versus deposition time, after sintering at 470°C, for TiO_2 and ZrO_2 membranes. Adapted from Larbot et al. [125].

A somewhat different technique consists in precipitating colloidal hydroxide or hydrous oxide particles, on a porous support. Such membranes, termed dynamic membranes are formed in place, in the medium to filtrate. They are made from salts

of Al, Fe and Zr. Such Zr membranes, often improperly termed Zr(OH)$_4$ dynamic membranes, are very efficient for many ultrafiltration applications [128].
Supported membranes can also be done by coating a fiber mat with the sol before gelation [129].

Catalytic sol-gel membranes

Because of their high specific surface area, sol-gel membranes can also be modified with additives to present interesting catalytic properties [130]. The additives can be inside the membrane pores or on top of the pores (Figure 9.6-4). A listing of such additives deposited in the gas phase or from a liquid phase is shown in Table 9.6-2.

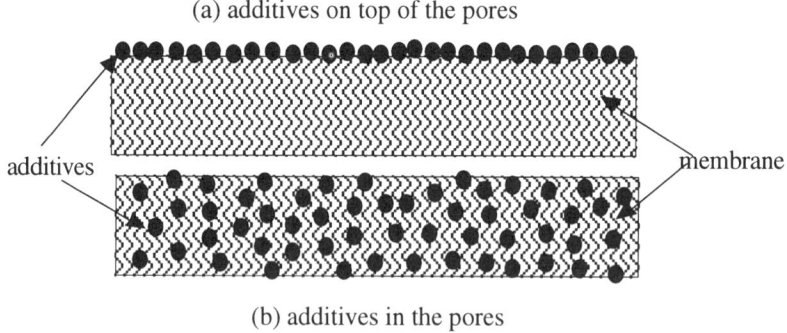

(a) additives on top of the pores

additives   membrane

(b) additives in the pores

Figure 9.6-4 - Possible structures of modified ultrafiltration membranes. Adapted from Burggraaf et al. [120].

Table 9.6-2 - Modified nanoscale ceramic catalytic membranes.
After Burggraaf et al. [120].

Membrane	additives	% additive	Pore size
γ-Al$_2$O$_3$	Fe or V oxide	5-10%	0.3 nm
	MgO/MgOH	5-100%	
	Al$_2$O$_3$/Al(OH)$_3$	5-100%	5 nm
	Ag	5-50%	> 10 nm
	CuCl/KCl		
	ZrO$_2$		
θ/α-Al$_2$O$_3$	ZrO$_2$ - Y$_2$O$_3$	1-100%	a few nm
Al$_2$O$_3$-TiO$_2$	V oxide, Ag		
TiO$_2$	V oxide		

9.7 - SOL-GEL CATALYSTS

Of the order of 90% of the industrial chemical processes use a catalyst [131]. Sol-gel ceramics have characteristics which make them good materials for such applications and review papers have been published on this subject [132-134]. Sol-gel derived ordered mesoporous materials offer themselves interesting applications. The synthesis of catalysts is an important domain of sol-gel processing which deserves special consideration.

CATALYSIS MATERIALS

Activity and selectivity of a catalyst

Catalysts can roughly be classified in four categories which are: metals, semiconducting oxides, insulator oxides and solid acids. They can be prepared as bulk solids or as colloidal particles which are dispersed on a support and it is possible to find examples of preparation of these four types of catalysts by sol-gel process [134].

In terms of kinetics, the efficiency of a catalyst can be characterized by its selectivity for a given product and its activity [135]. The activity of a catalyst refers to the rate at which it makes a reaction proceed towards chemical equilibrium. It can be expressed in several way such as $\frac{\text{(molecules reacted)}}{\text{(time)(area)}}$ or the turnover number given by :

$$\text{turnover number} = \frac{\text{(molecules reacted)}}{\text{(time)(active site)}} \qquad (9.7\text{-}1)$$

Turnover numbers depend on many conditions such as the temperature and pressure and they are not often provided in publications. Instead, it is often more convenient to indicate the temperature for which a given percent of reactant molecules is converted, for a given feed composition and pressure. For instance, T_{10} represents the temperature where 10% conversion occurs.

On the other hand, the selectivity measures the extent to which a catalyst accelerates the formation of a given product. It is defined as the percentage of the consumed reactant which is transformed to the desired product. Depending on the type of reaction considered, a catalyst may be useful for its activity, its selectivity or both. If several products can possibly be formed, the selectivity is often the more important property.

Oxides' active sites

Most sol-gel materials are oxides and in terms of catalysis mechanisms, the important aspects such as their surface sites which participate in the catalysis reaction, termed active sites, are very subtle, controversial and subject to important changes due to apparently minor modifications in the materials preparation methods. This is due to the fact that heterogeneous catalysis is essentially governed by local conditions around the active sites, which largely depend on the material preparation procedure.

An oxide catalyst may participate in a chemical transformation by acid-base reactions; in this case its acid or basic sites are important. An acid site may be of the Brönsted or of the Lewis type and its origin in materials is subject to controversy. Pure silica gels have a very weak surface acidity while pure alumina is amphoteric with both acidic and basic properties. Surface terminal hydroxo M-OH or aquo groups M-H_2O can liberate a protons, hence they can be Brönsted acid sites. Hydroxo and oxo M-O groups can capture a proton and act as basic Brönsted sites. On the other hand , terminal metal atoms can have missing electron pairs and constitute Lewis acid sites, while terminal oxo groups have unshared electron pairs which make them Lewis basic sites. Different types of Brönsted or Lewis acid sites may exist in a catalyst.

According to Pauling, the acidity of a catalyst can largely be tailored by adding cations with a different valence state to a host oxide. For instance, if a trivalent Al^{3+} cation substitutes for a tetravalent Si^{4+} cation in SiO_2, in such a way that it occupies a tetrahedral Si site, a net negative charge is created which must be balanced by a nearby positive charge. This can be an H^+ cation from water molecules which dissociate to fix a hydroxyl OH on the Al^{3+} cation. In terms of Kröger-Vink notation the overall reaction writes as in (9.7-2). The resulting structure corresponding to the right hand side of (9.7-2) may behave as a Brönsted acid as it can give up a proton according to reaction (9.7-3) or it may transform to a Lewis acid by losing a water molecule according to reaction (9.7-4)

$$Al_2O_3 + H_2O \Leftrightarrow \underbrace{Al'_{Si} + OH^{\circ}_O + H^+ + 3\,O^x_O}\qquad (9.7\text{-}2)$$

$$\begin{bmatrix} & H & \\ & | & \\ & O & \\ & |^- & \\ O - & Al & - O \\ & | & \\ & O & \end{bmatrix} + H^+ \Leftrightarrow \begin{bmatrix} H\diagdown \quad \diagup H^{\delta+} \\ O^{}_{\delta-} \\ | \\ O - Al - O \\ | \\ O \end{bmatrix} \qquad (9.7\text{-}3)$$

$$\text{Brönsted base} \qquad\qquad\qquad \text{Brönsted acid}$$

$$\begin{array}{c}\text{empty}\\\text{electronic}\\\text{case}\end{array} \qquad \Updownarrow \atop \text{Lewis acid}$$

$$H_2O \quad + \quad \begin{bmatrix} \Box \\ O - Al - O \\ | \\ O \end{bmatrix} \qquad (9.7\text{-}4)$$

An acid site can itself react with a hydrocarbon to form a carbenium ion of the type RCH^+CH_3 which is adsorbed on the Al site [135]. The latter can either give a proton as a Brönsted acid, to $RCH=CH_2$, or fix a H^- ion from RCH_2CH_3 on an electrophilic Al atom.

The strength of an acid site can be increased by replacing a hydroxyl group by halogens anions, such as fluoride or chloride anions or by sulfate anions. The acid strength of a solid can be determined by its ability to change a neutral organic base B adsorbed on the solid, into its conjugate acid form BH^+, in the case of Brönsted acidity. This acid strength is determined by the Hammett indicator H_0 defined as

$$H_0 = pK_a + \log \frac{[B]}{[BH^+]} \text{ for a Brönsted acid} \qquad (9.7\text{-}5)$$

$$H_0 = pK_a + \log \frac{[B]}{[AB]} \text{ for a Lewis acid} \qquad (9.7\text{-}6)$$

where pK_a is derived from the usual equilibrium dissociation constant of the acid, [B] is the organic base concentration in equilibrium on the solid surface, $[BH^+]$ its conjugate acid concentration for the Brönsted case and [AB] the compound formed with the base for the Lewis case. The lower H_0 the more acidic the solid surface.

An oxide may also participate in redox catalysis reactions, especially when it contains transition metal oxides linked to oxygen atoms which can easily be transferred out of the structure to produce a non-stoichiometric oxide. For the oxidation of a hydrocarbon R, the mechanism involves two steps, first oxidation of the hydrocarbon R by the catalyst oxygens (9.7-7), secondly re-oxidation of the catalyst by oxygen gas molecules (9.7-8).

$$\text{Cat-O} + \text{R} \Rightarrow \text{RO} + \text{Cat} \qquad (9.7\text{-}7)$$

$$2\text{Cat} + \text{O}_2 \Rightarrow 2 \text{ Cat-O} \qquad (9.7\text{-}8)$$

The catalyst oxygens can be network oxygens for an oxide, or chemisorbed oxygens for a metal.

Special characteristics of sol-gel oxides

The definition of the turnover number points out to the importance of increasing the number of active sites, hence the specific surface area of a catalyst. Gels and particularly aerogels have a very high porosity, for instance a specific surface area up to 1000 m^2/g for alumina aerogels, which corresponds to a pore volume up to 98% of a sample volume, and an apparent density as low as 0.03 g/cm^3 and currently of the order of 0.5 g/cm^3. Moreover their texture presents a good stability at relatively high temperature, at least in the intermediate temperature range according to the terms used in chapter 7, sometimes even in the presence of water vapor.

Their application in catalysis concerns two large domains: the synthesis of commercial organic products with a high added value and the protection of the environment. In both domains, sol-gel materials can participate in catalysis at two levels, either as catalytic active materials themselves, or as support of a catalytically active phase. Very often this active phase is a noble metal such Pt, Pd or a transition metal element. An active phase must remain dispersed as much as possible, hence the second possible action of sol-gel as catalyst support. A support with a large specific surface area maximizes the contacts between the reactants and the active phase, in so far as the active phase can be maintained finely dispersed at the atomic scale. Sol-gel processes can eventually hinder the sintering of this active phase by blocking it in an original fashion on the support.

An interactive effect between the support and the active phase, such as an active metal, can also enhance the activity of this active phase. The interaction can be textural, that is to say it modifies the types of pores and solid surfaces on which the active phase can be adsorbed, or structural due to the formation of a particular support phase such as a transition phase in alumina. For instance, amorphous or crystalline transition phases always tend to be more active than stable crystalline phases. Moreover, we mentioned before that the acidity of an oxide can be increased by dissolving cations with a different valence in the sites of the host cation oxide: sol-gel processes offer the potential to realize such substitutions without inducing phase segregation much more easily than conventional processes.

SYNTHESIS OF HIGH VALUE ORGANIC COMPOUNDS

For this type of applications, catalysts usually need to operate at a temperature $T < 400°C$, which is a temperature range where aerogels keep a high specific surface area. Aerogels are particularly interesting to magnify all catalytic reactions because of their very high surface area. They are often better than the xerogels because either the active phase is better dispersed, or the textural properties are better developed.

Types of chemical reactions catalyzed

The types of chemical reactions involved include partial oxidation, nitroxidation and hydrogenation. Examples are provided in Table 9.7-1 with the nature of the aerogel catalyst, and the selectivity, that is to say the percent of the desired compound formed, amongst all compounds formed other than the reactants. Sol-gel materials often have a high selectivity. However the activity is usually much lower.

Both single and mixed oxides of the transition elements present selective oxidation properties. On the other hand, selective hydrogenation is typically due to metal catalysts and sol-gel are mostly efficient as supports. Nitroxidation reactions corresponds to the formation of nitriles (A or ϕ)-C$\equiv$N , where A designates an aliphatic hydrocarbon such as propene, isobutylene, isobutane or propane and ϕ an aromatic hydrocarbon such as toluene, the xylenes and the monotolunitriles. These nitriles are obtained by reaction of alkenes with $NH_3 + O_2$, or of hydrocarbons with NO. In this case Ni^{2+} surface cations blocked in a spinel $NiAl_2O_4$ environment seem to be the active centers, and basic oxide additives such as MgO improve the catalyst activity. In the last example of Table 9.7-1, NH_3 is adsorbed on Fe^{3+} and Cr^{3+} oxidized site which are partially reduced and NO_2 is chemisorbed and decomposed on these sites.

Table 9.7-1 - Types of chemical reactions catalyzed by sol-gel

reactant	compound formed	aerogel catalyst	selectivity [reference]
Selective oxidation			
isobutylene	methylacroleine	$NiO-Al_2O_3$	67% [136]
isobutylene	acetone	$NiO-Al_2O_3$	25% [136]
isobutane, propane	acetone	$NiO-SiO_2-Al_2O_3$	100% [137]
Nitroxidation			
Hydrocarbons (aliphatics, or aromatics)	nitriles	$NiO-Al_2O_3$	80 to 90 % [132]
Selective Hydrogenation			
Cyclopentadiene	cyclopentene	$Cu-Al_2O_3$	100 % [138]
Toluene	methylcyclohexane (storage of H_2 for vehicles)	$Ni-SiO_2$	100% [139]
Benzene	cyclohexane	$Ni-MoO_2$	[140]
Nitrobenzene	aniline	$Pd-Al_2O_3$	100% [141]
CO & CO_2	(Fischer-Tropsch) methanol	$Fe_2O_3-SiO_2$ $Fe_2O_3-Al_2O_3$	300 x unsupported Fe_2O_3 [142,143]
Selective Reduction			
$NO + NH_3$	$N_2 + 2H_2O$	$Fe_2O_3-Cr_2O_3-Al_2O_3$	[144]
Polymerization			
ethylene	high molecular weight polyethylene	$TiCl_4-Al_2O_3$	[145]

Aerogels
Simple oxide aerogels are generally not active materials, so that they are used as supports. However they enhance the activity of a catalyst by a controversial mechanism known as spillover. This consists in a creation of active hydrogenation sites when in a H_2 atmosphere, by transfer of hydrogen from the active phase to the support. Spillover by oxygen transfer is also possible.

Silica is one of the most important support because of its inertness. Alumina is widely used in catalysis because of its acidic properties and as support of other active phases in bifunctional catalysts. Zirconia is interesting because it develops both redox and acid-base functions [146], hence it is used as a catalyst and as a support [147]. Titania is a reducible support used in photocatalysis [148,149]. A third oxide component can eventually be added to develop new functionalities.

Active metal particles supported on oxide aerogels can be prepared by impregnation of aerogels with a metal salt or by direct fabrication of a multicomponent aerogel. To obtain the metal particles, supercritical drying must be performed in a reducing environment, that is to say either in alcohol, or by replacing N_2 with H_2 before heating the autoclave or by flushing the autoclave with H_2 at $\approx 200°C$. In these conditions, NiO, CuO, PbO, V_2O_5 are reduced to finely divided metal particles with a size a few nm. A list of such supported metal particles made in situ in aerogels is given in Table 9.7-2.

Table 9.7-2 - Supported metal particles in aerogels made by in situ reduction of oxide aerogels. Adapted from Pajonk [132]

metal-aerogel	Specific surface area at $20°C$ (m^2/g)		[reference]
	total	metal	
Pt-Al_2O_3	450	120	[150]
Ni-SiO_2	620	356	[139]
Ni-Al_2O_3	150-650	7-53	[151]
Ni-SiO_2-Al_2O_3	480-730	5-68	[151]
Ni-MoO_2	1-15	0.7-15	[152]
Cu-Al_2O_3	660	30	[153]

Catalysis process engineering with aerogels
To be used industrially, aerogels can be encapsulated in honeycombs, Rasching rings, screens, mineral wool, boiling stones and glass or metal tubes. The reactant gases can be forced to flow through a catalyst fluidized powder bed.
Aerogel powders behave as type C beds as illustrated in Figure 9.7-1a. In the packed state represented in this figure, reactant gases do not pass easily. They can either push the powder bed as a plug illustrated in Figure 9.7-1b, or pass through channels as in Figure 9.7-1c. A fluidized bed type of type A is reached for a much higher reactant laminar velocity than with conventional powders, because strong Van der Waals attraction keep the aerogel particles together. This fluidized bed state is illustrated in Figure 9.7-1d. It corresponds to a break of the initial powder pack into clusters of aerogel powders with a size of the order of 1 mm.

PROTECTION OF THE ENVIRONMENT
The most important research in this field concern the reactions of:
- Air depollution , in particular the elimination of nitrogen oxides exhaust gases according to the reaction

$$NO_x \Rightarrow \frac{1}{2}N_2 + \frac{x}{2}O_2 \qquad (9.7-9)$$

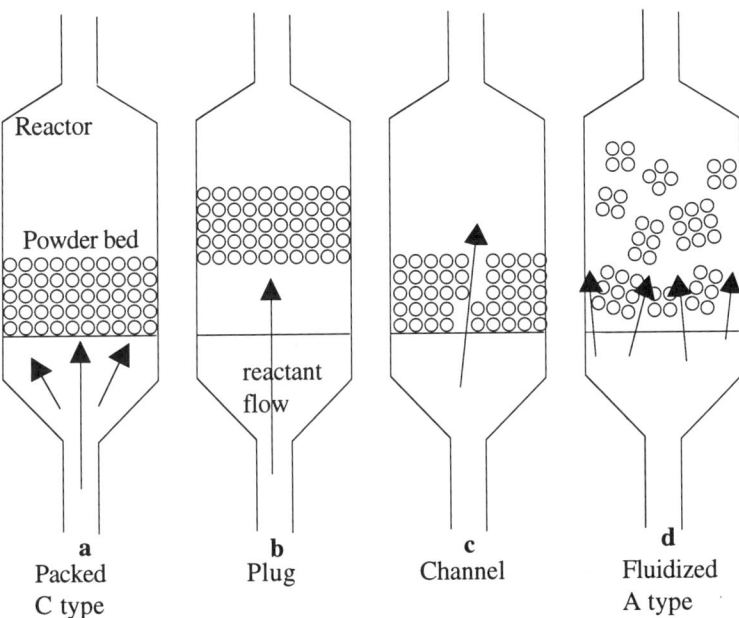

Reactor

Powder bed

reactant
flow

a
Packed
C type

b
Plug

c
Channel

d
Fluidized
A type

Figure 9.7-1 - Types of behavior of aerogel powders in a fluidized bed. Adapted from Pajonk [132]

- Clean catalytic combustion of hydrocarbons, so as to only reject CO_2 and H_2O . For instance with methane :

$$CH_4 + 2O_2 \Rightarrow CO_2 + 2\,H_2O \qquad (9.7\text{-}10)$$

- Catalytic post combustion depollution of automobile exhaust gases, known as three-way catalysis to eliminate CO , NO_x and HC rejects
- Catalytic combustion of diesel engine soot.

The catalysts which are needed often operate at a high temperature from $T \approx 900°C$ to 1200°C. At these temperatures, most aerogels lose their high specific surface area. In particular their sintering is promoted by H_2O present in engine exhausts. Chemical reactions between the supports and active phase also commonly occur, as well as poisoning of the active sites. Consequently, the advantages of sol-gel materials by comparison with conventional materials is not obvious.
However, as sintering can be accelerated by selecting the appropriate initial pore texture or additives, it should also be hindered by selecting other appropriate additives or initial pore texture. Hence research keeps going on in this field and some significant success has been achieved with phases derived from cubic alumina. Transition aluminas have a structure which is derived from a cubic close packing of

oxygen anions, and they are the single oxide materials able to keep the highest specific surface at the highest temperature, over 100 m^2/g up to 1000°C.

The best success so far has been obtained with materials derived from the hexaaluminate $BaAl_{12}O_{19}$ doped with Mn, Cr, Fe, Ni or Co by Machida et al. [154,155]. This compound has a structure derived from cubic spinel where densification is attenuated by the formation of lamellar structures. A few properties of such compounds are given in Table 9.7-3, including T_{10} and T_{90} which are the temperatures where the clean combustion of respectively 10% and 90% of methane is achieved.

Table 9.7-3 - Properties of hexaaluminate catalysts in the combustion of methane. Data from Machida et al. [154,155]

| Temperature (°C) | Specific surface area (m^2/g) | |
	$BaAl_{12}O_{19}$	$BaMnAl_{11}O_{19}$
1200 °C	50	
1300 °C		13.7
1600 °C	11	
	Catalytic combustion of CH_4	
	$BaAl_{12}O_{19}$	$BaMnAl_{11}O_{19}$
T_{10} (°C) ignition	700	490
T_{90} (°C) complete combustion	760	750

Other additives have been reported to enlarge the existence range of θ-Al_2O_3 towards higher temperature and hence to increase the high temperature specific surface area of Al_2O_3 based material. For instance, Y which could increase the residual specific surface area to 88 m^2/g at 1200°C [156]. Also, the resistance to sintering and transition to α-alumina seems to be improved by selecting a sol-gel process with organic complexing additives which produce a somewhat different initial pore texture as well as exotic transition alumina phases [157]. However such results need to be confirmed and the catalytic properties of these materials studied.

APPLICATION OF ORDERED MESOPOROUS CATALYSTS

Large pore mesoporous materials presented in chapte 6.3 can be used as supports for catalytic active phases such as heteropolyacids, amines, transition metal complexes and oxides. It is also possible to introduce Bronsted acid sites on the surface of the pores, or transition metals so as to make redox catalysts. Eventually, new bifunctional acid/metal oxide catalysts can be designed. By exchanging the alkaline cations for the Bronsted acid sites protons, the basicity of the conjugated base can also be increased. In summary, these mesoporous materials extend the catalytic possibilities of zeolites to larger molecules.

As acid catalysts, mesoporous type materials could be used for cracking or precracking [158], in association with zeolites, of large gas-oil molecules. However the thermal stability, and more particularly the hydrothermal stability during steaming regeneratioin at T ≈ 800°C of these MCM type materials is low, an inconvenience which can be attenuated by making pores with thicker walls. On the other hand, MCM type materials appear to be good acid catalysts for reactions requiring a lower hydrothermal stability, such as hydroisomerization, diesel production in mild hydrocracking reactions (MHC), demetallization and olefin oligomerization. NiMo supported on MCM-41 based catalysts performed better than zeolites from the point of view of hydro-desulfurization (HDS), hydrodenitrogenation (HDN) and MHC [159]. A three component NiW/MCM-41/zeolite has also been developed for deep hydrocracking [160]. A NiW/MCM-41 can be used to hydrocrack heavy waxes into lubrication oils [161]. MCM-41 supported bifunctional catalysts are useful for the isomerization of normal paraffins into isoparaffins [162]. MCM-41 based catalyst have also been studied to upgrade the disproportionation of larger olefins such as propylene.

The most promising domain of applications of ordered acid mesoporous catalysts is in the field of organic synthesis, to synthesize fine chemicals. Friedel-Crafts alkylation and acylation are of general use for the production of fine chemicals. In traditional processes, these reactions are catalyzed by stoichiometric proportions of $AlCl_3$, which induce waste disposal problems and must now be replaced by solid acids. MCM-41 type catalysts appear to be very efficient [163]. Such applications concern in particular the synthesis of fragrances and pharmaceuticals which involve the cylation of aromatic ketones [164]. Other mild acidic reactions are the acetalyzation of large size aldehydes [165], the Beckman rearrangement of cyclohexane oxide [166] and aldol condensation [167].

As base catalysts, much work needs to be done with the alkaline exchanged MCM-41 materials. The Na exchanged ones are active in the Knoevenagel condensation of benzaldehyde with ethyl cyanoacetate [168].

As redox catalysts, Ti-MCM-41 is more efficient than Ti-zeolites to catalyze the epoxidation of large olefins molecules by reactions with H_2O_2, because these molecules can diffuse in the large mesopores of the MCM based material [169]. Organic hydroperoxides can also be used as oxidants. Ti mesoporous materials made with a nonionic surfactant, termed Ti-HMS, appear to be good in the liquid-phase peroxide oxidation of methyl methacrylate, styrene, and 2,6-di-tert-butylphenol [170]. The oxidation of amines, interesting in some chemical and pharmaceutical synthesis can also be catalyzed by Ti-MCM-41 and Ti-HMS materials [171]. A Ti-MCM-41 catalyst containing both Ti^{IV} oxidation sites and acidic H^+ ones associated with Al^{IV} sites has been shown to be able to achieve at once the oxidation of linalool to cyclic furan and hydroxy ethers by using tert-butyl hydroperoxide [172]. When compared to TiO_2-SiO_2 aerogels, it appeared that the activity was higher per Ti atom in a Ti-MCM-41 than in an aerogel. However the aerogel contained more Ti^{IV} centers and had a very open mesoporous texture, so that it was globally more active with bulky olefins and organic hydroperoxides than the Ti-MCM-41 materials.

Other catalyzed reactions include the ring opening polymerization of lactic acid in Sn-HMS molecular sieves, synthesized with a nonionic surfactant. The Sn atoms were incorporated in the catalyst network instead of being mixed in the final products, which eliminated toxicological problems [173]. Tuel et al. also prepared ordered mesoporous silicas where the silica was amorphous and incorporated Zr cations. The obtained Zr-MS materials were active and selective for the oxidation of a large variety of substrates [174].

As supports, ordered mesoporous materials can be used to make stronger solid acids and bases. In particular heteropolyacids (HPA) such as $H_3PW_{12}O_{40}$ can be chemisorbed and used for reactions such as paraffin isomerization and isobutane/butene alkylation [175]. Strong basic catalysts can also be made by dispersion of amines in MCM-41. Such supported catalysts are useful to prepare monoglycerids from 2,3-epoxy alcohols and fatty acids [176]. Noble metals such Pt or Pd could also be finely dispersed by using a neutral $PT(NH_3)_2Cl_2$ precursor which goes in the hydrophilic part of the micelle [177]. The obtained catalyst performed well in the hydrogenation of benzene, naphtalene, phenanthrene and olefins. A modified MCM-41 material containing Fe and Pd allowed to control the formation of NO_x, CO and hydrocarbon emissions generated by oxygen rich combustion processes [178].
Selective oxidation catalysts can also be obtained by grafting organometallic complexes on the inner walls of mesoporous MCM-41 , or by functionalizing surface silanol groups with silanes followed by anchoring of a transition metal complex. Such procedures have been used to obtain Ti- containing or Mn-containing active sites [179,180]. The catalysts obtained showed good enantioselective properties, that is to say they catalyzed the formation of a certain chiral form of a compound [181].

9.8 - APPLICATIONS OF HYBRID ORGANIC INORGANIC MATERIALS [182]

In a hybrid material, the organic component can significantly modify the mechanical properties of the inorganic component [183-185]. It can make easier to process films and fibers as well as near shape molding [186].

Hybrids comprising polymers weakly bonded to an amorphous oxide sol-gel matrix can be made transparent, colorless, homogeneous and with good mechanical properties. The porosity can also be controlled as well as the hydrophilic/hydrophobic balance [187,188].
Materials with new optical [189,190] or electrical properties [191] can be tailored. Some materials can display new electrochemical reactions [192] as well as special chemical or biochemical reactivity [193-196]. For instance, dyes embedded in a sol-gel matrix can be luminescent, such as rhodamine 6G, rhodamine 640 and coumarin 4; photochromic such as spirospyrans or have non linear optics (NLO) properties

such as phtalocyanine. Such hybrids can be used as luminescent solar concentrators, dye lasers and photochromic or NLO devices.

It is also possible to synthesize hybrid materials with NLO properties by the technique of simultaneous gelation of a modified silicon alkoxide and an organic monomer. Two interpenetrating gel networks are formed, a polymer network which is confined in a silica gel network.

Ordered weakly bonded hybrids made by intercalation of organic conductive layers in between semiconducting V_2O_5 layers, yield mixed conductors comprising but their conduction properties are rather poor so far [197].

The properties of hybrids with strong organic-inorganic bonds can be outstanding for mechanical, rheological, optical, conductivity, or membrane selectivity applications. In particular, the mechanical properties depend on the size of the heterogeneities and on their interface with the matrix. In the case of hybrids from PDMS-TEOS or PTMO-TEOS the mechanical properties depends on the amount of acid when an acid catalyst used to gel TEOS. The Young's modulus decreases and the fracture strain increases as the molar ratio $\frac{[acid]}{[TEOS]}$ increases. Addition of titania [198] or zirconia [199] to PDMS-TEOS or PTMO-TEOS is possible after complexing $Zr(OR)_4$ and $Ti(OR)_4$ alkoxides with chelating groups such as the acetylacetonato (acac) ions so as to slow down their hydrolysis rate. This improves the mechanical properties of the hybrid.

Polyimide -SiO_2 composites made by reaction of TEOS with the macromonomer in figure 2.6-1 (Chapter 2) also have good mechanical properties, but moreover they can have NLO properties.

Polyamide-silica transparent hybrid selective membranes with a controlled pore size and hydrophilic/hydrophobic balance can be made by cocondensing polyoxazoline macromolecules and TEOS.

Several hybrid materials are being investigated for their ionic conduction properties. These include Si-O-PEG (polyethylene glycol) hybrids which can solvate small cations such as Li^+ but which slowly deteriorates during aging [200]. AMINOSILS made by condensation of $Si(OR)_3R'$ precursors where $R' = -(CH_2)_n-NH_2$ can solvate anions such as $CF_3SO_3^-$ and give good proton conductors. The conductivity can reach a value $\approx 1.4 \times 10^{-5}$ S. cm^{-1} in materials based on $Si(OR)_3(CH_2)_3NH_2$, $(CF_3SO_3H)_{0.1}$. Poly-(benzylsulfonic acid) siloxane (PBSS) hybrids have a high proton conductivity up to 10^{-2} S. cm^{-1} and are stable up to 300°C [201]. Electrochromic devices can be made by incorporating transition metal oxides in these hybrid materials but the oxide layer has a tendency to deteriorate due to the high acidity. The most stable hybrid ionic conductor is so far a three component complex material comprising a solvating polymer, a proton source and a proton vacancy inducer. All organic groups are anchored to trialkoxy silanes and the conductivity is of the order of 10^{-5} S. cm^{-1}.

Trifunctional alkoxysilanes $R'Si(OR)_3$ where R' is a vinyl, epoxy or methacrylate group, can form hybrids with special mechanical, optical or electrical properties. For instance, N-3 trimethoxy-silyl-propyl-pyrrole yield transparent sols which transform by polymerization of pyrrole to black xerogels with electronic conduction properties [202]. Bifunctional alkoxysilanes R'R"Si(OR)$_2$ where R' and R" are methyl or phenyl groups can be used to provide some flexibility to oxide chains networks [203]. Besides, methyl and phenyl groups are hydrophobic; polar solvents such as water or alcohol are not retained by the gel. Polyfunctional alkoxysilanes can be used to provide a variable spacing between silicon atoms, yielding materials with a variable rigidity. Their porosity depends to a great extent on the nature of the catalyst used to polymerize the alkoxysilane and the solvent. Their specific surface area can go up to 1262 m^2/g.

9.9 - REFERENCES

1 - Elmore W. C., Phys. Rev. 54 (1938) 309-310
2 - Charles S. W., Popplewell J, "Ferromagnetic Materials", vol. 2, Ch. 8, North-Holland, Amsterdam (1980) 510-559
3 - Michaud M., Leroy M.C., Livage J.,   Mater. Res. Bull. 11 (1976) 1425-1432.
4 - Livage J., Henry M., Sanchez C., Progress in Solid State Chemistry, 18 (1988) 259-341.
5 - Colton R.J., Guzman A.M., Rabalais J.W., Acc. Chem. Res. 11 (1978) 170-176 .
6 - Murawski L., Chung C.H., Mackenzie J.D., J. Non-Cryst. Solids 32 (1979) 91-104.
7 - Guestaux C., Kodak Pathé, French patent 2429252; 23 june 1978.
8 - Fritzsche H., "Switching and memory in amorphous semiconductors", in "Amorphous and Liquid Semiconductors", Ed. Tauc J., Plenum New-York (1974) 313-359.
9 - Adler D., Henisch H.K., Mott N.F., Rev. Mod. Phys. 50 (1978) 209-220.
10 - Adler D., Shur M.S., Silver M., Ovshinsky S.R., J. Appl. Phys. 51 (1980) 3289-3309.
11 - Aldebert P., Baffier N., Gharbi N., Livage. J., Mat. Res. Bull., 16 (1981) 949-955.
12 - Morineau R., Chemseddine A., Livage J., French Patent 2 527 219; 25 Nov. 1983.
13 - Judeinstein P., Livage J., Zarudianski A., Rose R.,   Solid State Ionics, 28-30 (1988) 1722-1725.
14 - Raman G., Gnanam F.D., Ramasamy P., J of Crytal Growth 75 (1986) 466-470.
15 - Fricke J., Journal of non-Crystalline solids 100 (1988) 169-173 .
16 - Poelz G., "Aerogels in High Energy Physics", in "Aerogels", ed. Fricke J.,   Springer Proc. in Physics, vol.6, springer Verlag Heidelberg, New-York (1986) 176-187.
17 - Schroeder H., Verres ref. 18 (1964) 89-97.
18 - Partlow D.P., Yoldas B.E., Am. Ceram. Soc. Bull. 59 (1980) 640-642.
19 - Bennett M.J., Houlton M., Price J.B., "New ceramic coatings for high temperature gas-cooled reactor materials protection", Oesterr. Forschunszent. Seibersdorf. Number OEFZS BER. No 4086 (1981) N1-N14.
20 - Dislich H., J. of Non-Crystalline Solids 57 (1983) 371-388.
21 - Nelson R.L., Ramsay J.D.F., Woodhead J.L., Cairns J.A., Crossley J.A.A., Thin Solid Films 81 (1981) 329-347.
22 - Sakka S, in "Treatise on Materials Science and Technology", Edited by M. Tomazawa and R.H. Doremus., Vol 22 (1982) 129-167.
23 - Schroeder H., "Oxide layers deposited from organic solutions", in "Physics of Thin Films ", Academic Press, New-York, Eds. Hass G., Thun R.E., vol. 5 (1969) 87-141 .
24 - Dislich H., Hinz P., J. Non-Crystalline Solids 48 (1982) 11-16.
25 - Dislich H., Hussman E., Thin Solid Film, 77 (1981) 129-139.

26 - Zelinski B.J.J., Uhlmann D.R., J. Phys. Chem. Solids 45 (1984) 1069-1090.
27 - Segal D.L., J. Non-Crystalline Solids 63 (1984) 183-191.
28 - Yoldas B.E., O'Keefe T.W., Appl. Opt. 18 (1979) 3133-3138.
29 - Brinker C.J., Harrington M.S., Solar Energy Mat. 5 (1981) 159-172.
30 - Dislich H., Angew. Chem. Internat. Edit. Engl. 10 (1971) 363-370.
31 - Nogami M., Moriya Y., Yogyo Kyokai-Shi 85 (1977) 59-65.
32 - Shinbo M., Tanzawa K., "Silica coating on glass products", Japan Kokai (patent) 18 Aug. 1976..
33 - Brinker C.J., Pettit R.B., "Sol-Gel derived antireflective coatings", Presented at " Ann. Meeting. Am. Ceram. Soc.", Chicago, Illinois (1983).
34 - Floch H., Priotton J.J., technical note CEL-V/D.LPP n° 214/88 (1988) of the "Commissariat A l'Energie Atomique", France.
35 - Spiers R.P., Subbaraman C.V., Wilkinson W.L., Chem. Eng. Sci., 29 (1974) 389-396.
36 - Yoldas B.E., Partlow D.P., Am. Ceram. Soc. Bull. 59 (1980) 640-642.
37 - Sakka S., Kamiya K., J. Non-Cryst. Solids 48 (1982) 31-46.
38 - Mahler W., Bechtold M.F., Nature (London) 285 (1980) 27-28.
39 - Kamiya K., Sakka S., Mizutani M., Yogyo Kyokaishi 86 (1978) 552-559.
40 - Kamiya K., Sakka S., Tashiro N., Yogyo Kyokaishi 84 (1976) 614-618.
41 - Kamiya K., Sakka S., Ito S., Yogyo Kyokaishi 85 (1977) 599-605.
42 - Kamiya K., Sakka S., Tatechimi Y., J. Mater. Sci. 15 (1980) 1765-1771.
43 - Kamiya K., Sakka S., Yogyo Kyokaishi 85 (1977) 308-309.
44 - Blakelock M.D., Hill N.A., Lee S.A., Goatcher C., Proc. Br. Ceram. Soc. 15 (1970) 69-83.
45 - Sowman H.G., German Patent 2 210 291 (U.S. Patent 3 795 524), 21 sep 1972 .
46 - Sowman H.G., German Patent 2 024 547 (U.S. patent 3 793 041) 19 Nov. 1970.
47 - Karst K.A., Sowman H.G., U.S. Patent 4047965; 13 sep. 1977.
48 - Sakka S., Kamiya K., J. Non-crystalline Solids 42 (1980) 403-422.
49 - Wills R.R., Markle R.A., Mukherjee S.P., Am. ceram. Soc. Bull., 62 (1983) 904-911.
50 - Verbeek W. Winter G., German Patent. 2,218,960, Bayer, A.G., Nov.8 (1973).
51 - Sakka S., Kamiya K., Proc. Int. Symp. Factors in Densification and Sintering of Oxides and Nonoxide Ceramics, Tokyo Institute of Technology, Tokyo, (1978) 101-109.
52 - Yoldas B.E., J. Mater. Sci. 14 (1979) 1843-1849.
53 - Souers P.C., Tsugawa R.T., Stone R.R., Rev. Sci. Instrum. 46 (1975) 682-685.
54 - Matsuyama I., Susa K., Satoh S., Suganuma T., Am. Ceram. Soc. Bull. 63 (1984) 1408-1411.
55 - Kamiya K., Sakka S., J. Mater. Sci. Lett. 15 (1980) 2937-2939.
56 - Ozaki Y., Hideshima M., Zairyo 26 (1977) 853-857.
57 - Nogami M., Moriya Y., Yogyo-Kyokai-Shi 87 (1979) 37-42.
58 - Yamane M., Aso S., Okano S., Sakaino T., J. Mater. Sci. 14 (1979) 607- 611.
59 - Yamane M., Okano S., Yogyo Kyokai-Shi 87 (1979) 434-438.
60 - Makishima A., Oohushi H., Wakakuwa M., Kotani D., Shimohira T., J. Non-crystalline solids 42 (1980) 545-552.
61 - Klein L.C., Garvey G.J., J. of Non-crystalline solids 48 (1982) 97-104.
62 - Yoldas B.E., J. Mater. Sci. 12 (1977) 1203-1208.
63 - Yoldas B.E., J. Non-crystalline Solids 38-39 (1980) 81-86.
64 - Hayashi T., Saito H., J. Mater. Sci. 15 (1980) 1971-1977.
65 - Hench L.L., Wilson R.L, Balaban C., Nogues J.L., in "Ultrastructure Processing of Advanced Materials", Uhlmann D.R. and Ulrich D.R., Eds., John Wiley & Sons, New-York (1992) 159-178 .

66 - Livage J., Lemerle J., Ann. Rev. Mater. Sci. 12 (1982) 103-122.
67 - Shoup R.D., in "Ultrastructure processing of glasses and ceramics", Eds. Mackenzie J.D, Ulrich D.R., John Wiley, New York, (1988) 347-354.
68 - Woignier T., Phalippou J., Zarzycki J., J. of Non-crystalline solids 63 (1984) 117-130.
69 - Decottignies M., Phalippou J., Zarzycki J., J. Mater. Sci. 13 (1978) 2605 -2618.
70 - Haas P.A., Clinton S.D., Kleinsteuber A.T., Can. J. Chem. Eng., December (1966) 348-353.
71 - Wymer R.G., Coobs J.H., Proc. Br. Ceram. Soc. 7 (1967) 61-69.
72 - Robbins J.M., Stradley J.G., 'Fabrication of sol-gel-derived thoria-urania by cold pressing and sintering", communication presented at the "69th Annual meeting, Amer. Ceram. Soc", New-York, May 2, 1967.
73 - Woodhead J.L., in " Sciences of Ceramics" Vol.4, Ed. Stewart G.H., Pub. Brit. Ceram. Soc., Stoke-on- Trent, (1968) 105-111.
74 - Fletcher J.M., Hardy C.J., Chem. and Ind., 13 , Jan (1968), 48 -51.
75 - Wheat T.A., J. Canad. Ceram. Soc. 46 (1977) 11-18.
76 - Blum J.B., Gurkovich S.R., J. of Mater. Sciences 20 (1985) 4479-4483.
77 - Colomban P., L'industrie Céramique 697 (1976) 531-535 .
78 - Lenfant P., Pias D., Ruffo M., Colomban P., Boilot J.P., Mater. Res. Bull. 15 (1980) 1817-1827.
79 - Haertling G.H., Land C.E., J. Amer. Ceram. Soc. 54 (1971) 1-11.
80 - Snow G.S., J. Amer. Ceram. Soc. 56 (1973) 91-96.
81 - Peercy P.S., Land C.E., Nucl. Intr. Method 209-210 (1983) 1167-1178.
82 - Brown L.M., Mazdiyasni K.S., J. of Am. Ceram. Soc. 55 (1972) 541-544.
83 - Mazdiyasni K.S., Ceramics International 8 (1982) 42-56.
84 - Mazdiyasni K.S., Dolloff R.T., Smith II J.S., J. Amer. Ceram. Soc. 52 (1969) 523-526.
85 - Smith II J.S., Dolloff R.T., Mazdiyasni K.S., J. Amer. Ceram. Soc. 53 (1970) 91-95.
86 - Perthuis H., Colomban P., J. Mat. Sci. Lett. 4 (1985) 344-346.
87 - Boilot J.P., Colomban P., Blanchard N., Solid State Ionics 9-10 (1983) 639-644.
88 - Colomban P., Perthuis H., Velasco G., in "Solid State Protonic Conductors II", Hindsgarle, Danmark, Eds. Goodenoogh J.B., Jensen J., Kleitz M., Odense Université Press (1983) 375-391.
89 - Perthuis H., Colomban P., Mat. Res. Bull. 19 (1984) 621-631.
90 - Barj M., Perthuis H., Colomban P., Solid State Ionics, 11 (1983) 157-177.
91- Perthuis H., Colomban P., Boilot J.P., Velasco G., Proceedings of the 5th Cimtec 14-19 June 1982, Lignano-Sabbiadoro (Italy), Ceramic Powders, P.Vincenzini ed., Elsevier, Amsterdam, Mat. Sc. Monogram 16 (1983) 575-582.
92 - Perthuis H., Velasco G., Colomban P., Jap. J. Appl. Phys. 23 (1984) 534-543.
93 - Boilot J.P., Colomban P., Gay A., Lejeune M., French patent n°8.306.934 of 27 april 1983.
94 - Mazdiyasni K.S., Lynch C.T., Smith II J.S., J. Am. Ceram. Soc. 48 (1965) 372-375.
95- Barringer E.A., Bowen H.K., Comm. Amer. Soc. (1982) c199-c201.
96 - Mazdiyasni K.S., Am. Ceram. Soc. Bull. 63 (1984) 591-594.
97 - Higuchi K., Naka S., Hirano S.S., Advanced Ceramic Materials 1 (1986) 104-107.
98 - Colomban P., L'industrie Céramique 792-3 (1985) 186-196 .
99 - Mazdiyasni K.S., Lynch C.T., Smith II J.S., J. Amer. Ceram. Soc. 50 (1967) 532-537.
100 - Sowman H.G., Leitheiser M.A., U.S. patent 4 314 827 (European Patent 24099) 25 Fev. 1981.
101 - Sane A.Y., European Patent 0115745- Eltech Systems Corporation, 15. 08 (1984).

102 - Lackey W.J., Angelini P., Laytin F.L., Stinton D.P., Vavruska J.S., Proc. Symp. Waste Management (1982) 391.

103 - Carturan G., Facchini G., Gottardi V., Guglielmi M., Navazio G., J. Non- crystalline Solids 48 (1982) 219-226.

104 - Roy R., J. Amer. Ceram. Soc. 52 (1969) 344-345.

105 - MacCarthy G.J., Roy R., Mackay J.M., J. Amer. Ceram Soc. 54 (1971) 639-640.

106 - Mukherjee S.P., Zarzycki J., Traverse J.P., J. mater. Sci. 11 (1976) 341-355.

107 - Keizer K., Burggraaf A.J., Sci. Ceram, 14 (1988) 83

108 - Hazlett J.D., Kutowy O., Matsuura T., Tremblay A.Y., Membrane Separations Symposium, 41st Canadian Chemical Engineerng Conference", Vancouver, October 6-9 (1991) 7-13.

109 - Eykamp W., Membrane Separations Symposium, 41st Canadian Chemical Engineerng Conference", Vancouver, October 6-9 (1991) 15-19.

110 - Chan K.K., Brownstein A.M., Am. ceram. Soc. Bull., 70 (1991) 703-705.

111 - Cot L., Larbot A., Guizard C., in : "Ultrastructure Processing of Advanced Ceramics", Mackenzie J.D. and Ulrich D.R., Eds., John-Wiley & Sons, New-York (1986) Chapter 34.

112 - Gillot J., "The developing Use of Inorganic Membranes. A Historical Perspective", in "Inorganic membranes. Synthesis, characteristics and applications", Bhave R.R., Ed, Van Nostrand Reinhold , New-York (1991)1-9

113 - Hoogland M.R., Fane A.G., Fel C.J.D.l, Proc. 1st Intl. Conf. Inorganic Membranes, Montpellier, 3-7 July (1990) 153-162.

114 - Hsieh H.P., "General Characteristics of Inorganic Membranes", in "Inorganic membranes. Synthesis, characteristics and applications", Bhave R.R., Ed, Van Nostrand Reinhold , New-York (1991) p 65

115 - Tanaka H., J. Non-Cryst. Solids, 65 (1984) 301

116 - Thomas M.P., Landham R.R., Butler E.P., Cowieseon D.R., Burlow E., Kilmartin P., J. Membrane Sci., 61 (1991) 215

117 - Raidel C., Spohr R., J. Membrane Sci., 7 (1980) 225

118 - Lee K.H., Khang S.J., Ceram. Engr. Sci. Proc., 8 (1987) 55

119 - Anderson M.A., Gieselmann M.J., Xu Q., J. Membrane Sci., 39 (1988) 243

120 - Burggraaf A.J., Keizer K., Van Hassel B.A., Solid State Ionics, 32/33 (1989) 771

121 - Kingery W.D., Bowen H.K., Uhlmann D.R., "Introduction to ceramics", Wiley-Interscience, , 2nd edition (1976) Chapter 10

122 - Brinker C.J., Scherer G.W., Sol-Gel science: The physics and chemistry of sol-gel processing", Academic Press, New-York (1990) Chapter 11.

123 - A. Larbot, J.P. Fabre, G, Guizard, L. Cot, J. Membrane Sci., 39 (1988) 209

124 - Okubo T., Haruta K., Kusakaba K., Morooka S., Anzai H., Akimaya S., J. membrane Sci., 59 (1991) 73

125 - Larbot A., Fabre J.P., Guizard C., Cot L., J. Am. Ceram. Soc., 72 (1989) 257

126 - Guizard C., Idrissi N., Larbot A., Cot L., Brit. Ceram. Proc., 38 (1986) 263

127 - Alcock D.S., McDowall I.C., J. Am. Ceram. Soc., 40 (1957) 355.

128 - Nakao S., Nomura T., Kimura S., J. Chem. Eng. Japan 19 [3] (1986) 221-226.

129 - Klein L.C., Giszpenc N., Am. Ceram. Soc. bull., 69 (1990) 1821.

130 - Cini P., S.R. Blaha, Harold M.P., J. Membrane Sci., 55 (1991) 199

131 - Haggin J., Chem. Eng. News, Sept. 3 (1990) 30

132 - Pajonk G.M., Applied Catalysis, 72 (1991) 217-266

133 - Ward D.A. and Ko E.I., I&EC Research 34 (1995) 421-433

134 - Cauqui M.A., Rodriguez-Izquierdo J.M., J. of Non-Cryst. Solids, 147&148 (1992) 724-738.

135 - Satterfield C.N., "Heterogeneous Catalysis in Industrial Practice", 2nd Edition, McGraw-Hill, New-York (1990).

136 - Astier M., Bertrand A., Bianchi D., Chenard A., Gardes G.E.E., Pajonk G., Taghavi M.B., Teichner S.J., Villemin B., in "Studies in Surface Science and Catalysis", vol. 1 "Preparation of Catalysts", Delmon B., Jacobs P.A., Poncelet G., Eds., Elsevier, Amsterdam (1976) 315

137 - Matis G., Juillet F., Teichner S.J., Bull. Soc. Chim. Fr., (1976) 1633.

138 - Taghavi M.B., Pajonk G.M., Teichner S.J., J. Colloid Interface Sci., 71 (1979) 45

139 - Klavana D., Chaouki J., Kusohorski D., Chavarie C., Pajonk G.M., Appl. Catal., 42 (1988) 121.

140 - Astier M., Bertrand A., Teichner S.J., Bull. Soc. Chim. Fr., (1980) 218.

141 - Armor J.N., Carlson E.J., Zambri P.M., Appl. Catal., 19 (1985) 339

142 - Blanchard F., Pommier B., Reymond J.P., Teichner S.J., J. Mol. Catal., 17 (1982) 171.

143 - Blanchard F., Pommier B., Reymond J.P., Teichner S.J., in " Studies in Surface Science and Catalysis", vol. 16 "Preparation of Catalysts III", Poncelet G., Grange P. and Jacobs P.A., eds., Elsevier, Amsterdam (1983) 395.

144 - Willey R.J., Lai H., Peri J.B., J. Catal. 130 [1991] 319

145 - Fanelli A.J., Burlew J.V., Marsh G.B., J. Catal., 116 (1988) 318.

146 - Paal Z., Menon P.G., "Hydrogen Effect in Catalysis", Marcel Dekker, New-York (1988).

147 - Nicolaon G.A., Teichner S.J., Bull. Soc. Chim. Fr., (1968) 388.

148 - Baker R.T.K, Tauster S.J., Dumesic J.A., ACS Symposium Series, "Strong Metal-Support Interactions", Am. Chem. Soc., Washington D.C., 1986, p 298.

149 - Paal Z., Menon P.G., "Hydrogen Effect in Catalysis", Marcel Dekker, New York (1988).

150 - Lacroix M., Pajonk G., Teichner S.J., in "Studies in Surface Science and Catalysis", vol. 7, Elsevier Amsterdam.

151 - Gardes G.E.E., Pajonk G.M., Teichner S.J., Bull. Soc. Chim. Fr., (1976) 1327.

152 - Astier M., Bertrand A., Teichner S.J., Bull. Soc. chim. Fr., (1980) 191.

153 - Pajonk G., Taghavi M.B., Teichner S.J., Bull. Soc. Chim. Fr., (1975) 983

154 - Machida M., Eguchi K., Arai H., Chem. Letters (Japan) (1986) 1993-1996.

155 - Machida M., Eguchi K., Arai H., Chem. Letters (Japan) (1987) 767-770.

156 - Ponthieu E., Grimblot J., Elaloui E., Pajonk G.M., J. Mat. Chem., 3 (1993) 287-293.

157 - E. Elaloui, A.C. Pierre and G.M. Pajonk, J. of Catalysis, 166 (1997) 340-346.

158 - Corma A., grande M.S., Gonzales-Alfaro V., Orchilles A.V., J. Catal. 159 (1996) 375.

159 - Corma A., martinez A., Martinez-Soria V., Monton J.B., J. Catal., 153 (1995) 25.

160 - Reddy K.M., Song C.S., Catal. Today 31 (1996) 137

161 - Apelian M.R., Degman T.F.Jr., Marler D.O., Mazzone D.N., U.S. Patent 5.227.353, (1993)

162 - Del rossi K.J., Hatzikos G.H., Huss A.Jr., U.S. Patent 5.256.277 (1993)

163 - Armengol E., Cano M.L., Corma A., Garcia H., Navarro M.T., J. Chem. Soc. Chem. Commun., (1995) 519.

164 - Gunnevegh E.A., Gopie S.S., Van Bekkum H., J. Mol. Catal., 106 (1996) 151.

165 - Climent M.J., Corma A., Iborra S., Navarro M.C., Primo J., J. Catal., 161 (1996) 786.

166 - Aucejo A., Burguet M.C., Corma A., Fornés V., Appl. Catal., 22 (1986) 187.

167 - Corma A., Chem. Rev. 97 (1997) 2373-2419.

168 - kloetstra K.R., Van Bekkum H., J. Chem. Soc. Chem. commun., (1995) 1005.

169 - Corma A., Navarro M.T., Pérez-Pariente J., J. Chem. Soc. Chem. Commun., (1994) 147.
170 - Iglesia E., Soled S.L., Kramer G.M., J. Catal., 144 (1993) 238.
171 - Sakane S., Tsubakino T., Nishiyma Y., Ishi Y., J. Org. Chem. 58 (1993) 3633.
172 - Corma A., Iglesias M., Sanchez F., J. chem. Soc. Chem. Commun., (1995) 1635.
173 - Abdel-Fattah T.M., Pinnavaia T., J. chem. Soc. Chem. Commun., (1996) 665.
174 - Gontier S., Tuel A., Appl. Catal., A , 143 (1996) 125.
175 - Kozheunikov I.V., Sinnema A., Jansen R.J.J., Pamin K., Van Bekkum H., Catal. Lett. 30 (1995) 241.
176 - Brunel D., Canvel A., Fajula F. Di Renzo F., Stud. Surf. Sci. Catal., 97 (1995) 173.
177 - Junges U., Jacobs W., Voigt-Martin I., Krutzsch B., Schuth F., J. Chem. Soc. Chem. Commun., (1995) 2283.
178 - Beck J.S., Socha R.F., Shihabi D.S., Vartulli J.C., U.S. Patent 5.143.707 (1992)
179 - Maschmeyer T., Rey F., Sankar G., Thomas J.M., Nature 378 (1995) 159.
180 - Burch R., Cruise N., Gleeson D., Tsang S.Ch., J. Chem. Soc. chem. commun., (1996) 951.
181 - Corma A., Iglesias M., del Pinto C., sanchez F., J. Chem. Soc. chem. Commun. (1991) 1253.
182 - Sanchez C., Ribot F., New J. Chem., 18 (1994) 1007-1047.
183 - Wilkes G.L., Orler B., Huang H.H., Polymer Prep., 26 (1985) 300
184 - Sur G.S., Mark J.E., Eur. polym. J., 21 (1985) 1051
185 - Morikawa A., Iyoku Y., kakimoto M., Imai Y., J. Mater. Chem. 2 (1992) 679
186 - Schmidt H., Popall M., in Sol-Gel Optics I, Mackenzie J.D. and Ulrich D.R., Eds., proc. SPIE 1328, SPIE, Washington, (1990) p. 249
187 - Izumi K., Tanaka H., Murakami M., degushi T., Morita A., Toghe N., Minami T., J. Non Cryst. Solids, 121 (1990) 344
188 - Chujo Y., Saegusa T., Adv. Polym. Sci., 100 (1992) 11
189 - Avnir D., Levy D., Reisfeld R., J. Phys. Chem., 88 (1984) 5956
190 - Dunn B., Zink J.I., J. Mater. Chem., 1 (1991) 903
191 - Kramer S.J., Colby M.W., Mackenzie J.D., Mattes B.R., Kaner R.B., in Chemical Processing of Advanced Materials, Hench L., West J.K., Eds., Wiley, New-York, 1992, p. 737
102 - Audebert P., Hapiot P., Griesmar P., Sanchez C., J. Mater. Chem., 1 (1994) 699
193 - Braun S., Rappoport S., Zusman R., Avnir D., Ottolenghi M., Mater. Lett., 10 (1990) 1
194 - Ellerby L.M., Nishida C.R., Nishida F., Yamanaka S.A., Dunn B., Valentine J.S., Zink J.I., Science 255 (1992) 1113
195 - Yamanaka S.A., Nishida F., Ellerby L.M., Nishida C.R., Dunn B., Valentine J.S., Zink J.I., Chem. Mater. 4 (1993) 495
196 - Audebert P., Demaille C., Sanchez C., Mater. Chem. 5 (1993) 911
197 - Kanatzidis M.G., Wu C.G., J. Am. Chem. Soc., 111 (1989) 4139.
198 - Parkhurst C.S., Doyle W.F., Silverman L.A., Singh S., Andersen M.P., McClug D, Wnek G.E., Uhlmann D.R., Mat. Res. Soc. Symp. Proc., 73 (1986) 769.
199 - Rodrigues D.E., Brennan A.B., Betrabet C., Wang B., Wikes G.L., Chem. Mater., 4 (1992) 1437
200 - Ravaine D., Seminel A., Charbouillot Y., Vincens M.A., J. Non Cryst. Solids 82 (1986) 210
201 - Sanchez J.Y., Denoyelle A., Poinsignon C., Polymer Adv. technol., 4 (1992) 99.
202 - Sanchez C., Alonso B., Chapusot F., Ribot F., Audebert P., J. Sol-Gel sci. Technol. 2 (1994) 161

203 - Babonneau F., Bois L., Maquet J., Livage J., in Eurogel 91, Vilminot S., Naβ R., Schmidt H., eds., Elsevier, Amsterdam, 1992, p. 319.

APPENDIX